HYDROGEOLOGY OF CRYSTALLINE ROCKS

Water Science and Technology Library

VOLUME 34

The titles published in this series are listed at the end of this volume.

HYDROGEOLOGY OF CRYSTALLINE ROCKS

edited by

INGRID STOBER

Geological Survey Baden-Württemberg,
Division of Hydrogeology,
Freiburg, Germany

and

KURT BUCHER

Institute of Mineralogy,
Petrology and Geochemistry,
University of Freiburg, Germany

SPRINGER-SCIENCE+BUSINESS MEDIA, B.V.

A C.I.P. Catalogue record for this book is available from the Library of Congress.

DOI 10.1007/978-94-017-1816-5

Cover illustration: Thermal spas have a long tradition in the crystalline basement of the Black Forest area. Shown is an illustration from Bad Wildbad (in: L. Phries, Strasbourg 1519). The bath is still in operation today. It utilizes 36 °C warm, upwelling mineralized groundwater from a deep reservoir in granite.

Printed on acid-free paper

Table of Contents

Chapter 4 Microbial Processes in Crystalline Rocks

Preface

Intense, multi-disciplinary research attempts to unravel the nature and behavior of water in the pore space of the continental upper crust. During the past 10 to 15 years it has been recognized that free water is almost universally present in fractured crystalline rocks. Continental deep drilling programs in Russia, USA and Germany confirmed the presence of saline fluids in the fracture pore space of the crystalline basement. National nuclear waste disposal programs of different countries and hot-dry-rock projects in various parts of the world added significantly to our general understanding of the hydrogeology of crystalline basement..

This has brought together diverse Earth-science disciplines that earlier had very little contact, communication and interaction. The interest of petrologists and geochemists in "Fluids in the crust" inspired research concepts and ideas that emerged from dealing with tectonically active crust where heat transfer controls dehydration, water production and migration. The time scale of the processes of interest is on the other of millions of years. Hydrogeologists deal with deep groundwater in the present day crust and most of the attention is paid to "normal", tectonically little active crust. Geophysicists are both, interested in geophysical signals of water present in the fracture porespace of the present day upper crust and in geophysical evidence of effects of fluids on rocks in the geologic past. At depth the deep groundwater hydrology and the fluid-related processes in the upper part of the continental crust of geophysists and petrologists become the prevalent research subject. We hope that this book inspires new inter-disciplinary research on this fascinating and important subject in the future.

Because of this steadily growing interest in hydrogeology of the crystalline rocks, the European Union of Geosciences organized a special symposium at EUG9, the biannual meeting in Strasbourg 1997 convened by the editors of this volume. This very successful symposium summarized the state-of-the-art of basement hydrogeology.

In this book, the reader will find a selection of papers about hydraulic, chemical, biological, and structural aspects of crystalline hydrogeology. Most of this research has been presented at EUG9. The first section of the volume highlights structural investigations on water conducting features and microstructural evidence of water flow in feldspars. The second part focuses on the hydraulic properties of crystalline rocks. Some aspects of water composition in the basement are treated in the third chapter. The recently recognized importance of microbial processes in deep groundwater environments is reflected in the final part of the book. At EUG10 in Strasbourg 1999, a special two-day symposium on "Hydrogeology of Crystalline Rocks" has drawn much attention and demonstrated the continuous interest of the scientific community in "water in the crust".

The editors would like to express their thanks to the authors for investing so much time and effort in this venture and for their willingness to share their ideas with the Earth science community. We also are very grateful to all colleagues who took their time and effort for constructively review the contributions and so helped to significantly improve the quality of the presentations. Without the generous effort of competent reviewers modern science would be impossible.

Reviews were provided by: Barbara John (Laramie), David Fountain (Laramie), Dirk Schulz-Mauch (Onalaska), Eric Frank (Würenlingen), Everett Shock (St.Louis), Helmut Wilhelm (Karlsruhe), Ingrid Stober (Freiburg), Jan Cramer (Trondheim), John Svenson (Laramie), Kurt Bucher (Freiburg), Ladislaus Rybach (Zürich), Hansruedi Maurer (Zürich), Simon Poulsen (Reno), Susan Swapp (Laramie), Tim Drever (Laramie), Tony Hoch (Boulder), Tullis C. Onstott (Princeton), Vala Ragnarsdottir (Bristol), Volker Dietrich (Zürich)

Ingrid Stober and Kurt Bucher

Freiburg, July 27. 1999

Chapter 1

Water conducting features in Crystalline Rocks

GEOLOGICAL AND HYDRAULIC PROPERTIES OF WATER-CONDUCTING FEATURES IN CRYSTALLINE ROCKS

MARTIN MAZUREK
Rock/Water Interaction Group (GGWW),
Institutes of Geology and of Mineralogy and Petrology,
University of Bern, Baltzerstr. 1, 3012 Bern, Switzerland
(mazurek@mpi.unibe.ch)

Abstract

Geological and hydrogeological field evidence from several sites (Grimsel Test Site, Äspö Hard Rock Laboratory, deep boreholes in northern Switzerland, various mines) shows that in spite of contrasting geological settings, evolutions and ages, several common characteristics of water-conducting features exist in crystalline basement rocks. Geometric and hydraulic properties of water-conducting features depend mainly on the mechanism of brittle deformation (*e.g.* faulting, jointing), on the nature and intensity of water/rock interactions (*e.g.* hydrothermal fracture sealing) and on rock type. Leucocratic rocks, such as aplite/pegmatite dykes, have higher fracture frequencies and transmissivities when compared to more basic rocks. Brittle deformation in most crystalline-rock environments occurred recurrently, and pre-existing structures (*e.g.* lithologic contacts, ductile shear-zones, older fault and fracture generations) were preferentially reactivated. Faults of different sizes, ranging from small cataclastic zones to regional lineaments, are the most important structures in which flow occurs. Due to the complex architecture of faults in directions parallel and perpendicular to strike, the spatial distribution of flow in faults is very heterogeneous. Hydrothermal alteration events lead to fracture sealing by mineral precipitation or to increased apertures due to the dissolution of pre-existing fracture infills, thereby enhancing the heterogeneity of the flowpaths on a small scale.

1. Definition and attributes of water-conducting features

Water-conducting features are zones with enhanced transmissivities within a rock body. They are the consequence of the hydraulic heterogeneity of the rocks and represent the dominant conduits for fluid flow through the formation. In fractured media, such as crystalline rocks, water-conducting features occur in structures generated by brittle deformation mechanisms, such as faults, joints and veins. The attributes of water-conducting features that are relevant for flow and solute transport are listed in Table 1 and can be grouped as follows:

- *Geometric/structural attributes* on a wide range of scales provide information on the characteristics of the fracture network within a volume of rock. They are the basis for the quantification of the connectivity of water-conducting features and of the flow field, including fluxes through single water-conducting features. Moreover, the spatial arrangement of flow-wetted surface and connected microporosity of the rock matrix determine the extent to which solute transport through fractures is attenuated by matrix diffusion (Neretnieks 1980). Geometric information is also required for upscaling procedures, such as the recalculation of

I. Stober and K. Bucher (eds.), Hydrogeology of Crystalline Rocks, 3–26.

transmissivities measured in individual boreholes to effective hydraulic conductivities of rock blocks with typical lengths of side of tens to hundreds of m.

- *Hydraulic attributes* determine the flow and advective/dispersive transport properties. Extrapolation of field measurements (up- and downscaling) requires information on the geometry and internal heterogeneity of water-conducting features.
- *Mineralogical/geochemical attributes* characterize the interactions between solutes and rocks by processes like mineral precipitation/dissolution, sorption and cation exchange. The residence times of groundwaters can be used to constrain the results of flow models (*e.g.* travel times).

TABLE 1. Attributes of water-conducting features

attribute	characterization technique	relevance
GEOMETRY AND STRUCTURE		
mechanistic principles and genetic aspects of fracture formation	macroscopic structural analysis	deformation mechanisms, stress-strain history, classification of fracture types, interpretation of fracture architecture and implications for transmissivity distribution
large-scale architecture	tunnel and surface mapping	characterization of structural heterogeneity, interpretation of hydraulic measurements
geometry: size, thickness, orientation frequency	core/tunnel logging, line counting, outcrop mapping, lineament analysis	interpretation and upscaling of hydraulic measurements, input to discrete fracture network models
age / stages of activity	dating techniques (e.g. isotope methods, electron spin resonance)	evolution of (recurrent) fault/fracture activity, recent movements
tectono-hydrothermal evolution (fracture creation and sealing)	petrologic and textural analysis	understanding the small-scale structure, structural and hydraulic heterogeneity, characteristics of flow-wetted surfaces and rock matrix domains
matrix porosity	gravimetric and injection methods	input to modelling matrix diffusion
HYDRAULICS		
local transmissivity / hydraulic conductivity, head	hydraulic testing, fluid logging	input for flow modelling
transmissivity distribution within a feature	small-scale hydraulic testing combined with mechanistic and structural understanding	hydraulic heterogeneity, channeling
flow porosity / fracture aperture	crosshole tracer tests	input to transport modelling
GEOCHEMISTRY		
fracture infill and wallrock mineralogy	petrography	input to modelling of water/rock interactions, e.g. sorption, cation exchange, dissolution/precipitation
groundwater composition	in-situ sampling and chemical analysis	input to modelling of water/rock interactions, *e.g.* sorption, chemical reaction
groundwater residence times	isotope geochemistry	constraints on paleo-flow rates and flow directions, consistency check to flow/transport model results

2. Geological and hydrogeological characteristics of sites discussed in this paper

The strategy adopted to characterize water-conducting features depends on the nature of the available dataset (*e.g.* tunnel- *vs* borehole-derived information) and on local geological characteristics. Whereas the hydraulic characterization follows common procedures that are often independent of a specific site, the steps taken to provide a geological characterization are much more site-specific. This contribution deals mainly with the following sites:

- the tunnel system at the Grimsel Test Site in the Swiss Alps (Bossart and Mazurek 1991, Frick *et al.* 1992, Vomvoris and Frieg 1992, Frieg and Vomvoris 1994);
- deep boreholes penetrating the crystalline basement of northern Switzerland (Thury *et al.* 1994, Mazurek 1998); and
- the tunnels at the Äspö Hard Rock Laboratory in southern Sweden (Rhén *et al.* 1997a,b, Stanfors *et al.* 1997, Mazurek *et al.* 1996).

Table 2 gives an overview of the relevant system characteristics of these sites. The sites cover a wide range of environments, from sparsely fractured, low-permeability systems (Grimsel) to highly fractured, higher-permeability systems with recurrent episodes of fracturing and hydrothermal[1] alteration/cementation (Äspö, northern Switzerland). Both, information derived from boreholes (northern Switzerland) and from drilled or blasted tunnels at different depth levels (Grimsel, Äspö) will be included.

At the Grimsel Test Site, only 10-20 weak inflow points were observed over a tunnel length of 700 m (average spacing: 35 - 70 m), and this number is subject to seasonal variation. In the tunnel system at Äspö, the average spacing between inflow points is 3 - 4 m measured along the tunnel. In the boreholes in northern Switzerland, 138 inflow points were detected, corresponding to an average spacing of *ca.* 42 m along hole. This number is not directly comparable to the tunnel data due to the different methodologies used to detect inflow points.

3. Methodology of investigating water-conducting features

According to the definition given above, water-conducting features are hydraulic anomalies that can be identified as water inflows in tunnels or by hydraulic test methods in boreholes. On a regional scale, major structures (such as fault zones) can be identified by surface-based methods (structural field mapping, geophysics), but their relevance as flow conduits at depth is generally inferred or based on extrapolation.

3.1 IDENTIFICATION OF WATER-CONDUCTING FEATURES IN TUNNELS

Water discharges can be observed directly as points/zones of moisture, drop or flow. Weak discharges will be identified only if the inflow exceeds a specific threshold value. This value is a function of the evaporation rate on the tunnel walls and thus of the tunnel ventilation system. At the Grimsel Test Site, about 1 ml water per second and

[1] Hydrothermal alteration is a term used to describe the interaction of rocks and fluids with T > 100 °C. Low-temperature alteration also occurs in many crystalline rocks but, in general, has limited consequences for the geologic and hydraulic properties of water-conducting features.

TABLE 2. Relevant characteristics of the sites discussed in this paper

	Grimsel Test Site	northern Switzerland	Äspö Hard Rock Laboratory
excavation technique	drilled tunnel system 700 m long	deep boreholes (total 5800 m of crystalline-rock cores from 6 boreholes)	blasted and drilled tunnel 3600 m long
overburden	400 m	200 - 2500 m	0 - 450 m
main rock types	granodiorite, granite	granites, gneisses	granitoids
tectonic environment	young Alpine system, currently uplifting	Variscan orogen, quiet tectonic regime	located at the intersection of Precambrian lineaments, presently uplifting due to post-glacial rebound
hydrogeological conditions	weakly fractured and low-permeability system, only few water inflows into tunnel (1-3 per 100 m)	higher-permeability system at shallow levels, low-permeability rocks only at depth, intensely fractured	high-permeability system, heavily fractured at all depths, 30-50 inflows per 100 m
geological structure (dekametric and smaller-scale)	simple	complex	very complex
degree of hydrothermal alteration	very weak	very intense	intense

m^2 of exposed rock surface is carried away by tunnel ventilation. The equivalent hydraulic conductivity to produce a visible moisture zone is *ca.* $3*10^{-8}$ m/s, and this value varies due to the seasonal variability of ambient air humidity.

In most cases, inflows into tunnels are concentrated in discrete points or zones within a planar structure intersected by the tunnel. However, the geometry of the inflow points is not necessarily representative of the conditions in the undisturbed formation. The presence of an underground construction, whether tunnel or shaft, affects the geologic and hydraulic properties of the adjacent rock. The main causes are:

- disturbance due to the excavation process (drilling, blasting)
- injection of grout or other sealing materials into the rock
- redistribution of rock stress and hydraulic heads (boundary conditions on tunnel wall: shear stress = 0, normal stress = 1 bar)
- possible development of an unsaturated zone.

The thickness of the excavation-disturbed zone is typically in the order of 1 tunnel diameter. The resulting skin effect may affect the hydraulic properties that are measured in the tunnel, such as inflow rates, inflow frequencies and spatial distributions of inflows. In addition, effects have also been identified on groundwater chemistry (Gascoyne and Thomas 1997).

In the abandoned mine at Stripa (Sweden), an experiment has been designed to explore the extent to which artifacts affect tunnel-derived hydraulic data (Olsson and Gale 1995, Olsson 1992). An array of parallel horizontal boreholes (each 100 m long) was drilled, and inflow rates were measured at different drawdowns. As shown in Table 3, there is a more or less even distribution of inflow into all boreholes, and the same is true for two fault zones that were penetrated by all boreholes (not detailed in Table 3). Subsequently, a tunnel 50 m long was excavated within the volume delineated by the

TABLE 3. Results of the SDE experiment in the Stripa mine (Sweden). Data from Olsson (1992)

INFLOW INTO BOREHOLE ARRAY BEFORE TUNNEL EXCAVATION				
drawdown		79 m	157 m	210 m
total inflow into boreholes		734 ml/min	1340 ml/min	1710 ml/min
thereof in	borehole D1	1.1 %	2.2 %	5.6 %
	borehole D2	18.8 %	19.1 %	21.3 %
	borehole D3	13.2 %	13.7 %	0.0 %
	borehole D4	23.6 %	23.1 %	17.7 %
	borehole D5	17.8 %	15.8 %	21.3 %
	borehole D6	25.5 %	26.1 %	33.3 %

COMPARISON OF INFLOWS INTO TUNNEL AND INTO BOREHOLES		
	into tunnel	into boreholes
total inflow	102 ml/min	876 ml/min
inflow from fault zone	101 ml/min	745 ml/min
inflow from rock outwith the fault zone	1 ml/min	131 ml/min

boreholes, and the quantity and spatial distribution of flow was measured. Table 3 shows that the total inflow into the tunnel is 8.5 times smaller than inflow into the equivalent borehole sections, suggesting the existence of a lower-permeability skin around the tunnel. More importantly, 99% of the total inflow into the tunnel were spatially focussed into one single fault zone, whereas the same fault zone accounted for only 85% of the inflow into the equivalent borehole sections. Inflow was highly focussed even within the fault zone itself, and one single fracture discharged about half the total inflow over a trace length of 1 m, with other sections of the same fracture having zero discharge. It is concluded that the existence of a tunnel modifies the spatial distribution of flow within individual structures in addition to discharge and transmissivity.

At Äspö, a completely dry tunnel-wall section was encountered between tunnel meters 2950 - 3000 at *ca.* 400 m below surface. Even though this section does not penetrate any major fault structures, it contains a network of relatively short fractures (<1 m trace length). Flow logging in five boreholes drilled from a niche into this zone yielded *ca.* 1 inflow point per m along hole, with typical transmissivities of $3*10^{-6}$ - $5*10^{-11}$ m^2/s (Winberg 1996). This example illustrates that in spite of the absence of inflow points into the tunnel, a hydraulically well-connected fracture network may exist in the rocks. The absence of moisture zones on the tunnel walls is either due to a high evaporation rate or to the presence of a hydraulic skin, as observed in Stripa.

3.2 IDENTIFICATION OF WATER-CONDUCTING FEATURES IN BOREHOLES

Water-conducting features in boreholes are identified by hydraulic packer tests or fluid logging techniques. The accuracy of localizing discrete features by packer tests depends on the length of the packer interval (generally a few m or more). Fluid logging comprises the acquisition of continuous temperature, electrical conductivity or vertical flow logs in the water column in the borehole. Differences in temperature and electric conductivity between the inflowing formation water and the borehole liquid define water inflow points. Uncertainties in the depth location are ±1 m under good experimental conditions, which allows generally unambiguous correlation with geological features in the cores. In cases of strong vertical water flow or non-optimum conditions (*e.g.* small contrast of electrical conductivity between formation water and borehole liquid), however, the resolution of depth location reduces to 2 - 4 m. Stacked electrical conductivity logs can also be used for the calculation of the transmissivities of the inflow points (Tsang *et al.* 1990, Paillet 1998). Fluid logging detects inflow points whose transmissivity is greater than *ca.* $5*10^{-10}$ m^2/s, but detection limits may be much higher in borehole sections with limited log quality (Tsang *et al.*, 1990). Packer tests have lower detection limits (northern Switzerland: $\leq 10^{-13}$ m/s) but are less accurate in the spatial resolution of inflows.

In drilling campaigns, water-conducting features are often used synonymous to inflow points of water into a borehole. However, it is a common observation that the frequency of inflow points in boreholes identified by hydraulic methods is smaller than the frequency of fractures identified in the corresponding core materials. This observation is consistent with either of the following hypotheses:

1. The rock formation contains different fracture generations with distinct geological characteristics (*e.g.* different orientations, infill materials) and therefore also different hydraulic properties.
2. Each fracture is heterogeneous in itself, *e.g.* contains transmissive and sealed segments. It is identified as an inflow point only if penetrated by the borehole in a transmissive segment.

Field evidence exists that genetically different fracture generations may have contrasting hydraulic properties, *e.g.* as a function of the orientation relative to the present-day stress field or the nature of the infill materials (Barton *et al.* 1995). In the Äspö Hard Rock Laboratory, a number of fracture sets, each with a distinct orientation, developed since the Proterozoic. Under the present-day stress regime, fracture sets with orientations perpendicular to the smallest compressive stress axis σ_3 are preferentially associated with inflow points because their orientation maximizes fracture apertures and therefore transmissivities (see Figure 1 and Munier 1993, 1995). These observations at Äspö support hypothesis 1. On the other hand, in core materials it is not generally possible to geologically distinguish fractures (of the same geological type and orientation) associated with inflow points from those without hydraulic signature. This is most probably due to the heterogeneous internal structure of each water-conducting feature on scales exceeding the core diameter, and this interpretation argues in favour of hypothesis 2. Most probably, both hypotheses play a role in explaining discrepancies between the inventories of structural and hydraulic discontinuities in boreholes. The consequence of hydraulic heterogeneity within and between fracture generations is that inflow points identified in boreholes record only a fraction of all water-conducting features.

3.3 GEOLOGICAL DATA ACQUISITION

Following the identification of water-conducting features by hydraulic methods or direct observation, a geological database can be compiled on the basis of the core materials or tunnel sections that are associated with the inflow points. The geometric/structural and geochemical attributes listed in Table 1 are addressed by investigating the *mechanistic principles of brittle deformation*, *structural elements* and *lithological domains* that are associated with the water-conducting features. The most relevant parameter groups are:

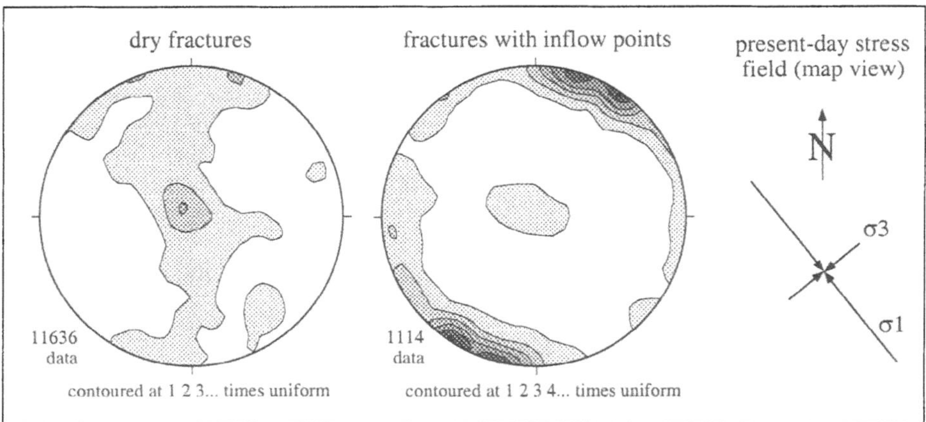

Figure 1. Lower-hemisphere equal-area plots of fracture orientations in the tunnels at Äspö, together with the orientation of the present-day stress field. Data from the SKB tunnel database (TMS).

- *Mechanism of brittle deformation:* Fractures through which water flows can be generated by faulting (brittle shear deformation) or jointing (dilation). The deformation mechanism affects size (= lateral extent), internal structure and heterogeneity of water-conducting features. The nomenclature of brittle discontinuities in rocks used here is consistent with the definitions used by the National Research Council (1996, ch. 2).
- *Pre-existing geometry:* The presence of older mechanical discontinuities in an otherwise homogeneous rock, such as dykes, ductile shear-zones/mylonites and cataclasites, focus the development of fractures. In many cases, the large-scale geometry of a water-conducting feature, namely size and orientation, are determined by pre-existing structural elements.
- *Internal structure:* Flow within water-conducting features may occur in single fractures or in complex networks of different types of discontinuities. Some types of water-conducting features, such as faults, consist of architectural components with contrasting hydraulic properties (*e.g.* fault core and damage zone, cf. Caine *et al.* 1996). The surface area of rock in contact with a unit volume of flowing water (flow-wetted surface) is a function of the internal structure and surface roughness of the water-conducting feature and determines the extent of chemical interaction between water, solutes and minerals.
- *Degree and type of hydrothermal alteration:* The alteration of wallrocks along water-conducting discontinuities results in changes of mineralogical composition and of matrix porosity. Many alteration products, such as clay minerals, have higher distribution coefficients for the sorption of solutes than magmatic or metamorphic minerals and so affect the degree of solute/mineral interaction along the flowpath. Matrix porosity determines the rates of diffusive mass transport between flowing water in fractures and stagnant water in the rock matrix.
- *Nature of fracture infills:* The presence and type of fault rocks (such as fault gouge) and mineralizations in fractures affect the hydraulic properties of water-conducting features. Gouges may act as barriers for flow (*e.g.* Forster and Evans 1991), and the irregular spatial distribution of fracture minerals may create small-scale hydraulic heterogeneity within fractures, with open channels (*e.g.* drusy

veins) and completely sealed segments. Moreover, fracture infills are in direct contact with solutes in the flowing water and thus affect the extent to which interactions between solutes and minerals take place.

- *Host-rock lithology:* Water-conducting features in different rock types may have contrasting structural, geochemical and hydraulic properties due to differences in mechanical properties, in mineralogical compositions and in large-scale geometries.

4. Relationships between structural, hydraulic and geochemical characteristics of water-conducting features

4.1 RECURRENCE OF DEFORMATION EVENTS AND THEIR ROLE FOR FRACTURE HYDRAULICS

In many (if not most) cases, the present-day network of water-conducting features in crystalline rocks consists of fractures that are products of a multiphase geological history. Stages of faulting and fracturing enhance permeability, whereas periods of cementation may seal existing flowpaths. Older structural elements, whether open or sealed, act as pre-existing heterogeneities for the development of younger structures. The overlay of all structural elements generated throughout the geological evolution defines the present-day network of water-conducting features.

The Migration shear-zone at Grimsel

Recurrence of deformation events and the role of pre-existing structural elements for younger stages of deformation can be demonstrated at the Grimsel Test Site. The so-called Migration shear-zone (Figure 2) in granodiorite has been used extensively for crosshole tracer tests, and its geological and hydraulic properties are very well investigated. Kralik *et al.* (1992) performed age datings of fault rocks from this shear-zone and obtained a spectrum of ages that were interpreted to reflect recurrent activity. The shear-zone discharges *ca.* 300 ml/min into the tunnel from a small number of discrete inflow points.

The Migration shear-zone originated as a ductile (mylonitic) shear-zone that was created during Alpine metamorphism some 30 Ma b.p. at temperatures of *ca.* 400 °C (Bossart and Mazurek 1991, Martel and Peterson 1991, Choukroune and Gapais 1983). The shear-zone is at least several dekameters long and acted as a fluid flowpath and conduit for mass transfer during ductile deformation, as indicated by the contrasting chemical compositions of the mylonite and the granodioritic protolith (Bradbury 1989). The matrix porosity of the mylonitic fault rocks of 0.8 % (Bossart and Mazurek 1991) is somewhat lower than that of the undeformed granodiorite (1.05 %), which is due to dynamic recrystallization during shear deformation and to the very dense, fine-grained fabric rich in sheet silicates. Because no discrete fractures developed during this first stage of deformation, the effects on present-day hydraulics are very limited. Given the presence of a planar fabric, hydraulic conductivity of mylonite in the ductile shear-zone is anisotropic but in all cases lower than that of the undeformed granodiorite under present-day conditions. Laboratory measurements yield mean values of $1*10^{-12}$ m/s, compared with $4*10^{-12}$ m/s for the undeformed rock (Tilch 1992). It is concluded that

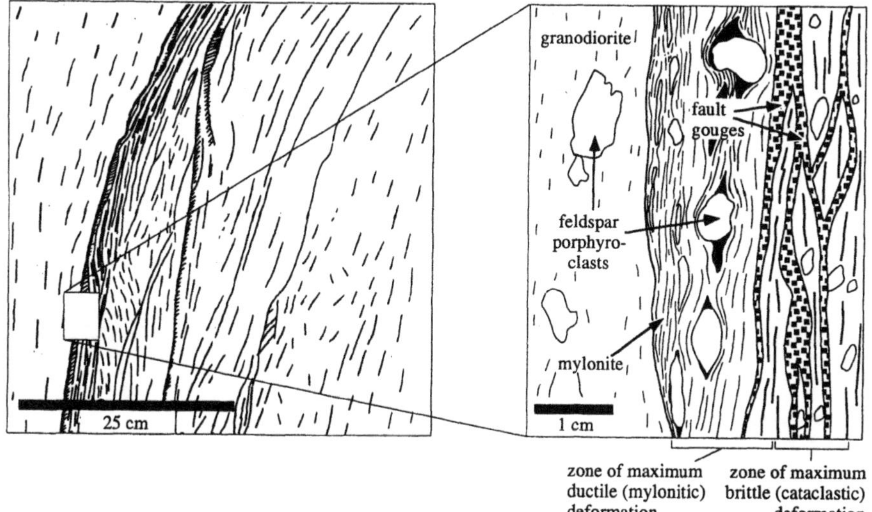

Figure 2. Architecture of the Migration shear-zone at Grimsel (vertical cross-sections) and relationship between ductile and brittle deformation. Adapted from Bossart and Mazurek (1991).

ductile shear-zones that were not affected by later deformations do not act as flow conduits but more likely slow down flow and diffusion through the rock matrix.

The major hydrogeological role of the ductile shear-zone was to act as a mechanical discontinuity and focus subsequent brittle deformation during differential regional uplift (Bossart and Mazurek 1991). The central parts of the mica-rich mylonites were reactivated as faults, namely in zones of rapid lateral transition from granodiorite to highly deformed and foliated mylonite (Figure 2). The regions where water flow occurs at present time contain a network of fault gouges. These consist of crushed wallrock material that, at Grimsel, was virtually unaffected by subsequent water/rock interaction and so reflects the mineralogical composition of the mylonite. The gouge materials have a high matrix porosity of 10 - 30 %, but due to the very small apertures of individual pores, flow through completely gouge-filled fracture segments is insignificant. Flow occurs either along microcracks between the gouges and the wallrock or in gouge-free channels that were generated by internal erosion (either due to natural flow or during excavation or hydraulic testing). In contrast, modelling of experimental tracer breakthrough curves indicates that the gouges are accessible for diffusion even within the short timescales of field experiments (Heer and Smith 1998).

The Migration shear-zone is a prototype structure at Grimsel and shares both structural evolution and hydraulic properties with a number of analogous features in the Test Site. Another distinct structure discharging water into the tunnel is located at the contact between a lamprophyre dyke and granodiorite. This contact had been affected by ductile deformation and was then reactivated by brittle faulting.

Water-conducting features at Äspö

The evolution of water-conducting features at Äspö has many common aspects with those described for Grimsel. Ductile structures (such as mylonitic shear-zones) have a focussing effect on the younger brittle structures. 31% of the water-conducting faults investigated by Mazurek *et al.* (1996) show evidence of ductile shear deformation in the adjacent wallrock, even though the volumetric proportion of mylonites is below 1 %. Differences to Grimsel include the presence of hydrothermal alteration and fracture sealing effects. Moreover, at least two stages of faulting can be distinguished. The older faulting event was followed by more or less complete cementation (epidote, quartz, chlorite, albite), such that the fault rocks are cemented cataclasites today. Subsequent stages of brittle deformation reopened these structures and produced fault breccias that are not fully cemented and so account for fracture permeability. Figure 3 illustrates the

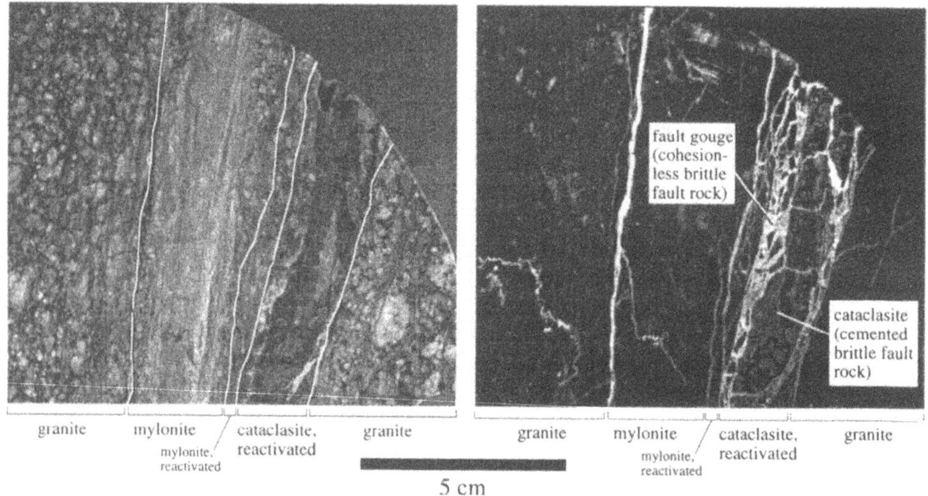

Figure 3. Small-scale fault architecture at Äspö. Left: Structural elements record recurrent activity (mylonite - cataclasite - fault gouge). Right: Core impregnated by fluorescent resin highlights present-day fluid pathways (UV light illumination).

4.2 BRITTLE DEFORMATION MECHANISMS AND WATER-CONDUCTING FEATURES

Brittle shear deformation (cataclasis, faulting)

At all sites investigated, faults are the most common type of water-conducting feature. In northern Switzerland, 43 % of all inflow points are related to faults and brittle shear-zones in granites and gneisses, and another 17% to aplite/pegmatite dykes affected by faulting (Mazurek 1998). Water-conducting features classified as faults include structures over a wide range of scales. Minor shear-zones consist of one single horizon containing cataclasite (*e.g.* a cemented horizon of fault rock a few cm thick), surrounded by a weakly developed damage zone. The other extreme are cataclastic zones related to major faults of several meters in thickness and disintegrated core material. What is common to cataclastic zones on all scales is the reactivation by fractures that follow cemented cataclasites or protocataclastic networks, and the common occurrence of vugs due to incomplete cementation or later dissolution of cataclastic matrices. At Äspö, major discharges into the tunnel occur almost exclusively from faults (Mazurek *et al.* 1996).

Similarly, on a regional scale of several km^2 and more, groundwater flow occurs mainly in major fault zones (Rhén *et al.* 1997a, Smellie *et al.* 1995).

Fracturing
Fractures and fractured zones are collective terms to describe zones with an increased frequency of brittle discontinuities. At least a part of the fractured zones are sets of joints (dilational structures), which are best seen in granitic rocks and are interpreted as tensile features that were generated during cooling or stages of regional tension.

Sets of hydraulically active fractures without shear deformation have been reported from the German KTB project (Durham 1997). In northern Switzerland, 32 % of all inflow points discharged from fractures and fractured zones in granites and gneisses, and another 6% from fractured but unfaulted aplites and pegmatites (Mazurek 1998). Fractures were generated in the course of different tectono-hydrothermal events, and younger fractures may reactivate or intersect older fracture generations. Vugs due to hydrothermal dissolution are often associated with water-conducting fractured zones.

At Äspö, faults are the dominant water-conducting features on dekametric to regional scales. In contrast, boreholes drilled into dekametric blocks delineated by large faults yielded fracture frequencies in excess of 1 m^{-1}, whereas only few structures clearly related to faulting were identified within such blocks. Fracture transmissivities are orders of magnitude lower than those of the faults. Thus at Äspö, networks of small fractures (with sizes mostly <1 m) are relevant water-conducting features on a small scale, whereas faults dominate the hydraulic properties in blocks with lengths of side larger than a few dekameters. The scale-dependence of water-conducting features and the importance of faults with increasing scale has also been observed by Caine *et al.* (1996).

4.3 EFFECT OF LITHOLOGY ON WATER-CONDUCTING FEATURES AND ON HYDRAULIC PARAMETERS

Granites and gneisses
In the crystalline basement of northern Switzerland, hydraulic conductivities of granitic rocks cannot be distinguished from those of gneisses. Depth below surface and degree of tectono-hydrothermal effects are more relevant factors (Thury *et al.* 1994, Voborny *et al.* 1994). In the adjacent Black Forest, Stober (1996) identified slightly higher conductivities in granites (log K [m/s] = -6.1 ± 0.8) when compared to gneisses (-6.9 ± 1.3), probably a consequence of minor differences in mineralogical compositions between these rocks types. Core logging and surface mapping showed that schistosity of gneisses (often weakly developed due to high-temperature recrystallization) is not overly relevant for the development of potentially transmissive structures and, at best, affects their orientations.

At Äspö, the three main lithologic units are all granitoid rocks. The average hydraulic conductivities have distinct values for each type of granite, in spite of substantial variability within each unit (Table 4). The leucogranite has the highest conductivities, whereas the more basic granodiorite has the lowest values. The systematic relationship between hydraulic conductivity and chemical/mineralogical composition reflects the tendency of more basic rocks to be more strongly affected by hydrothermal reactions and therefore fracture sealing. The more basic granitoids contain higher proportions of calcic plagioclase and biotite, which are commonly unstable under hydrothermal conditions and react to clays or other reaction products that may precipitate

TABLE 4. Geometric and hydraulic characteristics of lithological units at Äspö. K values from Rhén *et al.* (1997b)

rock unit (local name)	lithology	arithmetic mean of log K (m/s) ± 1σ	detailed-scale fracture frequency, m^{-1}
Äspö diorite	granodiorite	-9.9 ± 1.7	2.5
Småland granite	granite	-9.2 ± 1.7	no data
Fine-grained granite	leucogranite	-8.6 ± 2.0	6.5

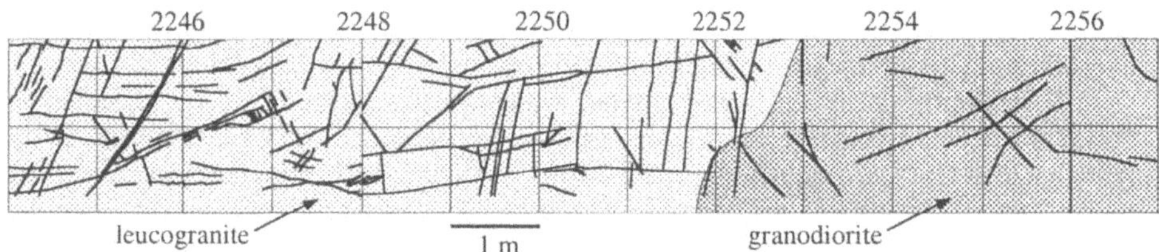

Figure 4. Fracture map (vertical tunnel wall) at Äspö highlighting enhanced fracture frequency in leucogranite when compared to granodiorite.

in the fractures. Moreover, leucocratic rocks tend to be more densely fractured than other rock types (Figure 4). Fracture frequencies in the leucogranite are 6.5 m^{-1}, compared to 2.5 m^{-1} in the other granitoids.

Leucocratic dykes: Aplites and pegmatites

Fracture density in aplites and pegmatites generally is much higher than in the country rocks (Daneck 1994). Whereas some fractures can be attributed to the cooling of the intruded magma, the more relevant characteristic is the very brittle behaviour during low-temperature deformation. In northern Switzerland, aplites and pegmatites focus faulting and fracturing and contain 23 % of all inflow points (compared to their volumetric proportion of only *ca.* 3.5 %). Fracture sealing by locally produced alteration products occcurs to a much lesser degree in leucocratic dykes when compared to all other rock types, and fractures tend to stay open over long periods of time. A clear dependence exists between dyke thickness and inflow points (Figure 5). About 10% of all dykes less than 1 m thick correlate with inflow points, whereas more than half the dykes thicker than 3 m contain inflow points.

The mechanical and hydraulic behaviour of aplites in northern Switzerland is analogous to that of the leucogranite at Äspö. At both sites, these leucocratic rocks represent the late stages of intrusive sequences, and the main difference is the depth at which intrusion occurred (Brisbin 1986). In northern Switzerland, dykes intruded at shallow levels into country rocks with brittle behaviour, resulting in regular shapes of the dykes. At Äspö, intrusion occurred at deep crustal levels into still plastic country rocks, resulting in irregular-shaped intrusive bodies with complex contacts and sizes in the range of meters to dekameters.

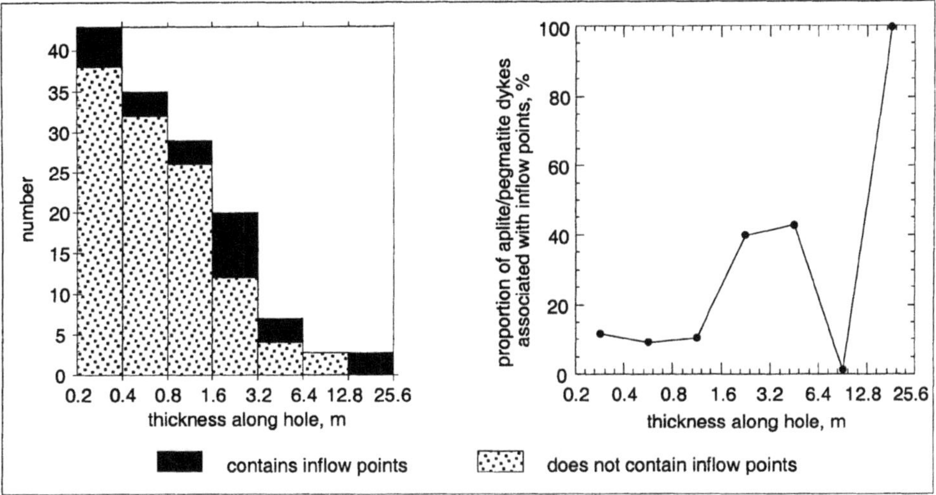

Figure 5. Thickness distribution of aplite/pegmatite dykes in boreholes of northern Switzerland and relationship to inflow points. From Mazurek (1998).

Basic dykes: Lamprophyres

In the Grimsel area, lamprophyre dykes focus both ductile deformation and brittle reactivation along the dyke contacts. In the Test Site, one of the few major water inflow points occurs at a lamprophyre contact.

In northern Switzerland, lamprophyres are very common features (54 dykes, total thickness of 100 m over a cored profile length in 6 boreholes of 5800 m), but none of them can be clearly correlated with inflow points. This is due to the strong effects of hydrothermal alteration in northern Switzerland, whereas alteration is largely absent at Grimsel. Lamprophyres contain basic minerals (pyroxene, amphibole, biotite) and calcic plagioclase that are unstable in most hydrothermal environments, such that abundant alteration products, mostly micas, chlorite and clay minerals, are produced and effectively seal available fracture openings.

Conclusions

Different rock types show contrasting behaviour in response to brittle deformation and alteration. Mineralogical composition and degree of alteration are factors of highest influence on hydraulic conductivity, whereas textural features (*e.g.* granites *vs* gneisses) play at best a second-order role. Rocks dominated by quartz and alkali feldspars (such as aplites, pegmatites, leucogranites) are more densely fractured than rocks containing abundant micas, and they often focus faulting. They are geochemically stable under most hydrothermal conditions, such that water/rock interactions and the sealing capacity are weak, resulting in conductivities higher than in other rock types. These findings are consistent with laboratory measurements of matrix permeabilities of samples from the Kola and KTB boreholes, where Morrow *et al.* (1994) report relative permeabilities of granodiorite > amphibolite > basalt.

Basic dykes, such as lamprophyres, are weakly fractured and may act as water flow paths only in systems with very limited water/rock interaction throughout geological evolution (such as Grimsel), whereas they are irrelevant for flow in environments that underwent extensive hydrothermal alteration (such as northern Switzerland).

4.4 DEPTH DEPENDENCE OF WATER-CONDUCTING FEATURES

In northern Switzerland, the geological characteristics of water-conducting features do not vary systematically with depth (observation interval: 200 - 2500 m below surface). The degree and nature of brittle deformation and the distribution of dykes are not systematic functions of depth. The only geological trend is the dominance of greenschist-grade alteration at depth (deeper than 800 - 1000 m below surface), whereas low-temperature alterations are more prominent at shallower levels.

Some degree of systematic distribution was recognized in the depth dependence of hydraulic conductivity. In some boreholes, conductivity substantially decreases with depth (*e.g.* Böttstein), whereas no depth dependence was observed in others (*e.g.* Kaisten, see Figure 6). All shallow occurrences of crystalline basement have higher permeability, while deeper parts of the basement contain higher-permeability sections in some boreholes and low-permeability sections in others. The overall trend of decreasing hydraulic conductivity with depth is in agreement with data obtained from several deep drilling programmes worldwide, consistent with a closure of pore apertures with increasing pressure. However, prominent exceptions to this trend exist, and the conductivity distribution at each specific site needs to be explored separately. In the German super-deep borehole (KTB), the distribution of hydraulic conductivities is discontinuous in the lower part of the drilled profile, with a zone of enhanced conductivity at 9.1 km below surface (Huenges *et al.* 1997).

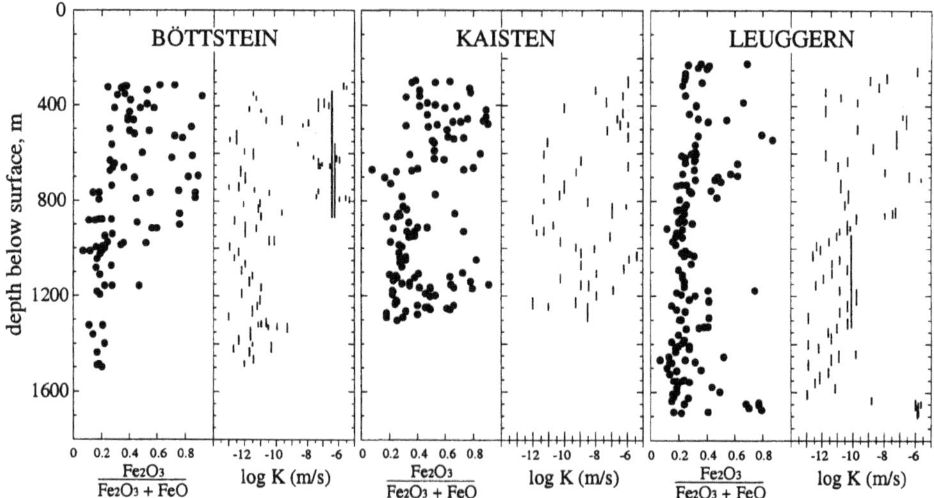

Figure 6. Degree of hydrothermal alteration (expressed by the degree of iron oxidation in the rock) and hydraulic conductivity in boreholes of northern Switzerland. Data from Mazurek (1998) and Küpfer *et al.* (1989).

In northern Switzerland, sections with higher permeability are strongly affected by brittle deformation and alteration. In contrast, tectono-hydrothermal effects are much less developed in sections with low permeability, and this is also reflected by the average frequency of inflow points (0.02 m^{-1} in low-permeability sections, 0.026 m^{-1} in higher-permeability sections). Figure 6 compares hydraulic conductivities with the weight ratio $\frac{Fe_2O_3}{Fe_2O_3 + FeO}$ of whole-rock samples. This ratio is a measure of the degree of hydrothermal alteration (and therefore of fracturing that created the flowpaths for the hydrothermal fluids) because all alteration phases were associated with oxidation of the wallrocks. There is an excellent correlation between low conductivities and low degrees of oxidation in the deeper parts of the Böttstein and Leuggern boreholes. Higher conductivities near the bottom-hole in Leuggern are also reflected by a high degree of iron oxidation. The conclusion is that zones with a high degree of fracturing and substantial hydrothermal oxidation have been relevant flowpaths in the past and still are at present. Hydrothermal alteration was intense but did not thoroughly seal the system.

4.5 HYDROCHEMICAL CONSTRAINTS ON FLOW THROUGH WATER-CONDUCTING FEATURES

Chemical and isotopic compositions of groundwaters provide information on recharge areas, average residence times and water/rock interactions along the flowpath. If sufficient data are available, the provenance and residence times of individual groundwater components can be derived, together with constraints on flow directions, flow velocities and mixing processes. Results of such investigations can be used to constrain structural/hydraulic conceptual models and flow models (*e.g.* initial and boundary conditions, travel times).

In northern Switzerland, Pearson *et al.* (1991) and Michard *et al.* (1996) identified four chemical groups of groundwaters. The youngest group, containing tritium and characterized by very low salinity, reflects recent recharge. The eastern and western groups have recharged during glacial and intraglacial periods, whereas the fourth, saline group evolved entirely within the crystalline rocks. The spatial arrangement and residence times of the groundwaters indicate that recharge took place in the Black Forest in southern Germany. Subsequent groundwater flow was to the SE, then turning SW and mixing with the saline waters. Chemical and isotopic data of the saline waters encountered in the deep parts of some of the boreholes indicate that the salinity is derived locally, *i.e.* by water/rock interaction within the crystalline basement. In deep groundwaters from the adjacent Black Forest, Stober and Bucher (1999) identified a saline, most likely marine component. These findings indicate that some of the deep groundwaters in the region are virtually stagnant, even over geological timescales.

At Äspö, Smellie *et al.* (1995) used structural and hydrochemical information from boreholes to derive a groundwater flow pattern. Chemical compositions of borehole-derived groundwater samples are consistent with mixing of surface-derived components (fresh/brackish water, seawater) and a deep saline groundwater component in different proportions. Combining the knowledge of the subsurface fault network at Äspö with limited hydraulic information (basically the distinction between recharging and discharging faults) yields a rough subsurface flow pattern. This conceptual model was substantially refined by integrating the chemical compositions of groundwaters. The relative proportions of the surface-derived and deep components in each groundwater sample were calculated, and the spatial distribution of mixing ratios was used to constrain the flow directions and the penetration depth of surface-derived waters. It was shown that in spite of the modest surface topography, the penetration depth of surface-derived waters is in excess of 500 m along some of the steeply dipping faults.

5. Evolution of water-conducting features and paleo-hydrogeology: Tectonics and hydrothermal activity in northern Switzerland

The presence of overpressured fluids can initiate brittle deformation by reducing the friction along discontinuities (Hubbert and Rubie 1959). On the other hand, fluid flow is often a consequence of brittle deformation due to the opening of new flowpaths and due to stress redistribution (*e.g.* Sibson 1975, Muir-Wood and King 1993). Coupled events of brittle deformation and fluid flow may result in hydrothermal or low-temperature alteration, and mineral dissolution/precipitation leads to enhancement, reduction or spatial redistribution of permeability. Combined structural, mineralogical and geochemical evidence can be used to unravel the paleo-hydrogeology of regions that experienced episodic deformation/fluid flow/alteration cycles. Recurrent stages of faulting, each stage linked with fluid circulation and distinct types of hydrothermal alteration, are illustrated taking the late- and post-Variscan evolution of the crystalline basement of northern Switzerland as an example. This case study also highlights the relationships between fracturing, chemical reaction and fluid flow over time.

5.1 GREENSCHIST-GRADE DEFORMATION AND ALTERATION

The first phase of regional brittle deformation post-dating the Variscan continental collision in northern Switzerland included faulting (cataclastic deformation) and fracturing on all scales (Meyer 1987). The fault rocks originally produced were fault breccias that acted as flow conduits, thus triggering fluid circulation and chemical reaction. Syn-genetic hydrothermal alteration is localized along faults and fractures and occurred at temperatures of 300 - 400 °C (Mazurek 1998). It included mainly the transformation of plagioclase to sericitic muscovite + albite and of biotite to chlorite + sericitic muscovite (Peters 1987a).

Deformation and alteration were linked to the shallow intrusion of Late Variscan granites (Diebold *et al.* 1991). The age of this event is constrained to the interval 290 - 320 Ma, *i.e.* to the Late Carboniferous (Mazurek 1998). This period was characterized by rapid basement uplift, erosion and tectonic unroofing at very high geothermal gradients. Maximum pressures of 300 - 700 bar derived from fluid inclusion studies correspond to depths of 3 - 7 km below surface (assuming hydrostatic conditions), which results in high geothermal gradients of 50 - 130 °C/km. Based on vitrinite reflectance data in the sedimentary rocks overlying the crystalline basement, Kempter (1987) calculated gradients around 100 °C/km for the Late Carboniferous.

The greenschist-grade tectono-hydrothermal event correlates with low-salinity Na-Cl waters identified in fluid inclusions. Within this group of fluid inclusions, a temporal evolution from high to low homogenization temperatures (400 → 140 °C) and decreasing salinities has been observed (Mullis 1987, Mullis and Stalder 1987). These trends are interpreted in terms of a progressive dilutilon of the formation waters by infiltrating meteoric waters. Meteoric signatures at depth are also recognized by the study of stable isotopes in minerals affected or produced by hydrothermal fluids (Mazurek 1992, Simon 1990, Simon and Hoefs 1987). The increase of the ratio

$\frac{Fe_2O_3}{Fe_2O_3 + FeO}$ in altered rock samples (*e.g.* due to biotite dissolution and hematite precipitation) also highlights the penetration of oxidizing, surface-derived waters to depths of several km, a process that requires a high permeability. Given the existence of very high geothermal gradients, thermal convection appears to be the most likely mechanism for the transport of surface-derived fluids into the crystalline basement.

Alteration of the wallrock penetrated centimeters to meters into the rock matrix, and the fault breccias were cemented by quartz and sericite to form cataclasites. By this process, the vast majority of all faults and fractures were fully healed (Figure 7). The plumbing system decayed, even though the driving force for flow (high geothermal gradients) outlasted this tectono-hydrothermal stage. A minority of all fractures were not fully sealed, leaving mineralized open channels. Provided they made part of a connected network of flowpaths, these channels have contributed to permeability throughout subsequent geological evolution. The effects of hydrothermal alteration on the hydraulic regime included mainly a reduction of permeability created by brittle deformation (see also Olsen and Scholz 1998) and the generation of channels, *i.e.* a patterning process resulting in hydraulic heterogeneity on a small scale.

5.2 LOW-TEMPERATURE DEFORMATION AND ALTERATION

Low-temperature faulting, fracturing and associated argillic alteration dominate by quantity over all other post-magmatic/post-metamorphic events and represent a major stage of regional crustal deformation (Meyer 1987). According to stratigraphic evidence

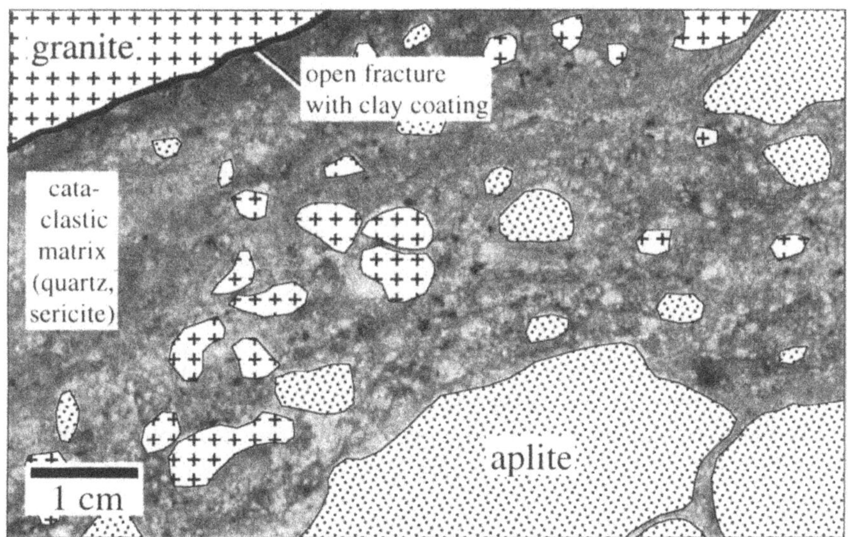

Figure 7. Contact between an aplite dyke and granite at 1502.2 m below surface in the borehole at Siblingen (northern Switzerland). The contact has been affected by cataclasis during the greenschist-grade deformation/alteration stage, and the fault rocks have been completely sealed by quartz-sericite-rich cement. A later reactivation during the low-temperature stage produced open fractures (coated with clay minerals) that account for present-day transmissivity of $4*10^{-7}$ m^2/s.

and K/Ar dates of illite, the age of this stage is constrained to 260 - 280 Ma (Early Permian). This time period is characterized by strong subsidence localized in deep Permo-Carboniferous troughs and is documented by thick piles of clastic sediments in northern Switzerland as well as in other troughs of Central Europe. The low-temperature hydrothermal phase and deformation are genetically linked to these crustal movements in a transpressive/transtensive tectonic framework (Arthaud and Matte 1977, Diebold *et al.* 1991).

Low-temperature alteration includes mainly the alteration of plagioclase to clay minerals and of biotite to chlorite at temperatures of 100 - 140 °C (evidence from fluid inclusions, Mullis 1987, Mullis and Stalder 1987). The heat source of this hydrothermal phase probably correlates with rhyolitic volcanism known in the Black Forest (Peters 1987b). Kempter (1987) postulates geothermal gradients of about 100°C/km for the Early Permian. Similarly to the greenschist-grade phase, low-temperature alteration occurred in a regime of thermally driven fluid circulation, consistent with meteoric signatures in alteration products (Mazurek 1992).

Cataclastic deformation was invariably accompanied by fracturing and jointing that typically reactivated pre-existing mechanical discontinuities, such as zones deformed (and sealed) in the high-temperature phase or dyke/wallrock contacts (Figure 7). Low-temperature cataclasis in itself was a multiphase, coupled process of deformation and hydrothermal activity, as demonstrated by the presence of cemented cataclasites as components in cataclastic zones. Late movements within this stage produced cohesionless fault rocks which, unlike most older deformation features, have not been healed by subsequent hydrothermal cementation and therefore still consist of crumbly rock fragments and unconsolidated fault breccias and gouges.

5.3 KAOLINITIC ALTERATION AND VUG FORMATION

A kaolinitic alteration and the generation of vugs/channels with mineralizations are the youngest rock/water interactions identified in the rocks, and these processes may still be continuing. Associated brittle deformation is weak and includes the reactivation of existing structures as joints. Alteration of the wallrock affects mainly plagioclase that is replaced by kaolinite, smectite and minor chlorite.

Open channels on a scale of millimeters to centimeters were generated in pre-existing structures, preferentially in fine-grained cataclastic matrices, by partial dissolution of the very fine-grained and porous gouge materials. Conspicuous, idiomorphic crystals of calcite, fluorite, baryte, siderite, quartz, celestite and minor ore minerals were deposited in the vugs. This dissolution/precipitation process enhanced the already existing heterogeneity of fracture apertures within any single structure. The open channels are of prime importance for the present-day hydrodynamics of the crystalline basement. Similar conclusions have also been reported from the geothermal borehole at Soultz-sous-Fôrets in the Rhine Graben by Komninou and Yardley (1997).

5.4 IMPLICATIONS FOR HYDRAULICS

The geometry and the hydraulic characteristics of water-conducting features in basement rocks evolve over geological timescales. Events of faulting and fracturing create permeability and, provided hydraulic gradients exist, enhance fluid circulation through the rocks. In regions with high geothermal gradients, convection cells may be activated. In such systems, events of fracturing, thermally driven flow and hydrothermal water/rock interaction are genetically linked and may occur in recurrent tectono-hydrothermal stages. Several phases of fracturing and related hydrothermalism are distinguished in northern Switzerland. Hydrothermal processes that follow brittle deformation have the following effects on the hydraulic properties of the rocks:

- Hydrothermal activity may result in cementation of fault rocks and in sealing of fractures. This leads to a reduction of permeability, and this process counteracts the enhancement of permeability by fracturing.
- On the other hand, dissolution of fracture infills and adjacent wallrocks may occur and results in permeability enhancement. Whether cementation or dissolution dominates is a function of the respective reaction volumes, fluid compositions, reaction rates and fluxes.
- Combined precipitation/dissolution reactions result in a redistribution of fracture aperture and permeability within a fracture, *i.e.* in heterogeneous (channel) flow.
- Alteration affects the mineralogy and matrix porosity of wallrock domains and thereby the extent to which solutes interact with the rock matrix (*e.g.* diffusion into the microporous matrix, sorption on mineral surfaces).

The existence of hydraulic heterogeneity within faults and fractures due to water/rock interactions or other processes has a bearing on the interpretation of hydraulic measurements in boreholes. A heterogeneous water-conducting feature will only be identified by hydraulic methods if the borehole penetrates a channel within the fracture, whereas no hydraulic response will be recorded if the borehole penetrates a cemented fracture segment. In order to evaluate the total number of (internally heterogeneous) water-conducting features in a borehole, additional information from the core materials is required. The proportion of the fracture area occupied by channels can be approximated by the ratio of inflow points to the total number of fractures identified by core logging.

6. Flow through faults

The hydraulic role of faults can be either as a conductor or as a seal (*e.g.* Forster and Evans 1991, Caine *et al.* 1996). Sandstone-hosted hydrocarbon compartments have been observed to have contrasting fluid pressures on either side of a fault separating them, indicative of the long-term sealing properties of such faults, most frequently caused by clay smears. It has also been reported that fault transmissivity can be anisotropic, with minimum values normal to the fault surface (Forster and Evans 1991). At all three sites investigated in this paper, faults are the structural elements that dominate the flow properties of the formation.

6.1 FAULT ARCHITECTURE, HETEROGENEITY AND SIGNIFICANCE FOR FLOW

Several authors investigated the anatomy of faults and provided mechanistic schemes for the evolution of faults in crystalline rocks (*e.g.* Segall and Pollard 1980, Sibson 1987, Martel and Pollard 1989). Increasing displacement along a fault results in the growth of the fault-surface area. Progressive shear leads to segment linkage, *i.e.* to the linking of adjacent faults into larger fault zones (*e.g.* Cartwright *et al.*1996, Cowie and Scholz 1992). The linkage is achieved in fault steps via parallel sets of dilational structures (splay cracks, dilational jogs; *cf.* Martel and Pollard 1989).

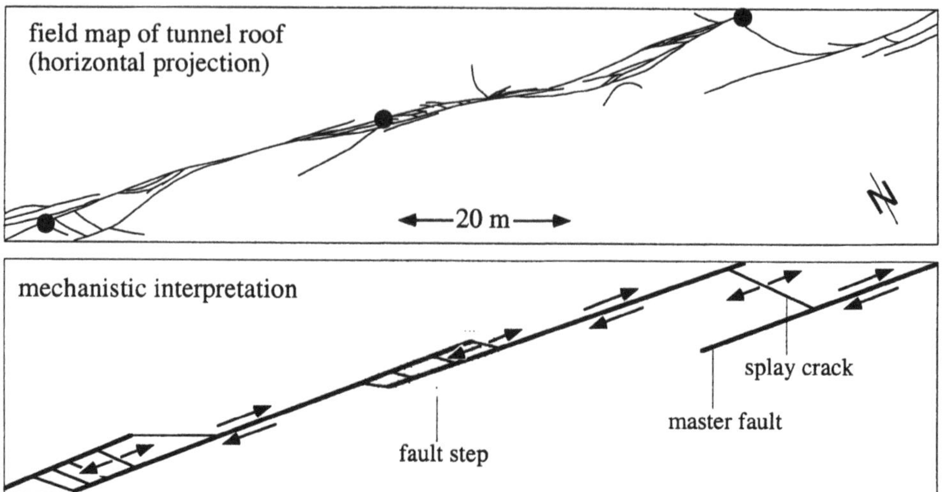

Figure 8. Trace map of a fault from the main tunnel at Äspö (adapted from Mazurek *et al.* 1996). Circles indicate main water inflows into the tunnel. A mechanistic interpretation of the observed fault architecture is given in the bottom part.

Figure 8 shows the trace map of a fault in the Äspö tunnel, together with the mechanistic interpretation. The structure is a CC type, *i.e.* characterized by clockwise shear-sense and clockwise arrangement of fault steps. A set of master faults and fault steps with connecting splay cracks can be recognized. The fault architecture is heterogeneous along strike, with segments consisting of one single master fault and other segments containing complex networks of master faults and splay cracks (fault steps). Whereas the entire structure is wet under local tunnel climate conditions, the three major discharges are all located within fault steps. In spite of the artificial redistribution of flow in the surroundings of tunnels (see above), such localization of flow is hardly a coincidence and indicates that structural heterogeneity of faults also has consequences for the distribution of flow. Shear deformation occurred along master faults, and these commonly contain fault gouges. In contrast, dilation occurred in splay cracks, and so fracture porosity was created in these structures. This, together with the lack of (potentially sealing) gouge materials, renders splay cracks relevant local flowpaths, at least as long as the stress regime that produced the fault system prevails and sealing by hydrothermal alteration products is not overly significant. On a larger scale, fault steps are essentially one-dimensional potential conduits, and their orientation is parallel to the intermediate regional stress axis. This means that in systems dominated by normal or thrust faulting, conduits are horizontal, whereas vertical fault steps occur in strike-slip fault systems (*e.g.* in the example shown in Figure 8).

In addition to heterogeneity along strike, faults are also heterogeneous in cross-section. They typically consist of networks of shear-planes (often containing gouge) that are embedded in a fault damage zone devoid of fault rocks (*e.g.* Caine *et al.* 1996). Gouge materials or hydrothermal effects may lead to an efficient sealing of the central parts of faults, and most of the fault transmissivity is accounted for by the damage zone.

6.2 SMALL-SCALE HETEROGENEITY

In addition to the heterogeneity related to fault architecture, small-scale heterogeneity occurs within single fractures that make part of the fault and is mostly related to hydrothermal effects. The action of hydrothermal activity results in an enhancement of the fracture-aperture spectrum, and the hydraulic consequence is, in the extreme case, the development of one-dimensional flow conduits within a planar fracture surface.

In northern Switzerland, hydrothermal water/rock interactions were very intense, and the small-scale fracture-aperture distributions are very heterogeneous. Only a fraction of the fracture area is unsealed and so has non-zero hydraulic aperture. At Grimsel, the scarcity of alteration results in much more homogeneous aperture and transmissivity distribution. Fracture apertures in both environments are illustrated and compared in Figure 9.

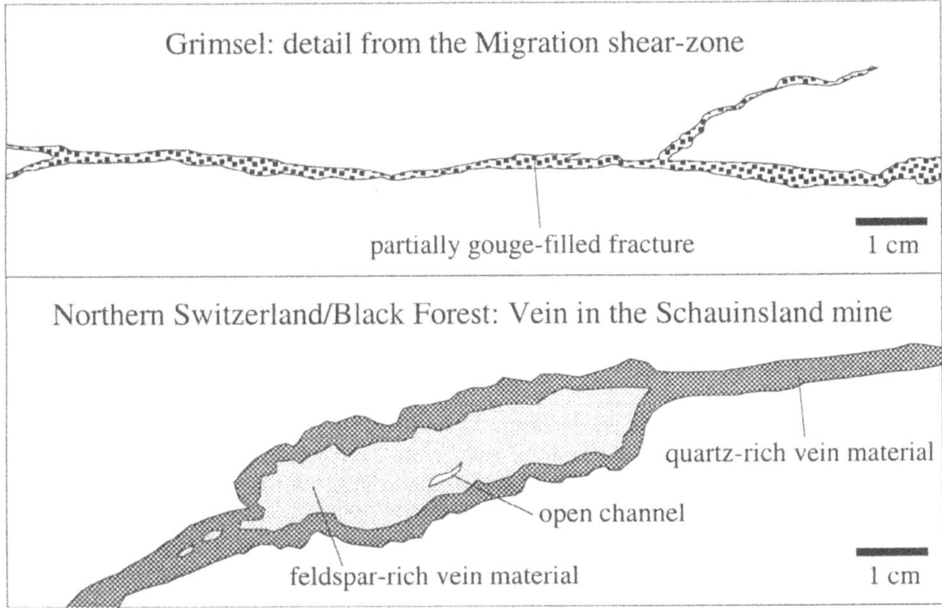

Figure 9. Fracture apertures in crystalline-rock environments affected by different degrees of hydrothermal water/rock interaction. Top: Fracture-aperture distribution in the Migration shear-zone at the Grimsel Test Site, unaffected by hydrothermal activity (adapted from Bossart and Mazurek 1991). Bottom: Quartz-feldspar-rich hydrothermal vein within a fault in the Schauinsland mine of the southern Black Forest (adjacent to northern Switzerland) with a localized open channel.

7. Summary and conclusions

Water-conducting features are hydraulic heterogeneities, and in crystalline rocks they are always related to brittle structures. Their identification in tunnels is complicated by the possible existence of hydraulic skins that result in redistribution of flow close to the tunnel walls. Point discharges are often observed but do not necessarily reflect natural conditions, and flow through the undisturbed rock body could be less strongly channeled.

In boreholes, inflow points generally occur only from a fraction of all faults and fractures that are present in the corresponding drillcore materials. This is explained by the existence of different fracture generations (each with different structural/hydraulic characteristics) and by spatial heterogeneity within each fracture.

In spite of the different ages and contrasting geological evolutions, water-conducting features have several common properties at all sites discussed. Faults are important conduits for water flow on regional and site scales. Water-conducting single fractures and fracture sets without a shear component also occur, but, at least in the case of Äspö, their hydraulic relevance is limited to small scales (metric to dekametric block sizes). The scale-dependence of water-conducting features and the growing importance of faults with increasing scale has also been observed by Caine *et al.* (1996, Fig. 3). The architecture of faults is heterogeneous along strike (master faults *vs* splay cracks) as well as in cross-section (fault core *vs* damage zone).

Faults and fractures preferentially develop along pre-existing discontinuities, such as dykes, lithologic contacts and ductile shear-zones. Lithologic control of the structure and transmissivity of water-conducting features was observed at Äspö and in northern Switzerland. Quartz-feldspar-rich rocks, such as aplite/pegmatites or leucogranites, are more densely fractured and more transmissive than basic rocks. Stages of brittle deformation and hydrothermal alteration are genetically linked and often recurrent. The hydraulic effects of hydrothermal alteration include a redistribution of flow channels within faults and fractures, thereby enhancing hydraulic heterogeneity.

Hydrochemical evidence indicates that regional flow through crystalline rocks and mixing of groundwater components of different origins occur. If hydraulic gradients are negligible (namely at great depth), stagnant, saline waters may reside in the formation over geologic timescales.

Acknowledgements

The author thanks Nagra (Swiss National Cooperative for the Disposal of Radioactive Waste) and SKB (Swedish Nuclear Fuel and Waste Management Co.) for providing all information available from their site characterization programmes and rock laboratories. Discussions and informal exchange of information with Tj. Peters (Uni. Bern), U. Frick, A. Gautschi, S. Vomvoris (all Nagra), O. Olsson, P. Wikberg (both SKB) and P. Bossart (Geotechnical Institute Ltd.) formed a basis for this paper and are greatfully acknowledged. Careful reviews of the manuscript were provided by U. Mäder, A. Matter Tj. Peters, H. N. Waber (all Uni. Bern), K. Bucher and I. Stober (both Freiburg).

References

Arthaud, F. and Matte, P. (1977) Late Paleozoic strike-slip faulting in southern Europe and northern Africa: Result of a right-lateral shear zone between the Appalachians and the Urals, *Bull. geol. Soc. Amer.* **88**, 1305-1320.

Barton, C. A., Zoback, M. D. and Moos, D. (1995) Fluid flow along potentially active faults in crystalline rock, *Geology* **23**, 683-686.

Bossart, P. and Mazurek, M. (1991) Structural geology and water flow-paths in the Migration shear-zone, Nagra Technical Report NTB 91-12, Nagra, Wettingen, Switzerland.

Bradbury, M. H., ed. (1989) Grimsel Test Site: Laboratory investigations in support of the migration experiments, Nagra Technical Report NTB 88-23, Nagra, Wettingen, Switzerland.

Brisbin, W. C. (1986) Mechanics of pegmatite intrusion, *Amer. Mineralogist* **71**, 644-651.

Caine, J. S., Evans, J. P. and Forster, C. B. (1996) Fault zone architecture and permeability structure, *Geology* **24**, 1025-1028.

Cartwright, J. A., Mansfield, C. S. and Trudgill, B. D. (1996) The growth of normal faults by segment linkage, in P. G. Buchanan and D. A. Nieuwland (eds.) *Modern development in structural interpretation, validation and modelling*, Geological Society Special Publication 99, London, pp. 163-177.

Choukroune, P. and Gapais, D. (1983) Strain pattern in the Aar granite (Central Alps): Orthogneiss developed by bulk inhomogeneous flattening, *J. Struct. Geol.* **5**, 411-418.

Cowie, P. A. and Scholz, C. H. (1992) Growth of faults by accumulation of seismic slip, *J. Geophys. Res.* **97**, 11085-11096.

Daneck, T. (1994) Platznahme und mechanisches Verhalten von Ganggesteinen im Grundgebirge des Südschwarzwaldes, Mitt. ETH Zürich, NF 296.

Diebold, P., Naef, H. and Ammann, M. (1991) Zur Tektonik der zentralen Nordschweiz, Nagra Technischer Bericht NTB 90-04, Nagra, Wettingen, Switzerland.

Durham, W. B. (1997) Laboratory observations of the hydraulic behavior of a permeable fracture from 3800 m depth in the KTB pilot hole, *J. Geophys. Res.* **B102**, 18405-18416.

Forster, C. B. and Evans, J. P. (1991) Hydrogeology of thrust faults and crystalline thrust sheets: results of combined field and modelling studies, *Geophys. Res. Lett.* **18**, 979-982.

Frick, U., Alexander, W. R., Baeyens, B., Bossart, P., Bradbury, M. H., Bühler, C., Eikenberg, J., Fierz, T., Heer, W., Hoehn, E., McKinley, I. G. and Smith, P. A. (1992) Grimsel Test Site - The radionuclide migration experiment - overview of investigations 1985 - 1990, Nagra Technical Report NTB 91-04, Nagra, Wettingen, Switzerland.

Frieg, B. and Vomvoris, S. (1994) Grimsel Test Site - Investigation of hydraulic parameters in the saturated and unsaturated zone of the Ventilation Drift, Nagra Technical Report NTB 93-10, Nagra, Wettingen, Switzerland.

Gascoyne, M. and Thomas, D. A. (1997) Impact of blasting on groundwater composition in a fracture in Canada's Underground Research Laboratory, *J. Geophys. Res.* **B102**, 573-584.

Heer, W. and Smith, P. A. (1998) Modelling the radionuclide migration experiment at Grimsel. What have we learned ? *Mat. Res. Soc. Symp. Proc.* **506**, 663-670.

Hubbert, M. K. and Rubie, W. W. (1959) Role of fluid pressure in mechanics of overthrust faulting, *Geol. Soc. Amer. Bull.* **70**, 115-205.

Huenges, E., Erzinger, J. and Kück, J. (1997) The permeable crust: Geohydraulic properties down to 9101 m depth, *J. Geophys. Res.* **B102**, 18255-18265.

Kempter, E. H. K. (1987) Fossile Maturität, Paläothermogradienten und Schichtlücken in der Bohrung Weiach im Lichte von Modellberechnungen der thermischen Maturität, *Eclogae geol. Helv.* **80**, 543-552.

Komninou, A. and Yardley, B. W. D. (1997) Fluid-rock interactions in the Rhine Graben: A thermodynamic model of the hydrothermal alteration observed in deep drilling, *Geochim. Cosmochim. Acta* **61**, 515-531.

Kralik, M, Clauer, N., Holnsteiner, R., Huemer, H. and Kappel, F (1992) Recurrent fault activity in the Grimsel Test Site (GTS, Switzerland): revealed by Rb-Sr, K-Ar and tritium isotope techniques, *J. Geol. Soc. London* **149**, 293-301.

Küpfer, T., Hufschmied, P. and Passquier, F. (1989) Hydraulische Tests in Tiefbohrungen der Nagra, *Nagra informiert* **11**, 7-23.

Martel, S. J. and Peterson, J. E. Jr. (1991) Interdisciplinary characterization of fracture systems at the US/BK site, Grimsel laboratory, Switzerland, *Int. J. Rock Mech. Min. Sci. Geomech. Abstr.* **28/4**, 295-323.

Martel, S. J. and Pollard, D. D. (1989) Mechanics of Slip and Fracture Along Small Faults and Simple Strike-Slip Fault Zones in Granitic Rock, *J. Geophys. Res.* **94**, 9417-9428.

Mazurek, M. (1992) Phase equilibria and oxygen isotopes in the evolution of metapelitic migmatites: a case study from the Pre-Alpine basement of northern Switzerland, *Contrib. Mineral. Petrol.* **109**, 494-510.

Mazurek, M. (1998) Geology of the crystalline basement of northern Switzerland and derivation of geological input data for safety assessment models, Nagra Technical Report NTB 93-12, Nagra, Wettingen, Switzerland.

Mazurek, M., Bossart, P. and Eliasson, T. (1996) Classification and characterization of water-conducting features at Äspö: Results of investigations on the outcrop scale, SKB International Cooperation Report ICR 97-01, SKB, Stockholm, Sweden.

Meyer, J. (1987) Die Kataklase im kristallinen Untergrund der Nordschweiz, *Eclogae geol. Helv.* **80**, 323-334.

Michard, G., Pearson, F. J. and Gautschi, A. (1996) Chemical evolution of waters during long term interaction with granitic rocks in northern Switzerland, *Appl. Geochem.* **11**, 757-774.

Morrow, C., Lockner, D. and Hickman, S. (1994) Effects of lithology and depth on the permeability of core samples from the Kola and KTB drill holes, *J. Geophys. Res.* **B99**, 7263-7274.

Muir-Wood, R. and King, G. C. P. (1993) Hydrothermal signatures of earthquake strain, *J. Geophys. Res.* **B98**, 22035-22068.

Munier, R. (1993) Segmentation, fragmentation and jostling of the Baltic Shield with time, Acta Universitatis Upsaliensis (Uppsala dissertations from the faculty of science) 37.

Munier, R. (1995) Studies of geological structures at Äspö. Comprehensive summary of results, SKB Progress Report PR 25-95-21, SKB, Stockholm, Sweden

Mullis, J. (1987) Fluideinschluss-Untersuchungen in den Nagra-Bohrungen der Nordschweiz, *Eclogae geol. Helv.* **80**, 553-568.

Mullis, J. and Stalder, H. A. (1987) Salt-poor and salt-rich fluid inclusions in quartz from two boreholes in northern Switzerland, *Chem. Geol.* **61**, 263-272.

National Research Council (1996) *Rock fractures and fluid flow,* National Academy Press, Washington D.C.

Neretnieks, I. (1980) Diffusion in the rock matrix: An important factor in radionuclide retardation ? *J. Geophys. Res.* **85**, 4379-4397.

Olsen, M. P. and Scholz, C. H. (1998) Healing and sealing of a simulated fault gouge under hydrothermal conditions: Implications for fault healing, *J. Geophys. Res.* **B103**, 7421-7430.

Olsson, O. (1992) Site characterization and validation - Final Report, Stripa Project Report 92-22, SKB, Stockholm, Sweden.

Olsson, O. and Gale, J. (1995) Site assessment and characterization for high-level nuclear waste disposal: results from the Stripa Project, Sweden, *Quarterly Journal of Engineering Geology* **28**, 17-30

Paillet, F. L. (1998) Flow modelling and permeability estimation using borehole flow logs in heterogeneous fractured formations, *Water Resources Res.* **34**, 997-1010.

Pearson, F. J., Balderer, W., Loosli, H. H., Lehmann, B. E., Matter, A., Peters, T., Schmassmann, H. and Gautschi, A. (1991) Applied isotope hydrogeology - A case study in northern Switzerland, Elsevier, Amsterdam and Nagra Technical Report NTB 88-01, Nagra, Wettingen, Switzerland.

Peters, Tj. (1987a) Hydrothermal alteration of a Variscan granite, magmatic autometasomatism and fault related vein metasomatism, in H. C. Helgeson (ed.) *Chemical transport in metasomatic processes,* Reidel, Dordrecht, pp. 577-590.

Peters, Tj. (1987b) Das Kristallin der Nordschweiz: Petrographie und hydrothermale Umwandlungen, *Eclogae geol. Helv.* **80**, 305-322.

Rhén, I., Bäckblom, G., Gustafson, G., Stanfors, R. and Wikberg, P. (1997a) Äspö HRL - Geoscientific evaluation 1997/2. Results from pre-investigations and detailed-scale site characterization. Summary report, SKB Technical Report 97-03, SKB, Stockholm, Sweden.

Rhén, I., Gustafson, G., Stanfors, R. and Wikberg, P. (1997b) Äspö HRL - Geoscientific evaluation 1997/5. Models based on site characterization 1986-1995, SKB Technical Report 97-06, SKB, Stockholm, Sweden.

Segall, P. and Pollard, D. D. (1980) Mechanics of discontinuous faults, *J. Geophys. Res.* **B85**, 4337-4350.

Sibson, R. H. (1975) Seismic pumping - a hydrothermal fluid transport mechanism, *J. Geol. Soc. London* **131**, 653-659.

Sibson, R. H. (1987) Earthquake rupturing as a mineralizing agent in hydrothermal systems, *Geology* **15**, 701-704.

Simon, K. (1990) Hydrothermal alteration of Variscan granites, southern Schwarzwald, Federal Republic of Germany, *Contrib. Mineral. Petrol.* **105**, 177-196.

Simon, K. and Hoefs, J. (1987) Effects of meteoric water interaction on Hercynian granites from the Südschwarzwald, southwest Germany, *Chem. Geol.* **61**, 253-261.

Smellie, J. A. T., Laaksoharju, M. and Wikberg, P. (1995) Äspö, SE Sweden: a natural groundwater flow model derived from hydrogeochemical observations, *J. Hydrol.* **172**, 147-169.

Stanfors, R., Erlström, M. and Markström, I. (1997) Äspö HRL - Geoscientific evaluation 1997/1. Overview of site characterization 1986-1995, SKB Technical Report 97-02, SKB, Stockholm, Sweden.

Stober, I. (1996) Hydrogeological investigations in crystalline rocks of the Black Forest, Germany, *Terra Nova* **8**, 255-258.

Stober, I. And Bucher, K. (1999) Deep groundwater in the crystalline basement of the Black Forest, *Applied Geochem.* **14**, 237-254.

Thury, M., Gautschi, A., Mazurek, M., Naef, H., Voborny, O., Vomvoris, S. and Wilson, W.E. (1994) Geology and hydrogeology of the crystalline basement of northern Switzerland, Nagra Technical Report NTB 93-01, Nagra, Wettingen, Switzerland.

Tilch, N. (1992) Strukturelle und hydraulische Aspekte von Kristallingesteinen des Grimselgebietes/Schweiz, Unpubl. diploma thesis, Univ. Clausthal, Germany.

Tsang, C. F., Hufschmied, P. and Hale, F. V. (1990) Determination of fracture inflow parameters with a borehole fluid conductivity logging method, *Water Resources Res.* **26**, 561-578.

Voborny, O., Resele, G., Hürlimann, W., Lanyon, W., Vomvoris, S. and Wilson, W. (1994) Hydrodynamic modelling of crystalline rocks, northern Switzerland, Nagra Technical Report NTB 92-04, Nagra, Wettingen, Switzerland.

Vomvoris, S. and Frieg, B. (1992) Grimsel Test Site - Overview of Nagra field and modelling activities in the Ventilation Drift (1988-1990), Nagra Technical Report NTB 91-34, Nagra, Wettingen, Switzerland.

Winberg, A. (1996) Descriptive structural-hydraulic models on block and detailed scales of the TRUE-1 site, SKB International Cooperation Report ICR 96-04, SKB, Stockholm, Sweden.

ALKALI FELDSPARS AS MICROTEXTURAL MARKERS OF FLUID FLOW

IAN PARSONS and MARTIN R. LEE
Department of Geology and Geophysics
The University of Edinburgh
West Mains Road
Edinburgh EH9 3JW
Scotland

ABSTRACT. Alkali feldspars provide an easily read microtextural record of fluid–rock interaction in the range from 450°C to diagenetic temperatures. The microtextures can provide unique insights into paths of fluid flow and mass transfer through crystalline rocks from the nanometre to the kilometre scale. The main thermodynamic driving force for the microtextural changes is elastic strain energy associated with coherency in strain-controlled microperthitic intergrowths and with tweed domain textures in orthoclase, both of which form at higher temperatures in cooling igneous and metamorphic rocks. Spontaneously strained feldspar dissolves in fluid films, reprecipitating as unstrained feldspar, a process which has been called 'unzipping'. This causes regular strain-controlled microperthites to coarsen to irregular patch and vein perthites, and orthoclase to recrystallize to tartan twinned microcline. Dissolution and reprecipitation around dislocation cores is an important part of these unzipping reactions, which lead to feldspars which are turbid, microporous and micropermeable. Crystal–fluid exchange reactions are driven by unzipping and the porous feldspars readily maintain alkali and isotopic exchange equilibrium with aqueous fluids down to ≤200°C. Intracrystal dissolution–reprecipitation is a process that has affected a high proportion of the alkali feldspar in the granitic upper crust of the Earth.

1. Introduction

Alkali feldspar, one of the most abundant minerals in the crust of the Earth, provides a unique microtextural record of water–rock interactions that have occurred from ~450°C to surface temperatures. This record can be read easily using electron microscopy, and interpreted in the light of established feldspar phase equilibria and phase behaviour (Brown and Parsons 1989). The reason for the exceptional reactivity of alkali feldspar is the inventory of defect microtextures that develop in all alkali feldspars of deep, high T origin during cooling. Exsolution of Na and K and ordering of Si and Al in the feldspar framework lead to the development of spontaneous intracrystal structural strains. Loss of this strain energy provides a thermodynamic driving force for dissolution–reprecipitation reactions in aqueous fluids, processes called 'unzipping' by Brown and Parsons (1993), which make the feldspar microporous and micropermeable,

I. Stober and K. Bucher (eds.), Hydrogeology of Crystalline Rocks, 27–50.

further enhancing its reactivity. Truly 'fresh' feldspar in plutonic felsic rocks is rare, and in most granitic rocks >20 vol%, and sometimes *all* the alkali feldspar, has been subject to dissolution and reprecipitation in aqueous fluids at T <450°C.

The microtextural changes are the underlying cause of the geochemical sensitivity of alkali feldspar to the passage of aqueous fluids. They are the reason why it is often the last mineral to close to ^{18}O–^{16}O exchange in multiphase assemblages (Giletti, 1985). In modern geothermal systems, aqueous fluids in granitic upper crust sampled directly in deep wells are usually in, or close to, equilibrium with feldspar assemblages. Redistribution of feldspar occurs in active hydrothermal systems with temperature gradients (Orville, 1963; Giggenbach, 1988) and in sedimentary basins (Saigal et al. 1988; Lee and Parsons, 1998). At T>200°C, feldspar–fluid equilibria are an important control of the composition of aqueous fluids in granitic crust, with departures from equilibrium becoming more marked, and kinetic controls more important, as surface T are approached.

2. Alkali feldspar phase equilibria

Because plagioclase and alkali feldspar together make up ~60% of the Earth's crust, the composition of fluids in the crust is strongly dependent on equilibria involving them. Recent summaries of feldspar phase equilibria are provided by Brown and Parsons (1989, 1994); we here define terminology and emphasise features relevant to low-T reactions. Much of the crust is composed of igneous granitic and granodioritic rocks, or feldspathic gneisses (simply called 'granites' below), containing two coexisting feldspar *phases* (plagioclase [PL] and alkali feldspar [AF]), so-called *subsolvus granites*. Granites in which only one feldspar phase grew from magma, *hypersolvus granites*, are more uncommon, usually forming small intrusions often associated with *syenites*, which are close to monomineralic alkali feldspar rocks. In subsolvus granites intercrystal equilibrium between feldspar pairs is attained in principle (but often not in practice) by reciprocal exchange through silicate liquid or aqueous fluid of the three main feldspar *components* $NaAlSi_3O_8$ (albite, Ab), $KAlSi_3O_8$ ('orthoclase', Or, called 'K-feldspar' below to distinguish the component from the special microtextural variety called orthoclase) and $CaAl_2Si_2O_8$ (anorthite, An). At equilibrium, the chemical potential of each of the three components must be same in both phases:

$$\mu_{Ab}^{PL} = \mu_{Ab}^{AF} \qquad \mu_{Or}^{PL} = \mu_{Or}^{AF} \qquad \mu_{An}^{PL} = \mu_{An}^{AF}$$

In P–T–X space this relationship defines the *ternary feldspar solvus* (Fig. 1), a dome-shaped surface which intersects the feldspar solidus at high T for some compositions. The intersection of the ternary solvus with the Ab–Or join gives the binary *alkali feldspar solvus*, and its intersection with the An–Or join causes almost complete immiscibility. Crystallization (either from melt or from aqueous fluids), for bulk compositions within the solvus, leads to the growth of two feldspars on a tie-line, the position of which depends on T and, to a smaller extent on P, forming the basis of the two-feldspar geothermometer (e.g. Fuhrman and Lindsley, 1988). The equilibrium

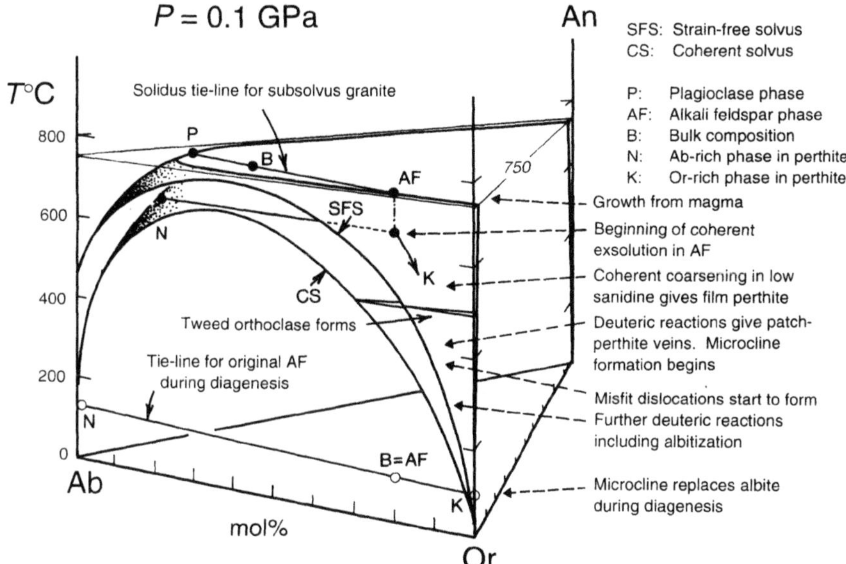

Figure 1. *T-X* prism for the system Ab–An–Or showing the ternary solvi. The SFS and CS are for equilibrium Si,Al ordering (Brown and Parsons 1989). The 750°C solidus isotherm is from Fuhrman and Lindsley (1988), for disordered feldspars. The labelled arrows at the right apply specifically to the Shap granite, as described in Section 5.2.

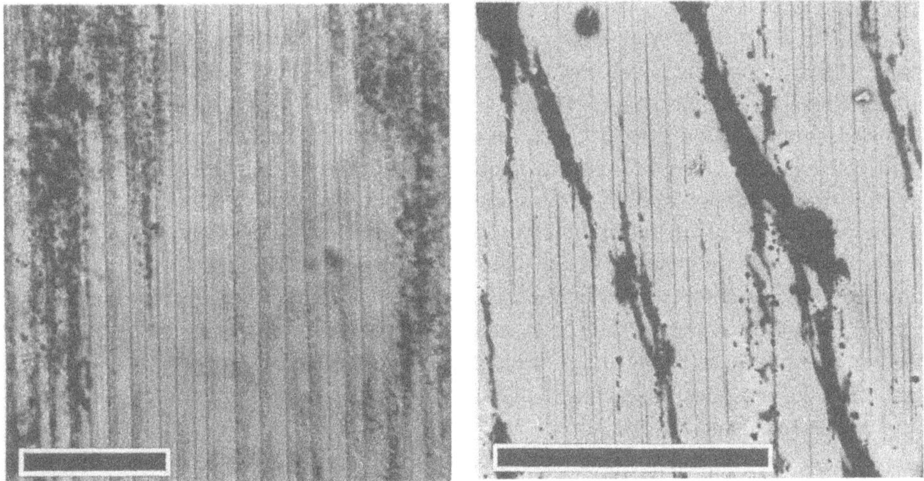

Figure 2. A. Ordinary light optical micrograph of an alkali feldspar from Shap. The clear areas are semicoherent lamellar film microperthites. In places the crystal is turbid, the turbidity partly following film lamellae. From Lee et al. (1995). B. BSE image, in an SEM, of similar feldspar. Albite appears dark, K-feldspar light grey. Ragged veins of patch perthite cut across areas of lamellar microperthite. Black dots are micropores. From Lee and Parsons (1997). Both scale bars 50 μm.

defined above involves only feldspar phases and components and is therefore independent of all other components in the system. Provided two feldspars occur in a system, their compositions are always defined, at equilibrium, by points on a solvus isotherm whatever the medium through which feldspar components are exchanged. In sediments or hydrothermal systems in which feldspar pairs may grow together or reach equilibrium via an aqueous phase, phases are always nearly pure end-member K-feldspar and albite, because of the low *T*. At the magmatic stage, most subsolvus granites crystallise an alkali feldspar solid solution (usually $Ab_{25-15}Or_{75-85}An_{<2}$) together with a plagioclase (usually oligoclase, $Ab_{90-70}Or_{<2}An_{10-30}$) which are in equilibrium with silicate liquid at $T \geq 700°C$. An important feature of the phase diagram is the large effect of the minor An component in alkali feldspar on its stability at low *T*. Although they are implicit in many models of fluid–rock interaction, and in two-feldspar geothermometry, phase equilibria of ternary feldspars at low *T* are poorly known. Si,Al order-disorder also affects the feldspar solvus, increasing order moving the solvus to higher *T*. Alkali feldspars may grow from magma with appreciable disorder, but the equilibrium forms will be fully ordered below ~450°C. For ordered alkali feldspars of the composition found in subsolvus granites, the solvus will be at ~100°C higher *T* than disordered. *P* also affects the solvus, with dT/dP ~200°C GPa^{-1}.

Cooling of a feldspar which grew on or above the solvus causes exsolution, leading to intergrowths known as *perthite* (Fig. 2). The compositions of the individual phases in a perthitic crystal are defined by the solvus surface but the bulk composition controls the proportions of the phases; in *perthite* sensu stricto, the matrix phase is Or-rich (as in subsolvus granites, Fig. 2), in *antiperthite* it is Ab-rich and in *mesoperthite* neither phase encloses the other (as in hypersolvus granites and syenites, Figs. 3, 6B). It is essential to state whether an analysis applies to the bulk composition of a crystal or to the composition of one or other of the intergrown phases. Perthites visible in an optical microscope (i.e. on the scale of >1 μm) are *microperthites*, sub-optical intergrowths are *cryptoperthites*. This distinction is important because it corresponds approximately with the upper size limit of intergrowths which are *coherent*, with a continuous Si,Al–O framework, and those which are discontinuous or *incoherent*. Modern microscopy has shown that most alkali feldspars from granites are both micro- and crypto-perthitic. Fully incoherent intergrowths are said to be 'strain-free', and compositions of the intergrown phases are defined by the *strain-free solvus* (Fig. 1). Coherent intergrowths involve spontaneous structural strain, and phase compositions are given by a *coherent solvus* which lies inside the strain-free solvus. The coherent or incoherent character of intergrowths provides an excellent marker of whether reactions between the feldspar and an aqueous fluid have occurred.

As well as exsolving, alkali feldspars also undergo Si,Al ordering, leading to the polymorphs *sanidine* (disordered, monoclinic) and *microcline* (ordered, triclinic). The common, partly ordered monoclinic form *orthoclase*, has a '*tweed*' domain texture on the scale of a few unit cells, geometrically similar to optical '*tartan*' twinning of microcline. Individual domains have triclinic structure but the average crystals are monoclinic. The ordering process in K-feldspar is slow, and even at low cooling rates the transition from sanidine to microcline becomes kinetically stranded when the tweed texture develops. The transition between the tweed and tartan microtextures provides a second type of marker of fluid–rock interaction. Ordering in albite is more rapid than in K-feldspar, because no symmetry change is involved, and in most low *T* rocks the albite is fully ordered. However, rapid growth at low *T*, for example during diagenesis, can lead to metastable growth of both Ab- and Or-rich disordered feldspars.

3. Two-feldspar–fluid equilibria

Important reactions in crustal rocks are governed by intercrystal equilibria in two-feldspar assemblages such as subsolvus granites and clastic sediments. These equilibria drive non-isochemical changes in feldspars which provide markers of the exchange of feldspar components through aqueous fluids and buffer their composition. In *principle*, as a two-feldspar granite cools, the PL and AF phases should remain as two homogeneous feldspars whose compositions are defined by points on the ternary solvus surface. Provided the bulk composition of the feldspar assemblage remains fixed, the phase compositions will lie on a tie-line passing through the bulk composition. The tie-line may rotate as *T* falls, by reciprocal exchange of Ab, Or and An. In *practice* intercrystalline equilibrium is virtually never attained. This can be inferred because the alkali feldspar in all granites is perthitic; it has left the ternary feldspar surface and cooled as an isolated phase, exsolving by processes we describe in Section 4.1. There are now three feldspars in the rock, a K-feldspar, albite in perthitic intergrowth, and (in most granites) an oligoclase, the original PL phase. The feldspar assemblage now violates the phase rule; we have entered the realm of phase behaviour rather than phase equilibria (Brown and Parsons 1989). As cooling proceeds phase behaviour may continue to increase in complexity as, for example, antiperthite forms in the plagioclase (we then have four feldspar phases).

Intercrystalline equilibrium may be reached in some rocks, such as granulites, which cool extremely slowly under essentially dry conditions (Brown and Parsons, 1988a; Kroll et al. 1993) and it is possible that intercrystalline exchange could occur at high *T* in normal granites. However, because growth temperatures estimated using two-feldspar geothermometry are often reasonable, it seems that high *T* intercrystalline exchange in cooling granites is unimportant, perhaps because water–rock ratios are generally low at this stage. Low-*T* intercrystalline equilibrium via an aqueous fluid was demonstrated directly, in a system with high water-rock ratios, by McDowell (1986). He showed that highly ordered authigenic feldspar pairs in fine-grained sandstones sampled in boreholes at measured *T* in the range 250–350°C, in the 16000 year old Salton Sea geothermal system, lie on the binary solvus for ordered feldspars (Bachinski and Müller, 1971). The feldspar compositions plot >150°C above the solvus for disordered feldspar pairs obtained by Smith and Parsons (1974), clearly indicating the importance of Si,Al ordering in feldspar reactions involving alkali exchange at low *T*.

The Na:K:Ca ratio of *liquids* coexisting with feldspar pairs depends on *P* and *T*, but only weakly on other components in the liquid or its physical state, whether aqueous fluid or silicate melt (Orville 1963). In alkali chloride solutions above the critical *T* of the alkali feldspar solvus, crystals and vapour change continuously in composition, but a narrow range of vapour compositions [near 25 mol% K/(K+Na)] is in equilibrium with a large range of feldspars. At lower *T* two feldspars are in equilibrium with vapour; the compositions of the feldspars define the solvus and the composition of the vapour is fixed irrespective of the proportions of the crystalline phases. The composition of the vapour in equilibrium with two feldspars decreases in K/(K+Na) with falling *T*, from 26 mol% K/(K+Na) at 670°C (just below the solvus critical *T*) to <16 mol% at 400°C. A hydrothermal fluid moving through and exchanging alkalis with a two-feldspar rock down a *T* gradient must react so that K replaces Na in the bulk solid assemblage, contributing Na to the cooling fluid. Reciprocal Na–K exchange through a pervasive fluid in a *T* gradient, provided it is in equilibrium with *two*-feldspars, will lead to K-enrichment in the bulk feldspar assemblage at the cooler end and Na-enrichment at the hotter end. Orville noted that solutions at 350°C with only 3–10 mol% K/(K+Na) are replacing plagioclase by K-feldspar in thermal springs at Wairaki, New Zealand, and at 200°C at Yellowstone, Wyoming. Fournier (1976) and Lagache and Weisbrod (1977) showed that similar relationships hold in the two-feldspars+liquid+vapour field, and Lagache (1984) summarized the affect of different salts in solution on partitioning of alkali metals.

Giggenbach (1988) discussed the relevance of feldspar reactions to geothermal systems, concluding that fluids in systems above 200°C were close to equilibrium with both Ab- and Or-rich feldspar. Thus the commonly observed K-metasomatism associated with major zones of upflow, and the Na-metasomatism associated with descending solutions, are ultimately controlled by these equilibria. Provided feldspar remains a stable phase and timescales of fluid movement are sufficiently long, they exercise an important control on fluids and feldspar assemblages even at lower *T*, for example during diagenesis in sedimentary basins (Sec. 5.2). In higher *T* regimes, such as contact metamorphic aureoles, porphyroblastic growth of K-feldspar may occur. It is important to note that petrographic evidence for K-metasomatism does not imply the passage of a K-rich aqueous fluid; an adequate volume of highly sodic but potassium-bearing fluid will replace Na in the feldspars by K at low *T*.

The experimental work shows that a number of different effects may control the composition of vapours and fluids in equilibrium with feldspars, but the buffering effect of two-feldspar assemblages is of great importance in late stage hydrothermal processes in all crustal rocks which contain feldspar pairs. Changes in *P*, *T* or fluid composition all cause bulk chemical changes in feldspars which provide a chemical driving force for microtextural and isotopic changes. In the next section we examine how spontaneous elastic forces in alkali feldspars can also make important contributions to their reactivity.

4. Intracrystalline reactions in alkali feldspars

4.1 STRAIN CONTROLLED MICROTEXTURES

The extreme reactivity of alkali feldspar in low-*T* aqueous solutions under *P–T–X* conditions in which feldspar is a stable phase is a consequence of the existence of a complex set of intracrystal microtextures which develop during cooling from growth *T*. Parsons (1978) reviewed the role of fluids in the development of these textures, and subsequent electron microscope work has provided insights into the mechanisms of change. Parsons and Brown (1984) showed that most crystals have a dual microtexture composed of *strain-controlled* (not involving a fluid) and *deuteric* (involving dissolution–redeposition in an aqueous fluid) regions. The former develop first, at higher *T*, and it is the latter which provide markers of fluid flow.

During cooling the alkali feldspar leaves the strain-free solvus (Sec. 3) and cools as a homogeneous phase (Fig. 1) until it intersects the coherent solvus. At some slightly lower *T*, coherent exsolution begins by diffusion of Na^+ and K^+ (and Ca^{2+} coupled to framework Al, although the amount of Ca is small) through a Si,Al–O framework which remains continuous. The resulting regular lamellar intergrowths share a common Si,Al–O framework but because the lattice dimensions of the Ab- and Or-rich phases are different the framework must be distorted, particularly near lamellar boundaries. Or-rich regions are under compression, and Ab-rich regions are under tension, in the plane of the coherent interface. The strain energy associated with these distortions leads to the coherent solvus, inside the strain-free solvus (Fig. 1).

The lamellar interface adopts an orientation which minimizes the strain energy associated with these *spontaneous elastic strains*. The orientations of such interfaces in alkali feldspars were calculated by Willaime and Brown (1974) using the unit cell parameters and elastic properties the feldspars; they vary with bulk composition and also with degree of framework order (Brown and Parsons 1988b). In alkali feldspars in subsolvus granites the lamellae form along planes approximately parallel to $\bar{6}01$–$\bar{8}01$ (Figs. 2, 4A), but in Ab-rich bulk compositions (as in hypersolvus granites), complex intersecting microcline lamellae form parallel to $\{\bar{6}\bar{6}1\}$, forming 'braid' perthite (Fig. 3), with lozenge-shaped columns of albite. The strain energy associated with coherent boundaries is relatively large, up to 2.4–5 $kJmol^{-1}$ in cryptoperthites (Brown and Parsons, 1993). One way to imagine the forces involved is to note that the difference in cell dimensions of the subordinate phase in a coherent cryptoperthite, from the relaxed values for the same material in a discontinuous intergrowth, may be as large as the variation brought about by a pressure change of 2 GPa. The free-energy associated with these elastic strains provides a thermodynamic driving-force for recrystallization leading to an unstrained microtexture composed of essentially the same phases.

During the early stages of exsolution of an alkali feldspar from a typical subsolvus granite, when planar lamellae of the volumetrically minor Ab-rich phase are small, it is possible for them to remain fully coherent. The lamellae have the form of relatively short, flat discs in three dimensions (Fig. 9B). As the feldspar cools the lamellae coarsen and lengthen (Fig. 4A). At the same time the entire structure stiffens and coherency forces become unsustainably large. There is a free energy advantage if coherency strain is decreased by the nucleation of edge dislocations (Fig. 4A). These form very flat elliptical loops which close around the exsolution lamellae, in two directions in the $\bar{6}01$ plane, one set with their major axis parallel to *b*, the other almost parallel to *c*. There is considerable energy in the core of edge dislocations, and also in the strained structure around them, and they play an extremely important role in the dissolution of feldspars in low *T* fluids, including during weathering (Lee and Parsons, 1995; Lee et al. 1998). They also provide a route by which water can gain access to the interior of crystals, leading to recrystallization giving unstrained feldspar (Figs. 4B, 8). Micas and clay minerals can nucleate on such dislocations (Brown and Parsons 1984, Fig. 3) and also following dissolution of highly strained structure at lamellar interfaces (Lee et al. 1997, Fig. 6c). Dislocations do not form directly in the more Ab-rich braid microperthites (Fig. 3), because coherency strains can be accommodated by the wavy interfaces and the short lateral extent of the Ab-rich lozenges (Brown and Parsons, 1984), but they can form when microtextures are modified near crystal margins by fluid-feldspar reactions (Lee et al. 1997).

The tweed microtexture of orthoclase provides an additional type of spontaneous strain which can drive reactions in the presence of a fluid. The alternating domains have partly ordered Si,Al distributions in which the Al is distributed with a left- or right-handed sense. Because of the change in orientation there is strain in the domain walls. Eggleton and Buseck (1980) showed that the lowering of free energy which would be achieved by further Si,Al ordering is matched by a build-up of strain energy in the walls of the domains. Thus a balance is reached and ordering stops; orthoclase can persist for billions of years although ordered microcline is the stable form below ~450°C (Brown and Parsons, 1989). The strain energy amounts to ~1.8–3.7 $kJmol^{-1}$; if fully ordered low microcline forms from the partly ordered orthoclase the total energy released is very large, ~5.5 $kJmol^{-1}$ (Brown and Parsons, 1993).

4.2 DEUTERIC MICROTEXTURES

In alkali feldspars that have cooled slowly, in the absence of external stresses or aqueous fluids, build-up of elastic strain energy stops coarsening of exsolution textures beyond ~1 μm and of tweed texture in orthoclase beyond a few tens of nm. Release of elastic strain energy provides a reduction in Gibbs free energy, and is the main thermodynamic driving force for the recrystallization of alkali feldspars during interactions with fluids at low *T*. Brown and Parsons (1993) called the processes leading to relaxation of elastic strain *'unzipping'* reactions; a strained microtexture is replaced by an assemblage of feldspars from which the constraint of coherency strain has been removed or much decreased. It is the existence of these types of reaction which leads to the extreme reactivity of feldspar in fluids at T <450°C, and makes feldspar an excellent marker of fluid–rock interaction. The converse, preservation of strain-controlled intergrowths which can develop *only* during cooling from relatively high *T*, provides a sure guide to volumes of crystals which have been preserved intact from magmatic or metamorphic growth *T*, and which may therefore preserve bulk compositions, trace element and isotopic signatures characteristic of those stages.

Once coherency strain is relaxed, Gibbs energy can be reduced by small changes in phase composition (as the Ab- and Or-rich phases move onto the strain-free solvus, Fig. 1), by decrease in surface energy as coarsening of exsolution lamellae proceeds, and by increase in Si,Al order as orthoclase transforms to microcline. Unzipping reactions usually lead to subgrain boundaries within individual crystals (Figs. 4B, 5), which make

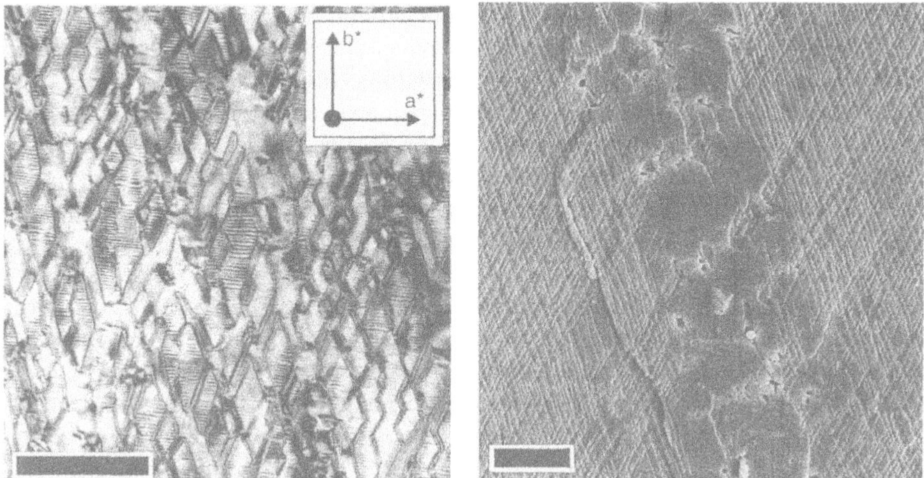

Figure 3. A. Bright-field TEM image of a braid cryptoperthite from the Coldwell syenite, Ontario. The diamond-shaped areas are Albite-twinned albite, zig-zag bands are microcline. From Waldron and Parsons (1993). Scale bar 500 nm. B. Secondary electron SEM image of HF-etched (001) cleavage surface of braid microperthite from the Klokken syenite. A band of deuterically coarsened, microporous patch perthite runs down the centre of the micrograph. From Waldron et al. (1994). Scale bar 5 μm.

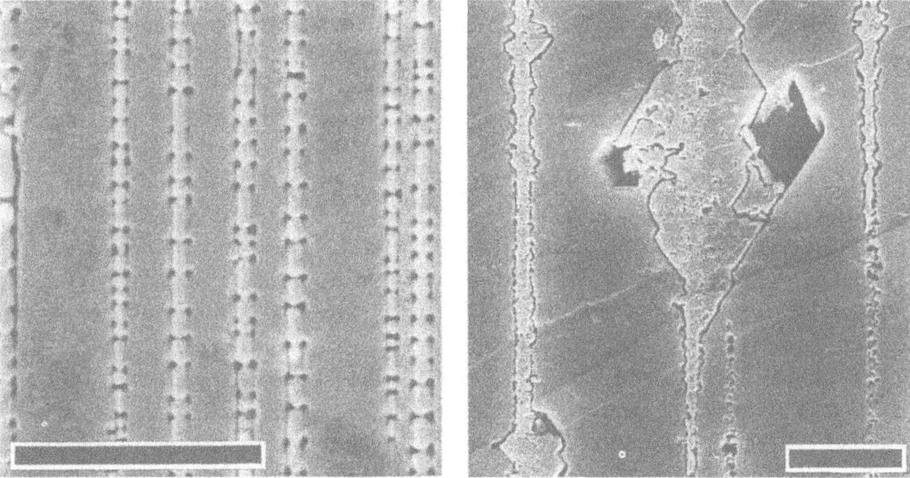

Figure 4. Secondary electron SEM images of etched (001) cleavage surfaces of Shap alkali feldspars. Both scale bars 5 μm. A. Semicoherent exsolution lamellae of albite parallel to *b* in orthoclase. Pairs of dots are etch pits at the outcrop of misfit dislocations. From Lee and Parsons (1995). B. Lamellae which have undergone unzipping reactions and albitization, and which have been replaced by rows of incoherent albite subgrains with the {110} habit. From Lee and Parsons (1997).

a positive surface energy contribution to the total free energy, but this is less than the lowering of free energy achieved by release of elastic strain, exsolution and ordering.

Unzipping reactions cause changes in hand-specimen and thin-section appearance of alkali feldspars which are familiar to petrologists (Fig. 2). Alkali feldspars which have escaped reactions with fluids are transparent and very dark blue, brown or bottle green in hand specimen; they are non-turbid or 'pristine' in thin section. Rocks with feldspars of this type are rarities in the exposed crust. The best-known example is the syenite known as larvikite, used as an architectural stone; in some varieties the massive rock is almost black in colour, with superimposed iridescence caused by optical diffraction by cryptoperthite lamellae. Granites in which the alkali feldspar is almost entirely pristine are extremely uncommon and have a largely anhydrous mafic mineralogy; striking examples are the nearly *black* rapakivi fayalite granites of the Prins Christian Sund pluton, South Greenland (Harrison et al. 1990).

In contrast alkali feldspar in most felsic plutonic rocks is translucent or chalky, pink or white. Crystals are variably translucent, often containing cross-cutting veinlets of more turbid feldspar. Turbidity in thin section is used routinely to distinguish feldspar from quartz, but the microtextural cause and significance were until recently largely overlooked. Montgomery and Brace (1975) investigated cloudy plagioclases using scanning electron microscopy (SEM) and found that they contained large pores on a scale of ~10 μm. They did not find pores in alkali feldspar. However, Worden et al. (1990), using SEM on fresh cleavage surfaces, showed that the variable turbidity of alkali feldspars from the Klokken intrusion was correlated with the presence of myriads of small (<1 μm) micropores (e.g. Figs. 3B, 5). Guthrie and Veblen (1991) showed the correlation of turbidity with micropores in alkali feldspars in Skye granites.

Walker et al. (1995) described the character, abundance and contents of pores in alkali feldspars from a range of rock types, including many two-feldspar granites. Although optical cloudiness of feldspar (Fig. 2) is often ascribed to clay minerals, most pores are empty, or may contain fluid. Clay minerals, metal-oxides and sulphides, noble metals and halite have all been reported (Worden et al. 1990; Guthrie and Veblen, 1991; Walker et al. 1995). Porosities in alkali feldspar reach 4.75 vol.%, and there may be 10^9 pores mm^{-3}. Walker (1990) showed that such feldspars are also micropermeable, by reacting porous feldspars with $H_2{}^{18}O$ for 75 h at 700°C, 0.1 GPa, and imaging the resulting ^{18}O distribution by ion microprobe. ^{18}O exchange occurred at least 50 μm inside the crystals. This extrapolates to fluxes of fluid *through* the crystals equivalent to transport over 5 km Ma^{-1}.

Profound microtextural changes accompany the development of pores. The most obvious is '*deuteric coarsening*' of perthite (Figs. 2B, 3B) (Parsons and Brown, 1984). Parsons (1978) drew attention to the apparently catastrophic character of the process, with a sudden breakdown of textures of great regularity to far coarser but chaotic intergrowths. In modern terminology, both strain-controlled perthitic intergrowths and tweed microtexture may be examples of self-organised criticality; intercrystalline water is the trigger which activates the step from microtextural order to irregularity. We will now examine the mechanisms by means of TEM images using examples from hypersolvus and subsolvus igneous rocks, and a granulite-facies metamorphic rock.

5. Unzipping reactions in alkali feldspars

5.1 HYPERSOLVUS ROCKS

Alkali feldspars in the Klokken syenite intrusion, south Greenland, have been the subject of numerous studies (e.g. Parsons 1978; Brown et al. 1983; Brown and Parsons 1984, 1988b; Worden et al., 1990). The remarkable feature of the intrusion is that it is a layered syenite in which some layers contain white feldspars which have been strongly affected by circulating deuteric fluids, interleaved with dark syenites which were impermeable, and in which pristine alkali feldspar with strain-controlled intergrowths predominate. These are braid crypto- and micro-perthites (Fig. 3B; similar to Fig. 3A) with bulk compositions near $Ab_{60}Or_{40}$ and almost devoid of micropores. During deuteric coarsening the strain-controlled microtexture gives way laterally, via a less regular zone of film microperthite, to an irregular '*patch perthite*' made up of Ab- and Or-rich subgrains (Figs. 3B, 5). Intergrowths coarsen by factors of up to 10^3. Within an individual thin-section the distribution of coarsened texture is irregular, some crystals being almost entirely recrystallized, others only in a few patches or veins (Fig. 3B) or at crystal edges (Fig. 6A). Subgrain walls link abundant micropores, which form voids where several subgrains are in contact (Fig. 5). Subgrains sometimes have irregular shapes but often are bounded by {010} and {110}, the *Adularia habit* characteristic of feldspars growing in low *T* hydrothermal veins and of authigenic feldspars. Diagenetic overgrowths are sometimes composed of {110} subgrains (Worden and Rushton 1992). Feldspar outgrowths in micropores often have the Adularia habit, giving pores characteristic shapes (Fig. 5B), and suggesting that reprecipitation of feldspar in voids is a step in the formation of subgrain mosaics.

In some of the dark syenite layers alteration is concentrated at crystal boundaries, forming complex textures called *pleated rims* (Fig. 6A) (Lee et al. 1997; Brown et al. 1997). Lee et al. showed that the *vein perthite* common in hypersolvus granites (Fig. 6B) forms from a pre-existing braid microtexture by advance of 'pleats' across entire crystals, leaving occasional relics of braid texture (Fig. 6B). The microtextural evidence is clear: much of the alkali feldspar (comprising perhaps 60 vol% of the entire rock) in most hypersolvus granites has undergone dissolution and reprecipitation in an aqueous fluid. The irregularity of microtextures in deuterically altered feldspars is in striking contrast to strain-controlled textures (compare Figs. 3A and 5A) and a reliable marker of fluid–rock interaction, in this case affecting crystals in their entirety.

Braid perthites contain low microcline which is stable only below ~450°C (Brown and Parsons, 1989) providing an upper *T* limit for the unzipping reactions. However the microporous texture, and small effective grain size, makes the crystals highly sensitive to subsequent reactions with fluids and it is likely that they continue to near-surface *T*. The marked turbidity of feldspar in most hypersolvus granites is thus a result of aqueous fluids acting on a mineral species in which strained microtextures provide a driving-force for pervasive intracrystal microtextural change.

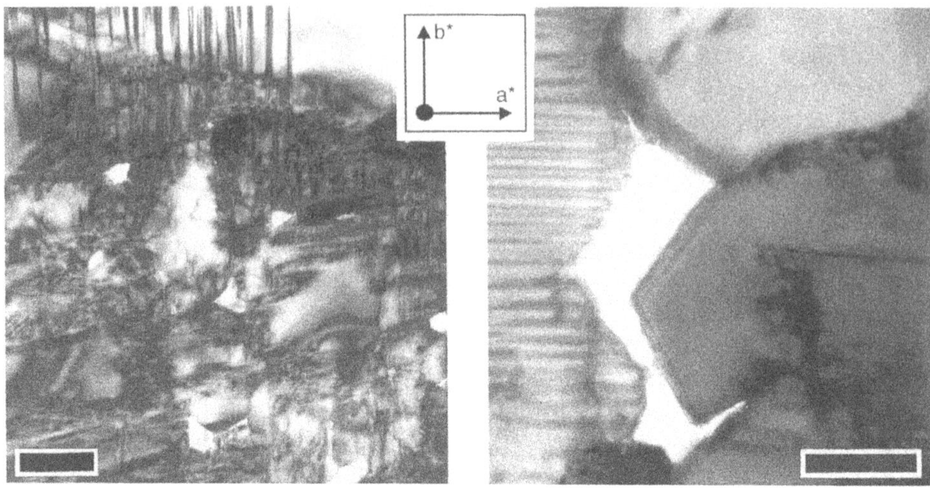

Figure 5. Bright field TEM images of patch perthites formed by deuteric coarsening, from the Klokken syenite. A. Irregular mosaic of subgrains of albite, tweed orthoclase and microcline. Some have the Adularia habit, and spaces between them define micropores. Scale bar 500 nm. B. Detail of a micropore between albite (left) and microcline (right), showing new microcline with the Adularia habit growing into the micropore. Scale bar 100 nm. Both previously unpublished.

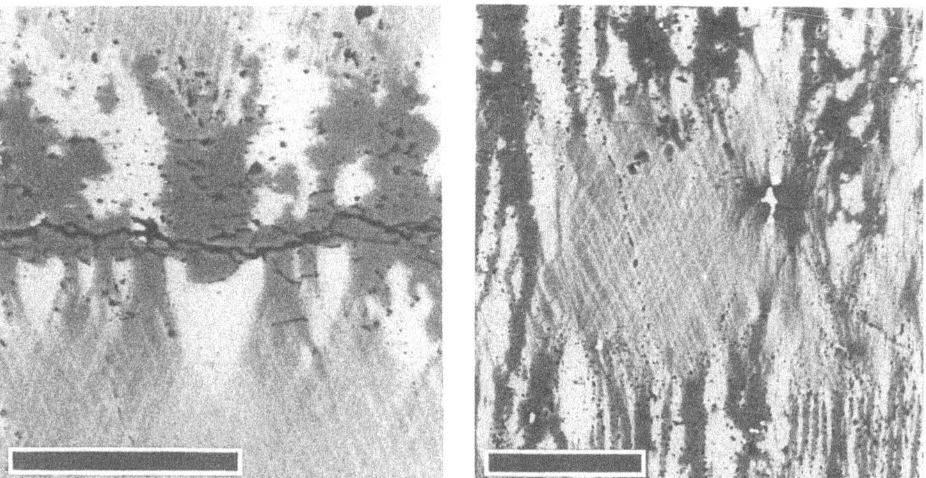

Figure 6. BSE SEM micrographs of microperthites from the Klokken intrusion. Albite appears dark, microcline light grey. Black dots are micropores. From Lee et al. (1997). Both scale bars 50 μm. A. Edge between two braid microperthite crystals showing pleated rims, from syenite. In the lower crystal the pleats are small and largely coherent, but in the upper crystal deuteric recrystallization has occurred. B. Relic of braid microtexture in the interior of a vein microperthite crystal from a quartz syenite.

5.2 SUBSOLVUS ROCKS

The microtextures in striking pink alkali feldspar phenocrysts in the Lower Devonian Shap granite, northern England, have been used to reconstruct the history of feldspar–fluid reactions from magmatic *T* (Lee et al. 1995), through replacement reactions late in igneous cooling (Lee and Parsons, 1997), to diagenetic reactions in the same feldspars following incorporation in an overlying Carboniferous conglomerate (Lee and Parsons, 1998). The events are summarized in Figure 7, and their *T* shown on Figure 1. By applying two-feldspar geothermometry (Fuhrman and Lindsley, 1988) we can suggest approximate *T* for these reactions. These are maximum *T*, based on feldspars with fully ordered Si,Al distributions; *T* for disordered feldspars would be ~100°C lower. In the relatively Or-rich feldspars (often ~$Ab_{20}Or_{80}$) characteristic of subsolvus granitic rocks, deuteric coarsening and development of microporosity is generally less pervasive than in feldspars from hypersolvus granites, probably because the contribution of strain energy to the crystal free energy is less. They have obvious dual microtexture, with relics of optically clear strain-controlled lamellar crypto- and micro-perthite, cut or surrounded by turbid, coarsened, microporous patch or irregular film perthite (Figs. 2, 4). Tweed orthoclase also simultaneously recrystallizes to microcline, illustrating a second type of unzipping reaction.

The Shap phenocrysts grew from melt at ~800–700°C, as low sanidine, cooling as homogeneous crystals until they intersected the coherent solvus at ~670°C. At some lower *T*, exsolution began by coherent nucleation, giving small, fully coherent lenses of nearly pure albite (Fig. 7A, 9B), coherency stresses being accommodated by nearly homogeneous strain of the minor phase. Coherent Albite and Pericline twins formed in the albite. The enclosing Or-rich feldspar ordered and adopted the tweed microtexture of orthoclase at ~500–450°. The exsolution lamellae became more abundant and coarsened by thickening and extending in length while remaining coherent, becoming film crypto- and ultimately micro-perthite (Fig. 7B). Up to this point the structure would be fully coherent, and no reactions involving external fluids would be involved.

Unzipping reactions began below 450°C, in the stability field of microcline (Fig. 7C). Irregular veins of turbid patch perthite, composed of slightly misorientated, incoherent subgrains of albite, tweed orthoclase and microcline, cut across the film lamellae (Figs. 2B, 4B). Within the veins subgrains vary from tens of nm up to > 1 μm, with sub-μm to nm micropores in spaces between them. Albite in the veins is richer in An (An_8) than that formed subsequently, and the An component must partly have been introduced from outside the phenocrysts. Two feldspar thermometry suggests *T* in the range 385–490°C, depending on degree of order. The veins are not affected by the coherent or semicoherent microtexture, which suggests that they preceded the formation of dislocations along the film lamellae (Fig. 7C; Sec. 4.1).

As the structure stiffened, dislocations formed on the broader film microperthite lamellae (Figs. 7D, 4A) and the morphology of textures formed by deuteric reactions changed. Semicoherent albite film lamellae coarsened and became microporous and irregular, by a dissolution–redeposition process which led to the development of bands of subgrains along the original lamellar direction (Figs. 7E, 4B). Small, fully coherent

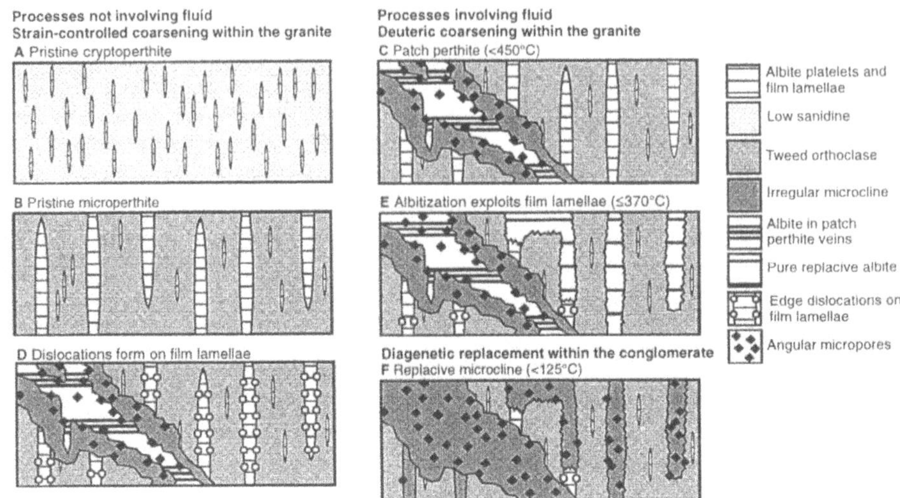

Figure 7. Diagrammatic illustration of intracrystal microtextural changes (in order, A to F) that have affected Shap alkali feldspar phenocrysts during during their igneous and diagenetic history. After Lee and Parsons (1997, 1998). Width of boxes ~50 μm.

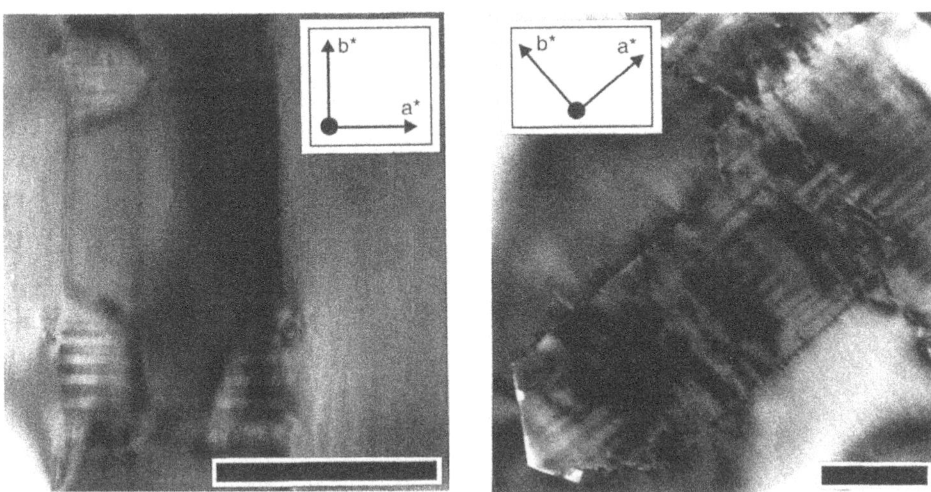

Figure 8. Bright field TEM micrographs illustrating the beginning of albitization in tweed orthoclase from Shap. From Lee and Parsons (1997). Both scale bars 200 nm. A. Semi-coherent albite film, not visibly twinned, running top to bottom, with pairs of new, incoherent, Albite-twinned albites forming on dislocations. B. Larger, semi-coherent peg-shaped crystal of replacive albite, terminated by planes with the Adularia habit. This leads to textures like Figure 4B.

lamellae were unaffected. This textural change may be sub-optical in scale and the accompanying microporosity probably causes much of the faint turbidity which characterizes plutonic K-feldspars (Fig. 2). Tiny albites nucleated on the spaced dislocations (Fig. 8A), and then grew laterally into the surrounding orthoclase (Fig. 8B). Deuteric reactions were thus guided by the dislocations, and boundaries defined by the film lamellae remain even when large areas have been replaced (Fig. 4B). This albite is close to pure albite, suggesting that it formed at low *T*, $<370°C$ and at *T* as low as those normally associated with diagenesis. Considerable volumes of K-feldspar were replaced by this generation of albite, leading to optically visible albitization similar to that seen in clastic sediments. This again suggests introduction of albite from outside the rock sample, probably by exchange of Ab and Or components in a hydrothermal system with a *T* gradient, as discussed in Section 3.

At some poorly defined low *T* a few semicoherent albite lamellae were replaced by microcline, so that what appears optically to be conventional perthitic intergrowths are in fact lamellar intergrowths of microcline in orthoclase. This continued during diagenesis after similar Shap phenocrysts were incorporated in overlying conglomerate (Figs. 7F, 9). In these crystals almost all albite has been replaced by microcline, both in film lamellae and patch perthite. Only those parts of the original crystals which had fully coherent albite lamellae or platelets were impervious to this replacement (Fig. 9B). Overall, the feldspars changed from bulk Ab_{28} to Ab_{9-12}. Apatite fission-track work (Green 1986) suggests that the conglomerate had reached 70–125°C during burial, the maximum *T* for the introduction of replacive microcline. Thus microtextures in the Shap feldspars provide a record of fluid–feldspar interactions occurring from $<450°C$ to $<125°C$. Shap is a typical two-feldspar granite and such microtextural diversity, and by implication fluid–feldspar reaction, is the norm.

5.3 UNZIPPING OF ORTHOCLASE

Relaxation of strain in the tweed microtexture of orthoclase, with the development of microcline, accompanies deuteric coarsening in Or-rich feldspars (Sec. 5.2) or drives unzipping in its own right. In more Ab-rich feldspars (such as the Klokken examples, Sec. 5.1) microcline formation usually precedes coarsening, for reasons discussed by Brown and Parsons (1984; 1988b). Fine-scale tweed frequently occurs within individual potassium feldspar crystals together with areas of coarsely tartan-twinned microcline. Crystals often show optically visible tartan twinning where turbid, whereas in clear areas the twinning dies away. TEM micrographs of these textures from granitic rocks are in Fitz Gerald and McLaren (1982) and McLaren (1984), and from metamorphic rocks in Bambauer et al. (1989). The latter authors recognized tweed and two types of microcline, 'regular' and 'irregular', in regional and contact metamorphic rocks. Tweed can grade or change abruptly into irregular microcline (Fig. 10A), and it is possible that irregular microcline is an intermediate stage in formation of regular microcline. Irregular microcline is often microporous, and Brown and Parsons (1993) noted that pores are less common in regular microcline, suggesting that they disappear by a healing process as the microtexture becomes more regular.

Figure 10A is a TEM micrograph of a sub-optical veinlet of irregular microcline in a mesoperthitic orthoclase from a granulite-facies metamorphic rock from the Adirondacks (Waldron et al. 1993). The vein contains a few micropores defined by microcline subgrains which often have the characteristic Adularia {110} habit. This is an example of a subtle fluid-feldspar reaction in this high-T metamorphic rock. Figure 10B shows a 100 nm micropore in a tweed orthoclase cryptoperthite from the Coldwell syenite intrusion (Waldron and Parsons, 1992) which is partly filled with new growth of feldspar and which appears to be associated with a trail of irregular microcline. It is possible that dissolution of strained structure at the leading edge of the pore has been followed by reprecipitation of unstrained, slightly more ordered microcline, at the trailing edge, leading to onward motion of the micropore through the structure.

6. Mechanisms and rates of reaction

Strictly speaking, evidence that deuteric coarsening and microcline formation reflect aqueous fluid-feldspar reactions is circumstantial. Deuteric microtextures have not been produced experimentally and microcline has not been synthesised from tweed orthoclase. The TEM work shows the coexistence of the tweed and tartan microtextures, but does not provide direct evidence for the mechanism of the transformation. There has, however, long been circumstantial petrological evidence that microcline development is related to the presence of water in the appropriate temperature range (e.g. Parsons and Boyd 1971; Parsons 1978; Tullis 1983).

Most readers will probably accept that the micrographs here and elsewhere speak for themselves; no process other than dissolution–reprecipitation could produce such textural changes. Correlation of textural and oxygen isotopic changes provides geochemical support. At low T dissolution–reprecipitation is a far more effective mechanism for ^{18}O–^{16}O exchange than volume diffusion because the driving forces for isotopic exchange are trivial (Giletti 1985). Guthrie and Veblen (1991), showed a correlation of micropore development and deuteric coarsening in alkali feldspars in Skye hypersolvus granites with low $\delta^{18}O$ measured by Ferry (1985), reflecting the circulation of large volumes of meteoric water through these highly altered rocks (Taylor and Forester, 1971).

O'Neill and Taylor (1967) studied mechanisms of oxygen exchange in feldspars experimentally and postulated reaction by a series of minute solution–redeposition steps in an advancing fluid film. The experiments were done with simultaneous Na–K exchange, providing a strong chemical driving force, ensuring that the reactions occurred on a laboratory time-scale. Deuteric coarsening involves changes which are overall isochemical and which take place in such a way that strain controlled intergrowths break down to a strain-free mosaic of sub-grains without affecting the external crystal morphology or, usually, its bulk composition. There is no strong chemical driving force, although there will be local exchange of, in particular, Na^+ and K^+. We would expect that small volumes of fluid advancing through the crystals would be very close to purely chemical equilibrium with the solid. It is because reaction rates

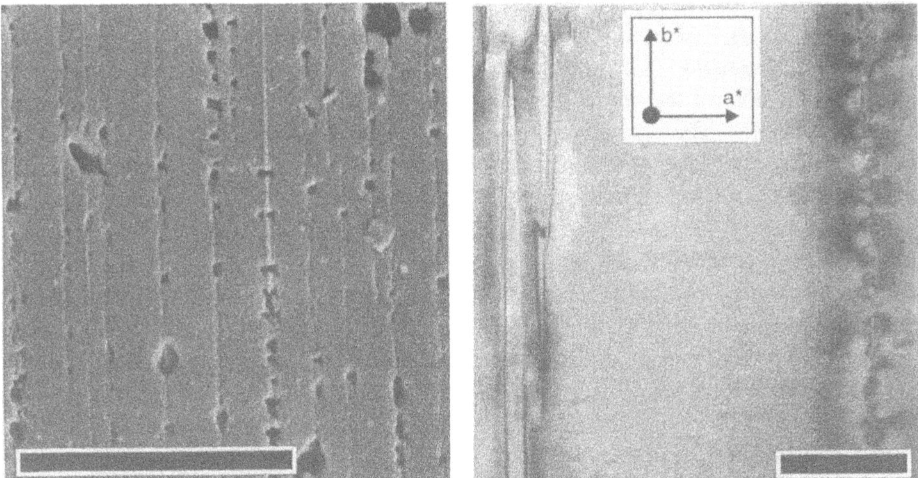

Figure 9. A. Secondary electron SEM image of an HF etched (001) surface of a Shap alkali feldspar occuring as a clastic fragment in Carboniferous conglomerate. Only one film lamella stands out in relief from the orthoclase matrix and is albite. The other lamellae form slight depressions, and are microcline. Scale bar 5 μm. B. Bright field TEM micrograph of a film lamella (right) replaced by microcline. Tiny fully coherent lenses of albite (left) are unaffected. Scale bar 200 nm. From Lee and Parsons (1998).

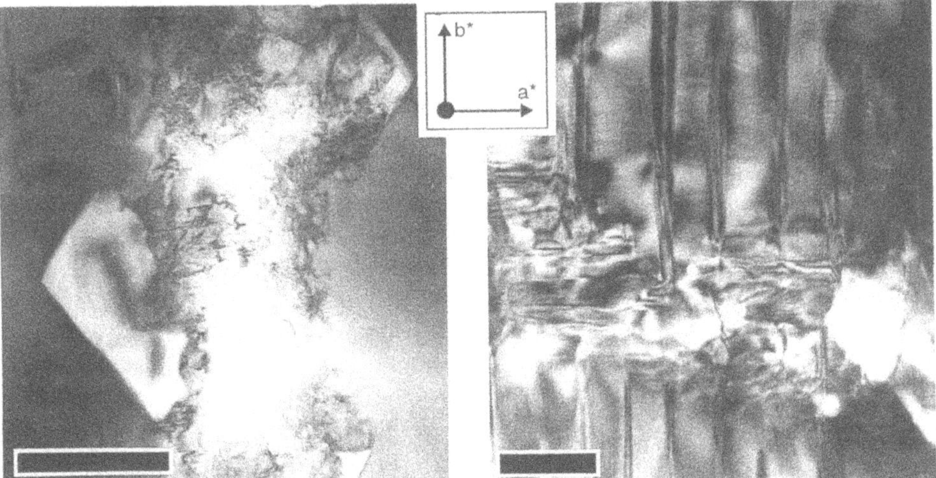

Figure 10. Bright field TEM images. A. Vein of subgrains of irregular microcline, often with the Adularia habit, cutting tweed orthoclase in a granulite from the Adirondacks. From Waldron et al. (1993). Scale bar 500 nm. B. Vein of microcline cutting a lamellar tweed orthoclase cryptoperthite from a Coldwell syenite. The microcline forms a trail behind a micropore which may have proceeded from left to right in the micrograph. From Waldron and Parsons (1992). Scale bar 200 nm.

under these circumstances are strongly dependent on chemical affinity (Velbel 1989) that the free-energy contribution of defects has such a powerful effect.

Since microtexture is central to this paper it is useful, if daunting, to consider the factors which control the rate at which a 'front' of deuteric coarsening might propagate into a crystal. This is an extremely complex problem, because the various microtextures, and the strains associated with them, evolve as T decreases. The reactions take place in confined spaces, in solutions very close to saturation, a situation which has been called the 'internal zone' in the weathering context (Hochella and Banfield 1995). Dissolution is likely to be the rate-limiting step, since feldspars >100μm can be grown in a few months at room T and P (Flehmig 1977). Dissolution rates based on conventional dissolution experiments (e.g. the 100–300°C experiments of Hellman, 1994, on Amelia albite) are of little help in the present context because they are carried out far from equilibrium. Lee et al. (1998) have shown that dissolution rates and mechanisms are very different in field and laboratory settings, being much more strongly defect dependent in nature.

By analogy with weathering, dissolution along dislocation cores is likely to be the fastest part of the deuteric coarsening process. Dislocation formation appears to be the precursor to coarsening of braid perthites (Lee et al. 1997) and for the most pervasive phase of deuteric reactions in film perthites (Lee et al. 1995; Lee and Parsons 1997). Figure 8 shows how, in deuteric fluids, new feldspar growth nucleates on these defects, which provide channels for water to enter from the crystal surface, providing a medium for loss of K^+ (from dissolving tweed orthoclase) and addition of Na^+ (forming albite). In confined spaces in soils advance of etch pits will be arrested when solutes can no longer diffuse away from the tip of the pit. Thus dissolution of etch pits during weathering and during reactions with cool hydrothermal fluids are likely to occur at broadly similar rates. During 5000 yr of weathering in soils at Shap (currently pH 3.4, at ~10±10°C, etch pits in the feldspars advanced at ~5 nm yr^{-1} (Lee et al. 1998).

For experimental silicate dissolution occurring far from equilibrium there is a good approximation to a linear Arrhenius relationship between log rate and $1/T$ (K) (Wood and Walther 1983). Dissolution rates of Amelia albite (Hellman 1994) show similar T dependence. Misfit dislocations in alkali feldspars such as those from Shap start to form in the range 350–400°C, depending on local strain fields, but their strain energies will increase further with cooling. If we assume that the dissolution rate of etch pits at 300°C is faster than that during weathering by the amount predicted by Wood and Walther's equation (by ~10^5) we obtain a dissolution rate for dislocations of 500 μm yr^{-1}. At 100°C the rate is 1.25 μm yr^{-1}. The latter value is rather close to the subgrain sizes in deuterically coarsened perthites (Figs. 3B, 4B, 5) supporting our contention that the low-T reactivity of feldspar is very much bound up with microtexture.

7. Mapping fluid flow paths on the nanometre to kilometre scale

In previous sections we have described how irregular patch and vein perthites within alkali feldspar crystals record the former presence of fluids. Going one step further, the spatial distribution of these perthite types can be used to map fluid flow paths. The distribution of patch perthites within individual grains records fluid movement on the nm to mm scale, whereas contrasts in the abundance of deuterically coarsened perthite between grains in the same rock sample, or between different samples or units in an intrusion, map fluid flow on the cm to km scale.

The volume of individual crystals affected by deuteric coarsening within a particular rock sample may vary considerably, even within a single thin section. For example, in the syenites from the Klokken intrusion (Parsons 1978; Brown et al. 1983; Worden et al 1990) it is sometimes the case that almost all crystals are coarsened, with only a few relics of strain-controlled microperthite remaining. Sometimes it can be inferred that coarsening is related to fractures (Fig. 3B) and crystal boundaries (Fig. 6A), but in other cases there seems to be no obvious relationship to pre-existing features; half a crystal may be affected pervasively, the other half not at all. Sometimes one crystal in a thin section is largely unaffected, while all others are coarsened. Brown et al. (1983) investigated whether this variability could be accounted for by variation in bulk composition, in particular whether there was a connection with An content, but found no correlation. Probably, the variation is in part related to the state of self-organised criticality which precedes unzipping (Sec. 4.2); once the process is triggered it tends to go to completion, and to advance pervasively through the crystal. Thus although the presence of deuteric microtextures indicates the presence of a fluid, on a local scale it may not necessarily be a good guide to the *volume* of fluid involved.

The most informative examples of localisation of the effects of fluids on the sub-mm scale are the 'pleated rims' (Fig. 6A) described by Lee et al. (1997) and Brown et al. (1997) in ~$Ab_{60}Or_{40}$ feldspars from the Klokken and Coldwell (Ontario) syenites. Detailed TEM study of these textures provided insights into the exact mechanism by which unzipping occurs. The largest pleated rims occur at grain boundaries (Fig. 6A), while smaller ones straddle subgrain boundaries and cleavage planes. Interaction of coherency strain fields in initial braid intergrowths with these gross structural features leads to a style of coherent coarsening which produces straight lamellae on which dislocations develop to relieve coherency strains. The dislocations permit ingress of water, leading to development of microporous vein microperthite. The presence of the coarsened pleats clearly shows which pathways existed at the sub-mm scale for water ingress during cooling below ~450°C, the upper stability limit of microcline. Ultimately, the pleats can propagate completely across crystals, leading to vein perthite (Fig. 6B), and the evidence for grain-scale pathways is lost.

It is significant that strain-controlled micro- and crypto-perthitic crystals, with variably developed pleated rims, are common in syenites, whereas the alkali feldspar in alkali granites is almost universally vein perthite. On petrological grounds the former crystallize from 'dry' magmas, the latter from more evolved, wetter magmas. Thus the microtextural evolution of the feldspars suggests that magmatic water was retained to T <450°C, and the magnitude of the deuteric changes reflect the build-up of water during normal igneous fractionation. Figure 6B shows a feldspar from a quartz syenite sheet cutting syenites largely composed of braid perthites in the Klokken intrusion. Because the two rocks shared the same cooling history this provides very clear evidence that fluid-feldspar reactions are the main factor controlling the microtextural evolution of the feldspars, rather than cooling rate.

Parsons and Becker (1986) showed a section through the Klokken syenite intrusion illustrating how layers of pale laminated syenite, in which the feldspars are microporous patch perthites (Fig. 5), are interleaved with dark coloured, finer grained syenite with a granular texture in which the feldspars are predominantly braid perthites (Fig. 3). They identified other, non-isochemical, mineralogical changes in the altered layers, particularly in pyroxenes, which showed that the laminated syenites acted as high-T aquifers. The feldspar microtextures thus map fluid-flow on the scale of several km. However, the oxygen isotopic composition of both syenite types is the same ($\delta^{18}O$ 3–8‰$_{SMOW}$), suggesting that the fluids were of magmatic origin (Parsons et al. 1991). Thus in this instance the feldspar microtextures are a more effective guide to fluid pathways than oxygen isotopes, which have been used so successfully elsewhere (Taylor and Forester, 1971; Ferry 1985). It is now technically possible, using the ion microprobe, to correlate microtexture and oxygen isotopic signatures in feldspars at scales down to ca. 50 μm, although this has not yet been done systematically.

In more Or-rich feldspars from subsolvus granites such as Shap, patch perthite microtextures indicate intercrystal fluid pathways on the scale of μm (Fig. 2). Textures reflecting fluid–feldspar interactions can be divided into two types (Sec. 5.2) reflecting two distinct phases of reaction with water, one occurring above the T at which dislocations form on the regular film microperthite lamellae (≤370°C), the other below (Fig. 7, stages C and E). The earlier phase (which formed at ~410°C) produced anastomosing veins of albite and microcline that cross-cut entire cm-sized phenocrysts. Although they were present at this time the exsolution lamellae exerted little control on the paths of fluid movement, because of the absence of dislocations. Probably the patch perthite shown in Figure 2B developed along fractures formed during cooling but all evidence is lost because of the dissolution-reprecipitation that has occurred. The second generation of patch perthite (Fig. 7E) is composed of compositionally pure albite that occurs only within 1 mm of grain boundaries. It has selectively replaced albite exsolution lamellae (with 8 mol% An), and surrounding orthoclase, by development and expansion of volumes of pure albite via a fluid which gained access to the crystal down the dislocations (Fig. 8). Thus the feldspar provides a marker not only of two phases of fluid-rock interaction which affected parts of the crystals only, but also of mass transfer.

Although variation in 'intensity of alteration' within granite bodies is often recorded using conventional microscopy by petrographers, and large-scale chemical changes have been intensively studied in the case of highly altered porphyry-type ore deposits, we are not aware of any work analogous to that described above on the Klokken intrusion, in which textures at the sub-mm scale have been systematically used to deduce fluid-flow pathways in an entire intrusion. It would be rewarding to establish whether the complex microtextures in the Shap granite (Fig. 7) are developed everywhere in the same way and to the same degree, and whether there is consistency of orientation of vein features such as those shown in Figure 2B in three-dimensions, but excellent exposure would be required to do this. Nevertheless, it is clear that there is great potential for mapping fluid pathways in granites at all scales using the microtextural methods described here.

8. Conclusions

Alkali feldspars have complex intracrystal microtextures which provide excellent markers of dissolution-reprecipitation reactions in aqueous solutions below ~450°C, continuing into the *T* range of diagenesis. The textures can be used to map fluid pathways on scales of nm to km. The particular suitability of alkali feldspars for this purpose arises because of the coherent microtextures (strain-controlled perthitic intergrowths and tweed domain textures in orthoclase) which develop in the feldspars during cooling from higher *T* before interactions with fluids begin. The main thermodynamic driving force for dissolution of feldspars of high *T* origin, in solutions with which they are close to equilibrium, is elastic strain energy associated with coherency in these microtextures. Feldspar dissolves, probably in minute fluid films, and unstrained feldspar grows, giving coarse, incoherent perthites and tartan-twinned microcline. Dissolution and reprecipitation around dislocation cores is an important part of these 'unzipping' processes, which lead to feldspars which are turbid. Because crystal–fluid exchange reactions can be driven and facilitated by unzipping, and because the resulting feldspar is microporous and micropermeable, feldspars readily maintain alkali and isotopic exchange equilibrium down to $T \leq 200$°C. Since granites in the broadest sense are so abundant, and turbidity is almost universal in feldspars collected at outcrop, we conclude that intracrystal solution–reprecipitation is a process that has affected a high proportion of the upper crust of the Earth. Granites in place in the deep crust, at *T* above those at which deuteric reactions occur, will be texturally, and perhaps geochemically, very different to those we see at the surface.

Acknowledgements

The ideas in this paper have developed during a succession of NERC Research Grants: GR3/6697, 8374 and 10290. We are grateful to our previous co-workers, Richard Worden, Kim Waldron, David Walker and Bill Brown, for the images and ideas they have provided.

References

Bachinski SW, Müller G (1971) Experimental determinations of the microcline–low albite solvus. J Petrol 12: 329-356

Bambauer HU, Krause C, Kroll H (1989) TEM-investigation of the sanidine/microcline transition across metamorphic zones: the K-feldspar varieties. Eur J Mineral 1: 47-58

Brown WL, Becker SM, Parsons I (1983) Cryptoperthites and cooling rate in a layered syenite pluton. Contrib Mineral Petrol 82: 13-25

Brown WL, Parsons I (1984) The nature of potassium feldspar, exsolution microtextures and development of dislocations as a function of composition in perthitic alkali feldspars. Contrib Mineral Petrol 86: 335-341

Brown WL, Parsons I (1988a) Inter- and intracrystalline exchange and geothermometry in granulite-facies feldspars. Terra cognita 8/3: 263

Brown WL, Parsons I (1988b) Zoned ternary feldspars in the Klokken intrusion: exsolution microtextures and mechanisms. Contrib Mineral Petrol 98: 444-454

Brown WL, Parsons I (1989) Alkali feldspars: ordering rates, phase transformations and behaviour diagrams for igneous rocks. Mineral Mag 53: 25-42

Brown WL, Parsons I (1993) Storage and release of elastic strain energy: the driving force for low-temperature reactivity and alteration of alkali feldspar. In: Boland JN and Fitz Gerald JD (eds) Defects and processes in the solid state: geoscience applications. The McLaren Volume. Elsevier Science Publishers, 267-290

Brown WL, Parsons I (1994) Feldspars in igneous rocks. In: Parsons I (ed) Feldspars and their reactions. NATO ASI Series C421, Kluwer Academic Publishers, 449-499

Brown WL, Lee MR, Waldron KA, Parsons I (1997) Strain-driven disordering of low microcline to low sanidine during partial phase separation in microperthites. Contrib Mineral Petrol 127: 305-313

Eggleton RA, Buseck PR (1980) The orthoclase–microcline inversion: a high resolution transmission electron microscope study and strain analysis. Contrib Mineral Petrol 74: 123-133

Ferry JM (1985) Hydrothermal alteration of Tertiary igneous rocks from the Isle of Skye, northwest Scotland. II Granites. Contrib Mineral Petrol 91: 283-304

Fitz Gerald JD, McLaren AC (1982) The microstructures of microcline from some granitic rocks and pegmatites. Contrib Mineral Petrol 80: 219-29

Flehmig W (1977) The synthesis of feldspars at temperatures between 0°C-80°C, their ordering behaviour and twinning. Contrib Mineral Petrol 65: 1-19.

Fournier RO (1976) Exchange of Na^+ and K^+ between water vapor and feldspar phases at high temperature and low vapor pressure. Geochim Cosmochim Acta 40: 1553-1561

Fuhrman ML, Lindsley DH (1988) Ternary feldspar modelling and thermometry. Am Mineral 73: 201-215

Giggenbach WF (1988) Geothermal solute equilibria. Derivation of Na-K-Mg-Ca geoindicators. Geochim Cosmochim Acta 52: 2749-2765

Giletti BJ (1985) The nature of oxygen transport within minerals in the presence of hydrothermal water and the role of diffusion. Chem Geol 53: 197-206

Green PF (1986) On the thermo-tectonic evolution of Northern England: evidence from fission track analysis. Geol Mag 123: 493-506

Guthrie GD, Veblen DR (1991) Turbid alkali feldspars from the Isle of Skye, northwest Scotland. Contrib Mineral Petrol 108: 398-404

Harrison TN, Parsons I, Brown PE (1990) Mineralogical evolution of fayalite-bearing rapakivi granites from the Prins Christian Sund pluton, South Greenland. Mineral Mag 54: 57-66

Hellmann R (1994) The albite-water system: Part I. The kinetics of dissolution as a function of pH at 100, 200, and 300°C. Geochim Cosmochim Acta 58: 596-611

Hochella MF Jr, Banfield JF (1995) Chemical weathering of silicates in nature: a microscopic perspective with theoretical considerations. In: White AF and Brantley SL (eds) Chemical weathering rates of silicate minerals. Min Soc Amer Rev Min 31: 353-406

Kroll H, Evangelakakis C, Voll G (1993) Two-feldspar geothermometry: a review and revision for slowly cooled rocks. Contrib Mineral Petrol 114: 510-518

Lagache M (1984) The exchange equilibrium distribution of alkali and alkaline-earth elements between feldspars and hydrothermal solutions. In: Brown WL (ed) Feldspars and feldspathoids: structures, properties and occurrences. NATO ASI Series C137, D Reidel Publishing Company, 247-280

Lagache and Weisbrod (1977) The system: two alkali feldspars–KCl–NaCl–H_2O at moderate to high temperatures and low pressures. Contrib Mineral Petrol 62: 77-101

Lee MR, Waldron KA, Parsons I (1995) Exsolution and alteration microtextures in alkali feldspar phenocrysts from the Shap granite. Mineral Mag 59: 63-78

Lee MR, Parsons I (1995) Microtextural controls of weathering of perthitic alkali feldspars. Geochim Cosmochim Acta 59: 4465-4488

Lee MR, Parsons I (1997) Dislocation formation and albitization in alkali feldspars from the Shap granite. Am Mineral 82: 557-570

Lee MR, Waldron KA, Parsons I, Brown WL (1997) Feldspar–fluid interactions in braid microperthites: pleated rims and vein microperthites. Contrib Mineral Petrol 127: 291-304

Lee MR, Parsons I (1998) Microtextural controls of diagenetic alteration of detrital alkali feldspars: a case study of the Shap conglomerate (Lower Carboniferous), Northwest England. J Sed Res 68: 198-211

Lee MR, Hodson ME, Parsons I (1998) The role of intergranular microtextures and microstructures in chemical and mechanical weathering: direct comparisons of experimentally and naturally weathered feldspars. Geochim Cosmochim Acta 62: 2771-2788

McDowell SD (1986) Compositional and structural state of coexisting feldspars, Salton Sea geothermal field. Mineral Mag 50: 75-84

McLaren AC (1984) Transmission electron microscope investigations of the microstructures of microclines. In: Brown WL ed: Feldspars and feldspathoids: structure, properties and occurrences. NATO ASI Series C, Reidel Publishing Co Dordrecht: 373-409

Montgomery CW, Brace WF (1975) Micropores in plagioclase. Contrib Mineral Petrol 52: 17-28

O'Neill JR, Taylor HP (1967) The oxygen isotope and cation exchange chemistry of feldspars. Am Mineral 52: 1414-1437

Orville PM (1963) Alkali ion exchange between vapor and feldspar phases. Am J Sci 261: 201-237

Parsons I (1978) Feldspars and fluids in cooling plutons. Mineral Mag 42: 1-17

Parsons I, Boyd R (1971) Distribution of potassium feldspar polymorphs in intrusive sequences. Mineral Mag 38: 295-311

Parsons I, Brown WL (1984) Feldspars and the thermal history of igneous rocks. In: Brown WL ed: Feldspars and feldspathoids: structure, properties and occurrences. NATO ASI Series C, Reidel Publishing Co Dordrecht: 317-371

Parsons I, Becker SM (1986) High-temperature fluid-rock interactions in a layered syenite pluton. Nature 321: 764-769

Parsons I, Mason RA, Becker SM, Finch AA (1991) Biotite equilibria and fluid circulation in the Klokken intrusion. J Petrol 32: 1299-1333

Saigal G, Morad S, Bjørlykke K, Egeberg PK, Aagaard P (1988) Diagenetic albitization of detrital K-feldspar in Jurassic, Lower Cretaceous, and Tertiary clastic reservoir rocks from offshore Norway, I. Textures and origin. J Sed Petrol 58: 1003-1013

Smith P, Parsons I (1974) The alkali feldspar solvus at 1 kilobar water-vapour pressure. Mineral Mag 39: 747-767

Taylor HP, Forester RW (1971) Low-^{18}O igneous rocks from the intrusive complexes of Skye, Mull and Ardnamurchan, Western Scotland. J Petrol 12: 465-498

Tullis J (1983) Deformation of feldspars. In: PH Ribbe, ed: Feldspar Mineralogy, Min Soc Amer Rev Mineral 2: 297-323

Velbel MA (1989) Effect of chemical affinity on feldspar hydrolysis rates in two natural weathering systems. Chem Geol 78: 245-253

Walker FDL (1990) Ion microprobe study of intragrain micropermeability in alkali feldspars. Contrib Mineral Petrol 106: 124-128

Walker FDL, Lee MR, Parsons I (1995) Micropores and micropermeable texture in alkali feldspars: geochemical and geophysical implications. Mineral Mag 59: 505–534

Waldron KA, Parsons I (1992) Feldspar microtextures and the multi-stage thermal history of syenites from the Coldwell Complex, Ontario. Contrib Mineral Petrol 111: 222-234

Waldron KA, Parsons I, Brown WL (1993) Solution-redeposition and the orthoclase-microcline transformation: evidence from granulites and relevance to ^{18}O exchange. Mineral Mag 57: 687-695

Waldron K, Lee MR, Parsons I (1994) The microstructures of perthitic alkali feldspars revealed by hydrofluoric acid etching. Contrib Mineral Petrol 116: 360-364

Willaime C, Brown WL (1974) A coherent elastic model for the determination of the orientation of exsolution boundaries: application to the feldspars. Acta Cryst A30: 316-331

Wood BJ, Walther JV (1983) Rates of hydrothermal reactions. Science 222: 413-415

Worden RH, Rushton JC (1992) Diagenetic K-feldspar textures: a TEM study and model for diagenetic feldspar growth. J Sed Petrol 62: 779-789

Worden RH, Walker FDL, Parsons I, Brown WL (1990) Development of microporosity, diffusion channels and deuteric coarsening in perthitic alkali feldspars. Contrib Mineral Petrol 104: 507-15

Chapter 2

Hydraulic properties of Crystalline Rocks

HYDRAULIC PROPERTIES OF THE UPPER CONTINENTAL CRUST:

data from the Urach 3 geothermal well

Ingrid Stober
Geological Survey of Baden-Württemberg,
Albertstr. 5, 79104 Freiburg, Germany
stober@lgrb.uni-freiburg.de

Kurt Bucher
Institute of Mineralogy, Petrology and Geochemistry,
University of Freiburg,
Albertstr. 23b, 79104 Freiburg, Germany
bucher@ruf.uni-freiburg.de

Abstract

The 4500m deep research borehole at Urach (South Germany) has been extensively used for hydraulic testing of the crystalline basement. The data permit a general interpretation of the hydraulic properties of crystalline continental upper crust. The typical granitic and gneissic basement contains an interconnected fluid-filled fracture system and behaves hydraulically like a confined fractured aquifer. Thus standard hydraulic well-tests can be used in the basement. The conclusions are based on data from the central part of the upper crust and are, therefore, believed to be characteristic and significant for the brittle upper continental crust in general.

The performed tests (including a > 500 hours long-term injection test) revealed a hydraulically effective porosity of the basement of typically 0.5 % and an average permeability of about 10^{-9} m/s. NaCl-rich brine with > 100 g/kg total dissolved solids (TDS) occupies the fracture pore space at depth. The basement can be best described as a homogeneous, isotropic aquifer and this characteristic hydraulic behavior persists to at least several hundred meters around the borehole. No evidence for hydraulic infiltration or the existence of impervious boundaries was found in the test data. The homogeneity of the aquifer, together

I. Stober and K. Bucher (eds.), Hydrogeology of Crystalline Rocks, 53–78.

with the highly saline water present in an interconnected system of abundant fractures appear to be characteristic of continental upper crust in general. Similar general aquifer properties were found in other deep boreholes into the crystalline basement of the Black Forest area, in the "Hot-Dry-Rock" well of Soultz-sous-Forêts (France), NAGRA deep wells (N-Switzerland), KTB wells (SE-Germany) and the Kola well (Kola, Russia).

1. Introduction

The continental crust consists, beneath a sedimentary cover of variable thickness, predominantly of diverse gneisses and granitic igneous rocks. The crust may be subdivided into an upper layer that is characterized by brittle deformation and a lower layer characterized by ductile behavior. The depth of the brittle-ductile transition zone depends chiefly on the thermal condition of the crust. However, 12 - 15 km are typical for stable steady state crustal areas with 25°C/km geothermal gradients (Wintch et al., 1995). The brittle upper part of the continental crust is characterized by on interconnected fracture network that provides the storage space for water and the conductivity for the water to flow (Mazurek, this volume). The existence of free water in a fracture pore space in the continental crust has been demonstrated by the German continental deep drilling program (KTB) to 9100m and by the Russian super deep well on the Kola peninsula down to 12000m. The flow of water in the crustal reservoir can be triggered by gravity (topographic flow), thermal or tectonic driving forces. The hydraulic properties of the crystalline basement of the upper continental crust such as permeability (hydraulic conductivity), porosity, storage coefficient, reservoir homogeneity and the dimensions of the reservoir (dimensions of fractures) can be explored by hydraulic well tests carried out in deep boreholes. The knowledge of these properties and parameters is crucial to the understanding of deep groundwater systems in the crystalline basement.

We report here on the planning, the testing and the data analysis of a number of well tests in the 4500 m deep research borehole Urach 3 (South Germany) in the crystalline basement of the Central European continental crust (Fig. 1). The tests included a unique long-term injection test. The Urach gneiss basement is a fractured hard rock aquifer and water is present on an interconnected fracture system to the final depth of the borehole. The Urach basement is a hydraulic conductor and its aquifer character is similar to the hydraulic behavior of the crystalline basement elsewhere, including the KTB borehole (Stober, 1986; 1995; Gustavson and Krasny, 1993). The aquifer behavior of the crystalline basement permits the utilization of well test techniques commonly employed in hydrogeology.

The term "basement" is used throughout this paper to describe a large volume of crystalline rock including its pore space, fracture and fault systems, voids, cavities and all structures that contribute to the hydraulic properties and behavior on a large scale. We prefer this expression over the word "rock" to avoid confusion with the properties of the solid rock matrix investigated in the laboratory. For instance, solid granite matrix has a typical permeability of 10^{-16} m s^{-1} whereas crystalline basement consisting of fractured granite may have a permeability of 10^{-6} m s^{-1}. The permeability of "rock" is measured in the laboratory, the permeability of "basement" is obtained in the field from well test data. For the fluid flow regime in the brittle upper crust the matrix permeability is a meaningless quantity, whereas large scale permeability of the basement as a whole is a quantity of fundamental importance that can be retrieved from hydraulic borehole tests. It is in this sense that we use the term "basement". The macroscopic large scale permeability represents a descriptive parameter that characterizes the conductive properties of the basement aquifer on a scale ranging from a few meters to thousands of meters.

Figure 1: Location of the Urach 3 research well in the state of Baden-Württemberg. The well was drilled into the crystalline basement of the Central European continental crust, the nearest surface outcrop of crystalline rock is the Black Forest to the west of Urach 3.

Hydraulic tests in boreholes give substantial information on aquifer properties controlling the patterns of fluid flow in the basement. Well test data analysis and interpretation provides a thorough insight into the hydraulic behavior of fluids in the crust. The data are evaluated by making use of thoughtful models that appropriately describe fluid flow in the tested portion of the crust. In turn, important aquifer parameters such as permeability and storage properties are computed from the test data based on the inferred model. Consequently, the hydraulic tests in boreholes provide important information on the overall size and structure of the water reservoir and its hydraulic homogeneity. The existence of hydraulic boundaries in an aquifer as well as their hydraulic properties can be detected. The derived aquifer parameter are model dependent. The frequency of fractures intersecting the borehole and their orientation in space can be estimated from well test data to some extent as well (Stober, 1986). The borehole data from Urach 3 also include temperature, electrical conductivity and water composition data. Temperature and electrical conductivity logs can be utilized in locating water inflow points in the basement and the logs yield information on the geothermal gradient and changes in water composition.

2. The research borehole Urach 3

The research borehole Urach 3 is located at the border of the Svabian Alb, SW Germany (Fig. 1). The earliest drilling activity dates back to the late seventies. The top of the crystalline basement of the central European crust is 1604 m below surface (b.s.). The thick sedimentary cover comprises the complete section from the Carboniferous, Permian, Triassic to the Jurassic. The borehole was drilled with research objectives and aims that changed considerably with time. Therefore, drilling took place at different time periods and sessions and with successively increasing "final" depths (Dietrich, 1982; Stober, 1986). A brief history of the Urach borehole follows:

* Initially and to a depth of 2500 m the borehole was used to calibrate various geophysical tools and techniques that were used in the exploration of the heat flow anomaly at Urach.
* Then the well has been deepened to 3334 m. After a series of leak-off-tests, it was used for frac-experiments. The frac-tests, inspired by similar tests in Los Alamos, were undertaken to generate artificial fractures for a single-well "Hot-Dry-Rock" (HDR) system. Artificial fractures were not generated, nevertheless the subsequent circulation tests were successful.
* The well has been deepened further to 3488 m. The temperature at bottom-hole was 147°C. A series of slug-tests, injection tests and circulation tests were carried out. To this date a total of non-recoverable 780 m^3 of water has been irreversibly injected into the crystalline basement.
* After a silent period in Urach, the well has been reactivated. The borehole was not deepened further, however, a new hydraulic research concept was established. The change in the testing and research philosophy was triggered by the rapidly growing evidence that water is universally present in the crystalline basement. The concept of "frac-tests in the crystalline basement" in its original meaning of artificially fracturing compact and intact basement rocks was gradually abandoned. In the new philosophy "frac-tests" expand a naturally present, water-saturated fracture system. Consequently the term "hot-dry-rock" experiments was deserted. The concept of a dry upper crust was replaced by a wet, water-saturated crystalline basement. For the first time ever, a long-term injection experiment of 3 weeks total duration was accomplished.
* Finally, the well was deepened again, now to a final depth 4444 m below surface. The new drilling was carried out in the context of an joint European

HDR-Project. This project at Soultz-sous-Forêts (France) explores the feasibility of the exploitation of geothermal energy using the so called doublet technique. This technique makes use of an injection and an extraction well. The research borehole Urach 3 was used in this context exclusively for the pre-testing of various geophysical techniques and instrumentation.

3. Geology of the Urach 3 drill site

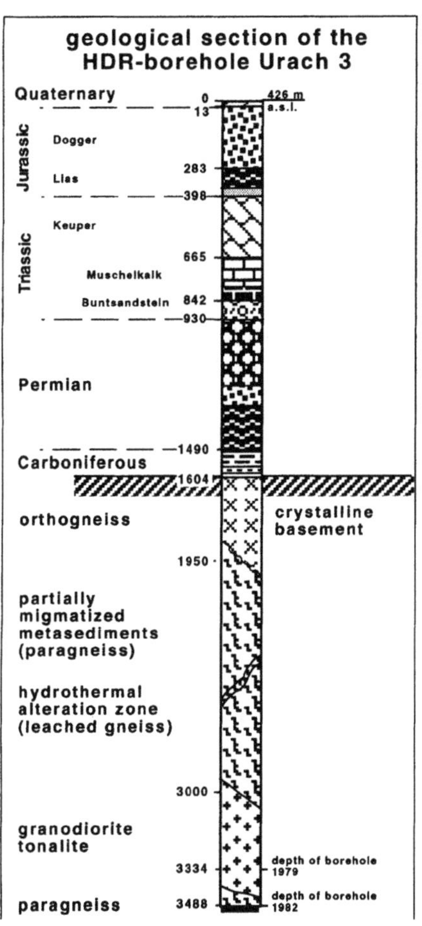

Figure 2: Geology of the well. The basement surface is 1604 m below surface and the basement consists of various rock types typical of upper continental crust. A total of 1884 m was drilled in the crystalline basement.

The crystalline basement at the Urach drill site belongs to the Moldanubian domain of the central European continental crust (Meissner, 1986). The last orogenic reworking of the crust occurred in the Carboniferous. At the drill site the basement is covered by 1604 m thick succession of sediments (Fig. 2). The inventory of crystalline rocks from the borehole includes a banded and interlayered sequence of biotite-amphibole gneiss, migmatitic gneiss, quartz-diorite, biotite-cordierite gneiss and similar rocks (Fig. 2). The rocks from the core-samples appear fresh and undeformed macroscopically. However, small scale hydrothermal alteration zones are frequent (Stenger, 1982). All rocks are strongly altered on a thin section scale (Bauer, 1987). Particularly plagioclase, biotite and cordierite are partly or completely replaced by secondary alteration products such as chlorite, white mica, carbonate and epidote. The alteration of the late Paleozoic mineral assemblages occurred predominately in the early Mesozoic (Lippold and Kirsch, 1994) but may continue until today.

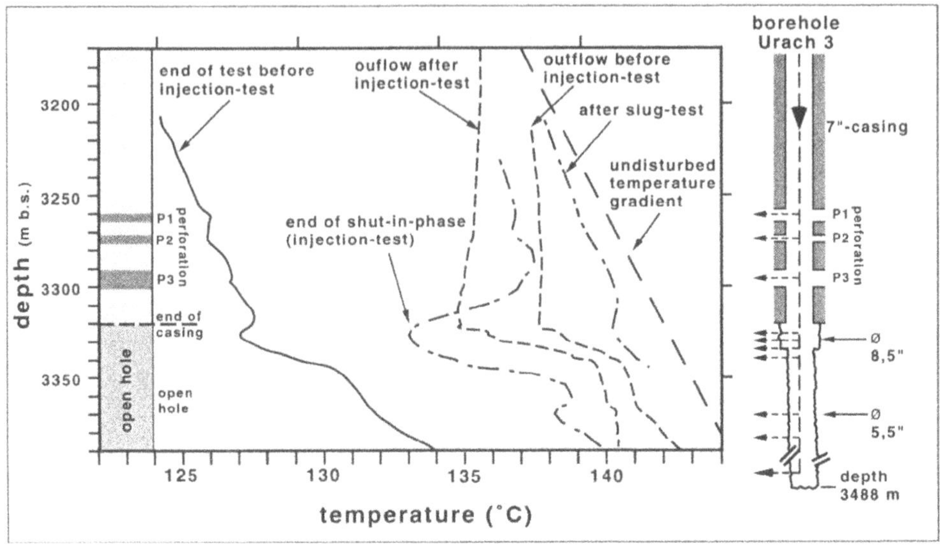

Figure 3: Temperature logs of the deeper part of the well measured before, during and after the different tests. The undisturbed temperature gradient shows the highest temperature, the low-temperature spikes on the logs are related to water outlet points to the basement at the levels indicated by horizontal arrows on the schematic well plan at the right hand side of Fig. 3. The low-temperature log on the left-hand side (solid line) represents the T-distribution prior to the long-term injection experiment. Note the distinct patterns of injection logs and outflow logs. The injected water is cold surface water.

The crystalline basement rocks contain very small intergranular pores. Consequently, the adhesive forces are generally large and hydraulic fracturing of the rocks is not a viable deformation mechanism. The rock matrix is nearly impervious ($k \ll 10^{-12}$ m s^{-1}). On a larger scale, however, the basement rock body is significantly more permeable, due to a ubiquitous fracture system of closed as well as open, interconnected fractures. This fracture system also provides pore space for a highly saline water (TDS > 100 g kg^{-1}). The fractures have variable orientations and dip in the full range of 0 - 90°. Slickensides are rare. Secondary mineral depositions on the fractures are common and species observed include: quartz, calcite (and other carbonates), barite, coelestine, chlorite, feldspars (mainly albite) and various sulfides (Stenger, 1982). Several cm thick leached zones can occasionally be observed in the rock matrix on both sides of an open fracture. In strongly hydrothermally altered migmatitic gneisses secondary minerals evolve from calcite to anhydrite. The system of water-conducting features is characterized by a high abundance of fractures, fissures and faults, 2 to 3 features per m is very common (Dietrich, 1982). The temperature distribution along the drill hole, measured during and shortly after the various hydraulic tests, revealed clearly distinctive zones such as fractures, fracture zones and other deformation zones that were capable of taking up water (Fig. 3).

4. Results from antecedent tests

Numerous hydraulic tests were completed in the crystalline basement of the deep research borehole. Some technical aspects and the concepts of the different tests will be briefly introduced below. Concept, approach and realization of the early tests performed in the borehole, such as leak-off-test, frac-test, originate from the oil and gas industry. After an extensive learning phase it was realized that the basement cannot be regarded as "dry rock". It can best be characterized as a thermal, fractured aquifer. Consequently tests typical of the groundwater industry were carried out in later work in the borehole, such as slug-tests or long-term injection tests.

4.1 Description of the well tests in the crystalline basement

Leak-off-test, frac-test and injection-test all use the same basic principle. A fluid (water) is injected with high pressure into sections of the borehole in the basement. During such tests, the pressure is recorded as a function of time and injection rate. This functional pressure-time-rate relationship is the central information for hydraulic modeling and interpretation (Stober, 1986).

Injection tests: Injection tests are the basic and fundamental tests used to explore the hydraulic properties and behavior of the basement. They are of extended duration and permit conclusions on the hydraulic properties of large volumes of the aquifer around the well. An injection test is usually performed like a well test (pumping test) with a constant injection rate, in contrast to leak-off- or frac-tests. The purpose of a long-term injection test is to characterize the effective hydraulic geometry of the bedrock and to measure the hydraulic and well specific parameters.

Frac-tests: The purpose of frac-tests is to create artificial fractures in order to increase the hydraulic conductivity of the rocks. The fractures are produced by applying very high hydraulic pressures during short test periods, thus irreversibly altering the rock material.

Leak-off-test: A test of short duration is similar to frac- and injection-tests, respectively. Leak-off-tests are often carried out prior to the latter tests in order to pre-test the hydraulic response of the basement, particularly its capacity to take up water.

All three types of tests stress the rocks and high hydraulic pressures ultimately will generate hydraulic fractures. In Urach 3 new fractures could not be generated. During initial pressure increase existing fractures were widened preventing the build-up of the high pressures necessary for new fractures to form. Table 1 lists typical test duration, injected volumes of water, and hydraulic pressures of the various tests run in Urach 3.

Type of test	injection rate Q (m^3/s)	well head pressure p (bar)	duration t (h)
Leak-off-tests	0.001-0.004	120-180	0.07-0.37
Frac-tests	0.002-0.020	315-640	0.12-1.00
Injection-tests	0.007-0.014	330-660	2.00-6.70

Table 1: Typical technical test parameters at the well Urach 3.

After the frac-tests at various depth in the borehole, so called circulation tests were carried out. Water was pumped to a section of the borehole that has been isolated with two packers. Short cut and leakage to other sections of the borehole occurred via an intersecting fracture system. Back flow was collected via a perforated casing in an inner pipe and the flow rate measured at the surface. Successful circulation requires that cool surface water that has been forced into a fracture system at over 3000 m b.s. could be recovered as hot deep water via the connected fracture network. The successful circulation in the basement was verified by a tracer test using Uranine (Stober, 1986).

The use of packers and cement bridges permits the examination and testing of discrete sections of the basement and the borehole (open hole, perforated sections). During and after tests the well is sealed at the surface. There the so called well-head pressure is measured, which is the most important test parameter. The pressure at the base of the borehole during the experiments can be estimated from the well-head pressure and the pressure of the water column in the well. Pressure increase is monitored during the injection phase. The instant when injection stops is called "shut-in". The pressure decrease during this shut-in phase is registered and monitored from this moment on. In order to rapidly dissipate the high pressure in the well after the shut-in test phase, the well-head is opened carefully and portions of the injected water flow out ("bleeding-out" of the well).

The viscosity and density of the water varies continuously at depth during the injection phase. The variations reflect the response to increasing pressure, decreasing temperature and changing hydrochemical properties of the water (usually cold surface water is injected). Decreasing temperature results in a increasing viscosity and density of the fluid and consequently in a decreasing permeability (K-value). In order to compute the physical properties of the injection fluid at depth, water temperature has been monitored at bottom-hole (Fig. 3). The temperature data permit a sensible estimate of the bottom-hole pressure which is given by the sum of the well-head pressure and the weight of the water column in the borehole. The pressure-temperature distribution in the borehole was always in the one-phase liquid field and boiling did not occur at any time.

The computation of the most critical parameters of interest, the transmissivity and the storage coefficient from a hydraulic model depend on reliable estimates of the density and viscosity of the water, which in turn can be computed from the bottom-hole pressure and the measured bottom-hole temperature.

4.2 Summary of important results of the hydraulic testing in the Urach 3 borehole

During the first test series in the crystalline basement in the late 1970ts, the 3334 m deep drillhole was cased down to 3320 m and the open-hole was 14 m long. The first leak-off- and frac-tests were performed in the open-hole. Later, the casing was perforated along three sections (3259-3264 m, 3271-3276 m, 3294-3299 m). Frac-tests were carried out in each of these perforated sections separately using a double packer. In accordance with the "Hot-Dry-Rock" concept of the late 70ts, a gel-substance with proppings was pressed under high pressure into the opened fractures behind the perforated casing after every frac-test in an attempt to keep the "newly generated" fractures open.

After the borehole has been inactive for some years, slug-tests, injection-tests and circulation-tests were performed in order to examine if the hydraulic properties of the basement have changed in the mean time. No significant changes could be observed, although newly formed calcite crystals coated fresh surfaces. Then the well has been deepened until the open-hole was 168 m long.

The computational data analysis of all tests (leak-off-, frac-, injection- and slug-test) revealed the following major results (Stober, 1986):

- The transmissivity (permeability) of the basement increased during each of the tests that exceeded a certain threshold pressure. The highest measured transmissivity at the end of pressure buildup was greater than $T > 10^{-6}$ m^2/s; the smallest value at the beginning of the same test was well below $T < 10^{-7}$ m^2/s. Towards the end of a pressure-decrease phase, the transmissivity approached its original value asymptotically.

- The storage coefficient showed an analogous behavior.
- The wellbore storage computed from the pressure data was consistently larger than that calculated from the well geometry data.
- At the beginning of an injection-test when the effects of wellbore storage ceased, hydraulically relevant fractures could be detected on the pressure data plots. The pressure-drop data immediately after shut-in showed either the influence of a fracture system or a negative skin effect, also diagnostic of a fracture system.

These test results suggest that the crystalline basement responds to increasing water pressure in the following way:

A set of primary and secondary openings connect the borehole with the rock matrix. The voids include borehole break-outs, open fractures and other cavities. They generate together with the volume of the borehole itself a wellbore capacity which is larger than the predicted value on the basis of the borehole-geometry data alone.

During the hydraulic tests, water enters the basement via the borehole and the related void space. The measured variations of the transmissivity and the storage coefficient shows that the bedrock behaves elastically during pressure buildup and pressure drawdown, respectively. With increasing water pressure, fractures widen and pore space and transmissivity increases (Fig. 4) and reversibly decreases as pressure drops. During pressure-buildup existing fractures expand and they collapse upon pressure decrease. This means that the basement takes up ever increasing amounts of water as pressure and injection rates are gradually increased.

From the leak-off-tests it is evident that the elastic behavior of the basement is restricted to pressures above a distinct threshold pressure. At least, elastic behavior was observed in all experiments with well-head pressures above p=176 bar.

A comparison of the hydraulic tests at 3320 m b.s., prior to the deepening to 3488 m, shows that the rocks in the open-hole section are much more transmissive compared to the perforated casing sections, probably because the open-hole section is much longer (14 m) than the basement behind the perforated casing sections where an additional resistance must be overcome. Perforation 1 was more transmissive than number 2 and 3, which were similar.

Propping material pressed into the widened fractures did not measurably change transmissivity. The emplacement of aperture supporting material to keep the hydraulically widened fractures open had no effect in the crystalline basement.

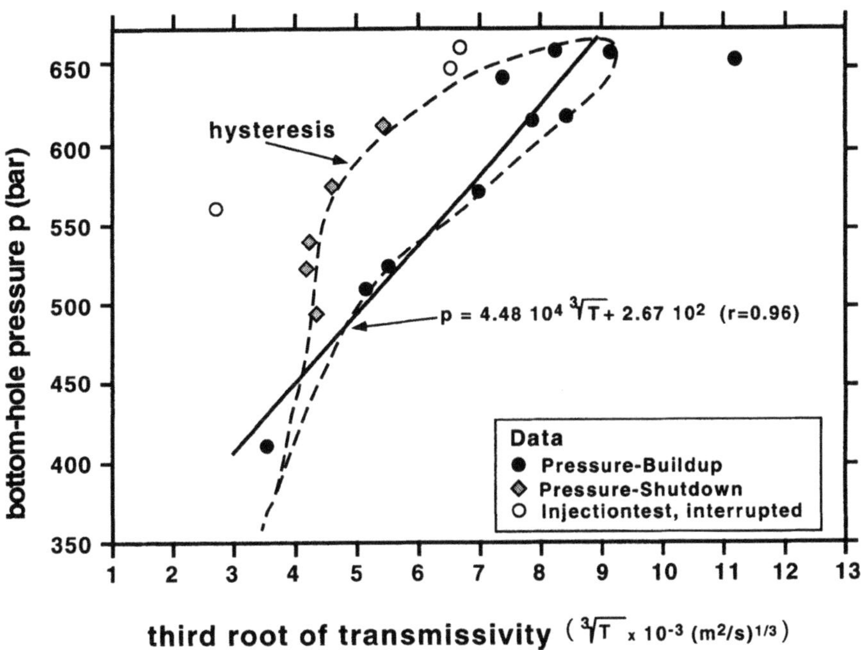

Figure 4: Bottom-hole pressure versus transmissivity relationship of an injection test in the Urach 3 well. The linear relationship between the third root of transmissivity and the bottom-hole pressure during the pressure-buildup phase of the test is followed by a hysteresis during pressure-shutdown.

Slug-tests were carried out just before drilling additional 154 m (Schädel and Stober, 1984a). The slug-tests indicated higher transmissivities than the injection-tests despite the much greater thickness of the hydraulic connection to the rock after the extra meters were drilled. This is an often observed effect of slug-tests which is related to the method. Because of the short test duration, the small volumes of injected fluid and the low pressure, slug-tests characterize the hydraulic properties of the rock close to the drillhole which is in general more fractured as a result of the drilling process and the decompression of the rock. Therefore, the rocks next to the borehole are usually more permeable than the undisturbed rock further away from the hole. After termination of the drill operations several injection tests were performed. They showed that the transmissivities determined from frac-experiments immediately before the deepening of the borehole were significantly lower than from the very first injection test after the drill operations. The original situation was gradually restored with each further injection test. The step-by-step increase of the transmissivity approaching the original value is related to the washout of drilling mud (bentonite) and to the dissolution of secondary minerals deposited during and after the earlier leak-off tests and frac experiments. The injection of water during the hydraulic tests moves the mud, rock fragments and other debris into the fracture system of the rock and spreads it there. The mud front is visible on pressure-time data plots of injection tests as a hydraulic boundary. This boundary moves continually further into the basement from one test to the next.

5. Results from the long-term injection-test

All previous hydraulic tests had the intention, according to the HDR concept at that time, to produce hydraulic fracs, respectively to open and widen the natural fracture system, in order to circulate in a single well system. Therefore, the experimental procedure did not necessarily meet the requirements of a hydraulic test with rigorous quantitative analysis of the data (see above). Particularly, the variable injection rates and the short duration of the tests made a quantitative analysis and interpretation of the data difficult. Because of all this, at Urach 3 a unique and innovative long-term injection-test of altogether 500 hours duration was finally carried out (Fig. 5). This test has been especially designed for geohydraulic research by the first author.

5.1 Planning and realization of the long-term injection experiment

Due to technical and financial restrictions it was not possible to test the three perforated sections of 5 m length each and the open hole of 154 m length separately. The test was performed in the borehole as it was at that time, that is with a 7"-casing (Fig. 3). The estimated bottom-hole pressure is related to the density of the fluid, which in turn depends on temperature, pressure and fluid composition (see above). Therefore, during any hydraulic tests the three variables, temperature, pressure at bottom-hole and the total of dissolved solids (TDS) should be measured continuously (TDS for instance via electrical conductivity). At Urach 3, the electrical conductivity of the injected water at the hydraulically conducting sections of the basement could not be measured for technical reasons. However, the injected water had generally a low TDS and no significant variation of TDS with time because large amounts of cold low-TDS water has been injected during earlier tests. The bottom-hole temperature was continuously measured during the test (Fig. 3).

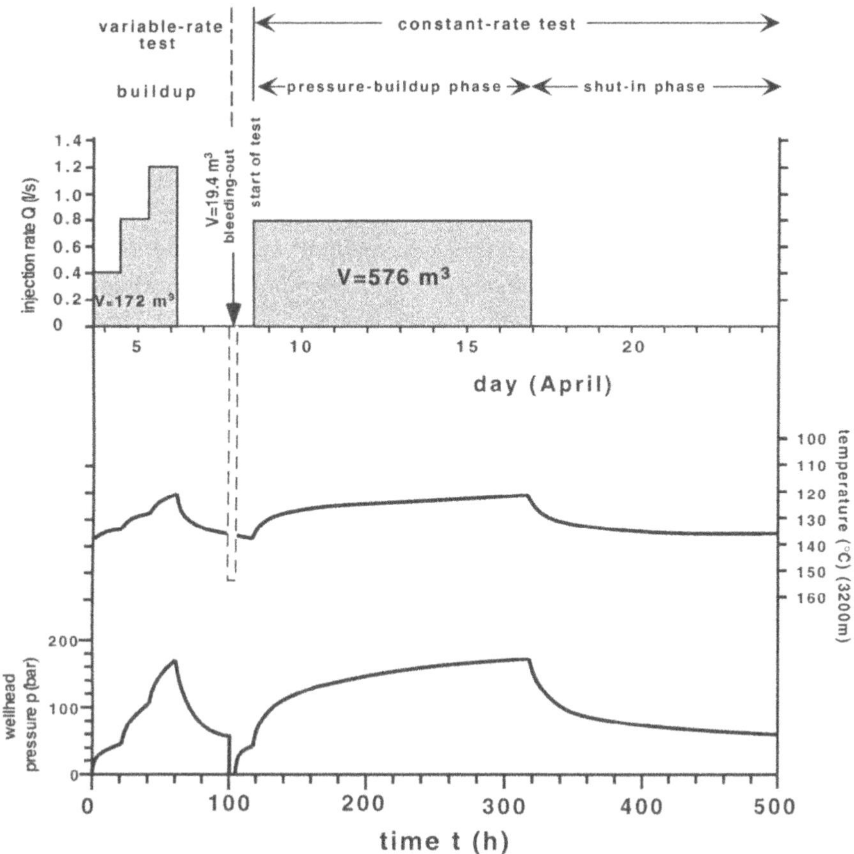

Figure 5: Technical data and technical program of the long-term injection experiment in the Urach 3 research well. Injection rates, temperature and well-head pressure during the experiment are shown. Shaded area indicates total amount of injected water. Initial step-test was run at three different Q.

With the new experiment, it was planed to avoid elastic deformation of the basement rocks. The threshold pressure for opening a fracture, which is the hydrostatic pressure in equilibrium with the lithostatic pressure should not be exceeded. The value of the threshold pressure depends on the geometrical orientation of the fracture in the stress field. In brittle crystalline crust, such as beneath Urach, the vertical pressure component typically exceeds horizontal pressure components which has the consequence that the opening pressure for vertical fractures is smaller than for horizontal fractures for which it is equal to the load of the overburden. Because open fractures have no shear strength one can estimate the opening pressure (p_{op}) from the vertical pressure component (p_v), the horizontal pressure component (p_h) and the dip angle of the fracture (α).

$$p_{op} = 0.5 \left[(p_v + p_h) - (p_h - p_v) \cos 2\alpha \right] \qquad (1)$$

The vertical pressure p_v at Urach can be calculated using a density of $\rho = 2.7\ 10^3$ kg m^{-3} to $p_v = 890$ bar. The horizontal component is estimated from a Poisson number of $\nu = 0.31$ to $p_h = 400$ bar. The most commonly observed dip angles of the open fractures of 20° to 30° result in calculated p_{op} in the range of 450 to 520 bar. The range of these values fit well with the observed opening pressure of 176 bar at the well-head measured in leak-off tests, corresponding to about 500 bar bottom-hole pressure.

The actual injection experiment has been subdivided into two test segments: (i) a step-test with three different injection rates ($0.4\ 10^{-3}\ m^3/s$, $0.8\ 10^{-3}\ m^3/s$, $1.2\ 10^{-3}\ m^3/s$) and 20 hours duration each, a pressure recovery phase and (ii) an aquifer test that has been running for 200 hours (!) with a constant rate of $Q = 0.8\ 10^{-3}\ m^3/s$ and a terminal pressure recovery phase (Fig. 5). The planned injection rates were calculated from the results of the earlier experiments in Urach 3 so that pressures at which elastic deformation occurs were just not reached. For the planning of the aquifer experiment, the results of the step-test immediately run before, were of great value. At the end of the aquifer test a well head pressure of 164 bar was reached. This pressure was still below the opening pressure of the fractures as required by the concept of the experiment. On the other hand, the full possible injection pressure without causing elastic deformation has been used.

Following the step-test, the pressure drawdown was measured during 40 h. Subsequently, the well was left bleeding-out at the well head. The aquifer test was not started before near stationary conditions were reached after renewed pressure buildup. At the end of the injection during the aquifer-test, the pressure of the shut-in phase was monitored for 180 h. Fig. 5 shows the injection rates and the resulting pressure-response at the well-head as well as the continuously measured temperature at bottom-hole.

Temperature measurements (Fig. 3) show that the basement takes up or releases much water along certain sections of the well. Other regions take up little or no water and their hydraulic connection to the drillhole is poor. The temperature logs also show that the fracture zone below the casing-shoe (3325 m b.s.) takes up about 40 % of the injected water. The perforated sections transmitted only minor amounts of water to the basement. The deepest part of the open hole, which is inaccessible for geophysical measurements with tools because of technical reasons (two different hole diameters), consumes 55 % of the injected water.

5.2 Analysis and evaluation of the long-term injection experiment

The quantitative analysis of the hydraulic test data from the research borehole Urach 3 relies on a single source of information on the response to the input signal: the measured pressure-time relationship at the well-head of the Urach 3 well. Separate observation wells are not available.

A first overview on the relationship between injection rate and pressure buildup is provided by the step-test. Fig. 6 shows the measured well-head pressures plotted against time for different, constant injection rates. Extrapolations were calculated from hydraulic computer programs appropriate for fractured aquifers (Stober, 1986). It can be seen that the well-head pressure has approximately doubled and tripled, respectively, from the first to the second and then to the third step. Therefore, rate and pressure are proportional. The take-up capacity of the basement remained unchanged when injection rates were increased by discrete increments. With increasing pressure transmissivities remained unchanged indicating that compression of the basement rocks was small. This in turn shows that the experiment, as desired, remained below the critical pressure for fracture opening.

The next step in the geohydraulic data evaluation involved the selection of an appropriate hydraulic model. Models for different infiltration geometry, spherical or radial symmetric (cylindrical) from the borehole, vary considerably. A comparison of the pressure-time data of the actual aquifer test and the mathematical formulation for a point sink (Carslaw and Jaeger, 1959) showed that the injected water did not infiltrate the basement spherically from the borehole.

The relationship of the pressure build-up and pressure shut-down data versus the logarithm of time during the step test and the aquifer test (p vs log t and log p vs log (t + Δt)) shows that this relationship becomes log linear after about 7 hours experimental time. Such a log-linear relationship is consistent with a continuous unlimited radial-symmetric dispersion of the injected water (Kruseman and De Ridder, 1991; Stober, 1986). The data do not indicate any form of approaching a capacity limitation. The basement apparently has an unlimited capacity to take up water. The basement behaves hydraulically like any other infinitely extended confined aquifer. The slope of the straight line relationship is related to the transmissivity, as a measure of the average permeability of the tested section of the basement: $T = 2.5\ 10^{-7}\ m^2\ s^{-1}$.

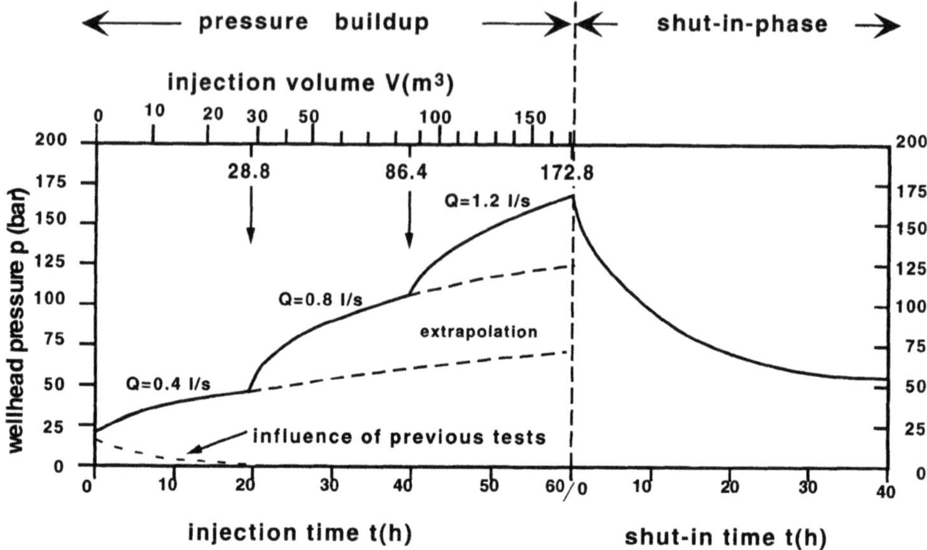

Figure 6: Pressure buildup and pressure shutdown data of the step-test run with three different injection-rates. Dashed curves show extrapolated pressure-time relationship.

The storage coefficient, a measure of the capacity of the basement to take-up water can be estimated from:

$$S = H \cdot \rho \cdot g \cdot \phi \cdot c_t \qquad (2)$$

The "thickness of the aquifer" (H), the total length of all individual hydraulic tests sections is H = 169 m; g is the acceleration due to gravity, ρ = density of water. The porosity ϕ (the fraction of open fracture space of the basement) in the area of the test sections of Urach 3 has been calculated from previous test data and ranges between $\phi = 0.002$ and $\phi = 0.005$ (0.2 - 0.5 %). For comparison, the range of porosities calculated from well test data of pumping tests in wells of the entire Black Forest basement is $\phi = 0.001$ to $\phi = 0.021$. The total compressibility c_t of the system can be approximated by the compressibility of water c_w ($c_w = 4.2\ 10^{-10}\ Pa^{-1}$) because the basement rocks are much less compressible than water. Calculated storage coefficients were between $S = 1.4\ 10^{-6}$ and $S = 3.4\ 10^{-6}$.

Prior to the pseudo-radial flow period (the linear behavior between p and log t) other, transient, flow conditions dominated water infiltration. Wellbore storage effects alone cannot be responsible for such long-lived effects (t = 6.9 h). Additional effects, like the influence of a few large fractures affected the measured pressure-time relationship during the early stages of the test.

The measured pressure data of the injection tests, particularly of the long-term aquifer test are in excellent agreement with the theoretical flow model of Gringarten and Ramey (1974), describing the flow behavior during a well-test in an aquifer with a radial symmetrical horizontal fracture of finite dimension (Fig. 7). The examination of the measured data using the theoretical Gringarten and Ramey model results in calculated transmissivities in radial direction (direction of the fracture) of $T_{(f)} = 7.3\ 10^{-7}\ m^2\ s^{-1}$. Furthermore, the product of the storage coefficient and the square of the radial dimension of the fracture is $S \cdot r_{(f)}^2 = 3.74\ 10^{-1}\ m^2$. The calculated radius of the fracture between $r_{(f)} = 332$ m and $r_{(f)} = 517$ m follows from the estimated value of the storage coefficient. From the parameter $h_D = T_{(f)}.H / r_{(f)}^2.k_{fz} = 10$ (Fig. 7) a vertical permeability, normal to the fracture of $k_{fz} = 1.1\ 10^{-10}$ m/s to $k_{fz} = 4.6\ 10^{-11}$ m/s can be computed.

The deduced fracture radii must be understood as hydraulically effective fracture radii. The model concept conceives one single fracture, which is continuously connected and filled with movable compressible water. In reality, many fractures are present, the fractures have variable and discontinuous apertures, there are supporting contact areas along the fractures with adhesive, fixed, stationary water. Flow occurs predominantly along discrete channels rather than continuously along the entire fracture surface.

The calculated total transmissivity $T = 2.5\ 10^{-7}\ m^2\ s^{-1}$ is referred to the hydraulically tested basement section, with several points acting as water inlets to the basement (Fig. 3). The inlet-points that take up water are fractures, that is structures of increased permeability that are being intersected by the borehole at a certain defined location. The fractures may intersect the borehole under various but unknown angles. In the mathematical-analytical model used for data analysis, all fractures that take up water were described by one single model fracture which is characterized by the averaged geometrical properties of all real fractures. The used horizontal model fracture represents the complete set of the predominantly flat lying real fractures in the basement that actually have been measured in the cores of the drillhole.

The data also provide some clues on the time-dependent flow behavior during an injection or shut-in phase in a fractured aquifer. During the first minutes of the injection experiment, the wellbore storage can just barely be seen. The wellbore storage here is determined from the borehole volume under test conditions and all breakouts and fractures present in the vicinity of the borehole. This flow period can be seen on Fig. 7 at about $t < 200$ s. During the subsequent five hours of the injection test one can identify the mentioned linear flow period on the log p versus log t plot (Fig. 7) given by a straight line through the data with a theoretical model slope $m = 0.5$.

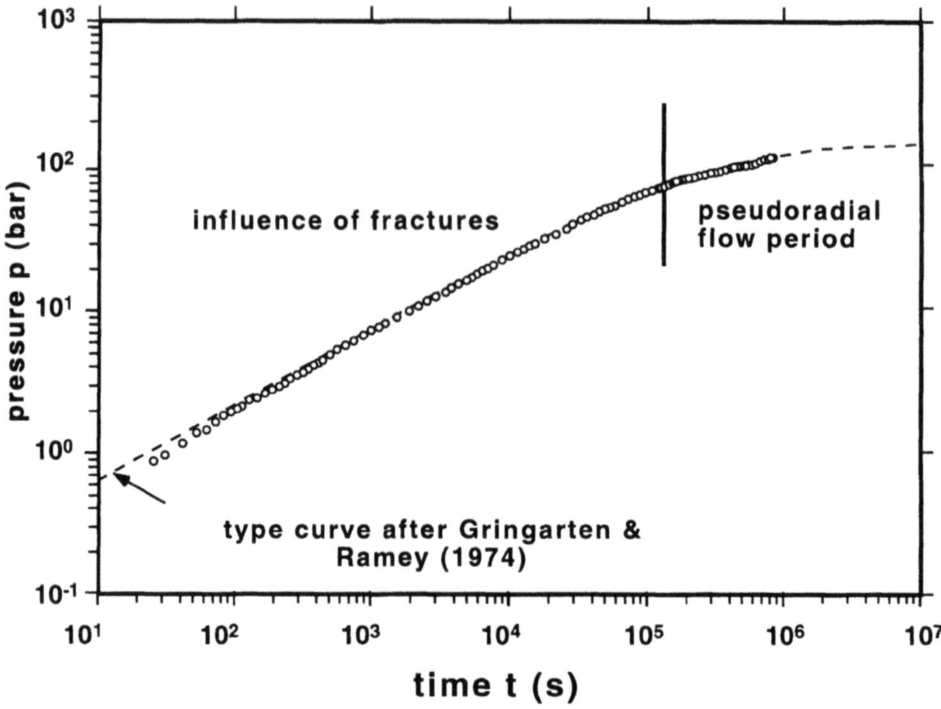

Figure 7: Logarithmic pressure versus time plot (Horner plot) of the long-term injection test. Wellbore storage effects can be barely seen at $t < 100$ s, later the flow regime is dominated by the influence of fractures, at $t > 10^5$ s the basement behaves like a homogeneous, isotropic aquifer (pseudoradial flow period).

5.3 Hydrogeological interpretation

The hydrogeological interpretation of the observed data pattern is, that the injected water enters from the borehole in different and variably inclined fractures but with an average overall property of a nearly horizontal fracture. From this network of intersecting fractures the injected water reaches greater distances from the borehole. This is equivalent in the hydraulic model to the transfer of water from the horizontal fracture vertical into the basement (linear vertical flow period). After a transition period of about 2 hours, the so-called pseudo-radial flow period is firmly established (Fig. 7). This means, with respect to the hydraulic model, that the horizontal fracture with its finite extension looses continuously its hydraulic importance. The average hydraulic properties of the entire basement gradually dominate the flow regime when continued injection of water reaches ever more distant volumes of the basement.

The characteristics of the pressure decrease during shut-in after the injection phase (Fig. 5) show that the injected water continues to infiltrate the basement until the original pressure situation is gradually approached.

If well tests are run long enough in the crystalline basement, pseudo-radial flow period conditions are established like in porous aquifers. The basement behaves, after the first short flow periods, like a homogeneous isotropic aquifer. This observation strongly suggests that water-conducting features with widely differing orientations, random and regular distribution and frequent intersections are present in the basement over large regions. The data did not indicate effects of either external infiltration or the presence of impervious boundaries. The basement-cover interface ca 1900 m above the open-hole cannot be seen in the data patterns. It is expected that major fault zones in horizontal direction from the borehole would be recognized first. The absence of boundary effects in the data set suggests that if major hydraulic barriers were present in the basement they would be at least several hundred meters away from the borehole.

It can be concluded from the long-term injection experiment that under the experimental conditions the crystalline basement is able to take-up surprisingly large, in fact unlimited, amounts of water. During injection, pressure increase with time diminishes ($d^2p/dt^2 < 0$). The injected water infiltrates the basement along an interconnected fracture system. The basement behaves like an aquifer. The injected water uses space that results from compression of the water in the fracture pore space and from the replacement of in-situ water present in the reservoir prior to the experiment. This hydraulic reaction is typical for confined aquifers and thus permits the operation of hydraulic well tests in basement drillholes. Such tests typical of the groundwater industry can be prognosticated and analyzed in crystalline basement wells like well tests in any other fractured aquifer.

6. Comparison with borehole data from the crystalline basement of the Black Forest area

Compared with permeability data from wells in the crystalline basement of the central Black Forest (Fig. 1), the data from the research well Urach 3 are relatively low (Stober, 1995; Stober and Bucher, 1998). The calculated transmissivity together with the length of the hydraulic test section results in the calculated permeability of T/H = K = 1.5 10^{-9} m/s for the gneissic basement at the Urach 3 location. The range of permeabilities of the Black Forest crystalline basement is K = 3.5 10^{-10} m/s and 8.7 10^{-5} m/s (!) with an average of 2.1 10^{-7} m/s. The individual values follow a log-normal distribution (Stober, 1995). Permeabilities of the basement may reach values that are normally typical of porous gravel aquifers.

Granitic basement is, on an average, more permeable than gneissic basement. The average for granitic parts of the Black Forest basement is 9.6 10^{-7} m/s, in areas dominated by metamorphic rocks the permeability is lower by a factor of 20 on an average, about 5.0 10^{-8} m/s (Stober, 1996a). The permeability of the basement beneath Urach is therefore considerably lower than the average of the Black Forest gneiss basement.

Gneisses have a relatively high modal content of micas and other sheet silicates that can be easily deformed and a layered structure with parallel oriented micas in bands or lenses. The response to tectonic stress is mainly by ductile deformation. The brittle deformation behavior of feldspar-rich granites explains their higher average fracture density and consequently the higher hydraulic conductivity of granites compared with gneisses in the Black Forest.

In contrast to granites, the permeability of metamorphic basement decreases slightly with depth because of diminishing near-surface effects such as slope creep, weathering, influence of the ice ages and smaller decompression effects, which in turn may explain the relatively low permeability at Urach.

The most significant water conducting features in the crystalline basement are, in general, related to faults and fault zones, along contacts between different lithological units and along alteration zones related to old flow systems of hydrothermal fluids (Stober, 1995; Mazurek, this volume). These highly conductive zones can be a few centimeters to several meters thick. Spatial orientation and frequency of the high permeability zones may vary considerably. Therefore, the high permeability zones in the basement form an interconnected network in which the actual water transport takes place. This network can be very dense and frequent, particularly in granitic basement, such that the flow behavior approaches that of a homogeneous isotropic aquifer with corresponding high permeability right from the beginning of a hydraulic test.

It has been concluded from the injection test, that the flow behavior of the Urach gneiss basement locally controlled by a relatively small number of important water conducting fractures, on a larger scale (with an increasing volume of infiltrated basement) flow behavior corresponds to flow in a homogeneous porous medium.

In addition to this hydraulic behavior observed in the Urach gneiss basement, two other types of flow behavior were found in the basement of the Black Forest: a) in granitic basement the early local fracture dominated flow phase could not normally be detected and flow corresponds to flow in a homogeneous isotropic aquifer also on a small scale. This is particularly the case in strongly fractured granites. b) in some wells the pressure-time relationships of the well-tests indicated the presence of a hydraulically effective boundary within the tested volume and (or) leakage in the vicinity of mineralized veins or major fault zones.

7. The continental upper crust: hydraulic properties and a summary of fluid properties

The hydraulic properties of the crystalline basement beneath Urach were deduced from the analysis of well test data as outlined above. In the upper crust of the continents brittle deformation prevails. Its lower boundary is given by the brittle-ductile transition zone. The depth of which in turn depends mainly on the local thermal state of the crust. The ductile crust below the brittle-ductile transition zone is typically devoid of free aqueous fluid phase (Frost and Bucher, 1994). Regular upper crust is about 12-15 km thick (Wintsch et al., 1995). The Urach upper crust may be considerably thinner because of the high geothermal gradient, the Urach heat flow anomaly (Schädel and Stober, 1984b; 1984c; Fuchs, 1986). The evaluation of seismic reflection data suggested the presence of a large low-velocity body in the middle and deep crust at Urach (Bartelsen et al., 1982).

The hydraulic properties of the crust deduced in this paper are based on data retrieved from 4.5 km depth and are significant for a tested volume of at least several hundred meters around the open-hole. Therefore, the derived properties from the middle of the brittle crust at Urach are believed to be characteristic and significant for the brittle upper continental crust in general. The validity of this extrapolation is supported by similar findings and conclusions by the KTB program (Emmermann and Lauterjung, 1997).

The permeability of the basement is rather low with a characteristic value of 1.5 10^{-9} m/s. Average permeability of gneisses in the Black Forest basement to the west of Urach is about 2.1 10^{-7} m/s (Stober, 1996b; Stober and Bucher, 1999a). The crystalline basement behaves like a confined aquifer and has an unlimited capacity to take-up water from an external source. The calculated porosity of the basement ranges from 0.2 to 0.5 % (Black Forest basement 0.1 to 2.1 %). This porosity is largely related to fractures and faults, the permeability relates to the connectivity of this fracture and fault system. Water-filled open and interconnected fractures characterize the hydrogeology of the basement at 4.5 km depth at Urach.

7.1 Water in the upper continental crust - a synopsis of data, observations and ideas

- The geodynamic state of the crust largely controls the overall aspect of fluid flow and fluid composition in the crust. Active geothermal areas, active continental rifts, areas of ongoing continental collision and present day lithosphere subduction are examples of tectonic settings characterized by ongoing vigorous and dynamic processes of fluid generation (e.g. dehydration), transport (e.g. thermal convection) and fluid consumption (hydrothermal alteration). However, the vast bulk of the continental crust is geodynamically inactive and the hydrogeological summary below refers to inactive brittle upper continental crust, that is the normal state of the crust.

- Open water-bearing fractures and fault zones with circulating water were encountered in all deep boreholes that were drilled into the crystalline basement of the continents so far, including the Kola borehole (12500 m) and the KTB borehole (9100 m) (Kozlovsky, 1984; Emmermann et al., 1995).
- We are unaware of any deep boreholes into crystalline basement that was 'dry'. The brittle upper crust is 'wet'. More precisely, the grain boundaries of the rock matrix are usually 'dry' but water is prevailingly present in the fracture pore space and other macroscopic water-containing features.
- The composition of water in the brittle upper crust is characterized by high amounts of total dissolved solids (TDS) and Na-Ca-Cl brines with TDS > 100 g/kg are typical (Frape and Fritz, 1987; Stober, 1995; Stober and Bucher, 1999a; Edmunds and Savage, 1991; Pauwels et al., 1993; Althaus, 1982; Bucher and Stober, this volume). The dominant anion is chloride, dominant cations are Na and Ca in varying proportions.
- The residence time of these waters is large, millions of years rather than thousands of years. Evidence comes from presence of preserved fossil seawater at depth in basement areas that had a marine cover millions of years ago (Stober and Bucher, 1999b). Rapid exchange with surface water does not occur due to the very low hydraulic potential gradients combined with low permeabilities.
- The permeability of the crystalline basement gradually decreases at great depth, also in brittle granitic basement. The abundance of open water-filled fractures decreases with depth and the mesh size of the network of intersecting fractures increases (Stober, 1995).
- The salinity of the water increases with depth (Gascoyne and Kamineni, 1993; Stober and Bucher, 1999a) and the brines may be in equilibrium with halite and other salts at great depth (Markl and Bucher, 1998; Stober and Bucher, 1999b).
- High salinity and high abundance of brine in the crust may locally lead to a high electrical conductivity of the crust that can be detected by magnetotelluric methods (Haak and Hutton, 1986, Jones, 1992). Abnormally high conductive crust requires NaCl-brine and very high porosity. The porosity (0.5 %), permeability (10^{-9} m/s) and chemical composition of the water (> 100 g/kg TDS) in the basement at Urach is insufficient for a "electrical conductivity anomaly". The crust is resistive. In general, magnetotelluric sounding does not "see" the wet crust.
- Water-filled fractures may also be the cause of seismic reflections in the crust (Smithson et al., 1979; Mair and Green, 1981; Jones and Nur, 1984). However, none of the water-conducting features described in this paper from Urach or from the Black Forest basement (Stober, 1995) is associated with a necessary zone of density contrast of sufficient thickness and extension to create a

detectable seismic reflector. The upper continental crust is seismically transparent (Warner and McGeary, 1987) and electrically resistive (Haak and Hutton, 1986) despite that it contains saline water in the pore space of an interconnected fracture system.

- The few seismic reflectors and electrically conductive zones that can be seen in the crystalline basement of the upper crust are probably related to graphite-bearing crustal shear- and fault-zones (Warner and McGeary, 1987; Emmermann and Lauterjung, 1997).
- The hydraulic potential (hydraulic head) in the Urach 3 borehole has been measured over a long period of time and at all stages of the year-long drillhole history at different depth of the borehole. The hydraulic potential decreases significantly with increasing depth. Also in the KTB main hole (9100 m) the hydraulic potential is significantly decreasing with depth and has not yet reached steady state (Schulze et al., this volume). The water system connected with increasing depth to the borehole is at a gradually lower hydraulic pressure. This means that water is actively used up by the basement at depth. The water consumption is probably related to the hydration of unstable mineral assemblages in crystalline basement rocks. The process contributes to the increasing salinity at increasing depth.
- The mechanism of water transport from higher-levels in the crust to the consumption area at depth is unknown at present, but we suggest here that tidal pumping analogous to seismic pumping could be a viable means of water transport to the depth. Tidal effects of about 12 cm amplitude has been reported from the Urach 3 well (Stober, 1992). Tidally induced water-table fluctuations have been observed also in the KTB pilot hole (4000 m) and the Kola borehole, respectively (Schulze et al., this volume).
- The open interconnected fracture system in the upper crust gradually looses its connectivity with depth. Extended fractures pinch off and neck-down leading to isolated water-filled regions. These regions are cut-off from any recharge and take a local, isolated evolution path. With increasing temperature at depth the water in these fluid pockets is consumed by hydration reactions until the system reaches equilibrium or all water has been used up (Frost and Bucher, 1994).

Acknowledgments

The technically and financially extensive well tests reported in this communication required the effort of many institutions and individuals. We thank all who contributed to the successful operation of the tests. The Urach 3 project has been financed by Bundesministerium für Forschung und Technologie (BMFT). The overall concept of the well tests at Urach 3 research well has been designed by the Geological Survey of Baden-Württemberg by the first author. The long-term injection test has been technically carried out by the Survey of Lower Saxony in Hanover (BGR) with their equipment and under the auspices of Dr. Jung. The temperature logs were taken by the Federal Geological Survey in Hanover under the guidance of Mr. Zoth. Responsible for technical aspects of the well tests was Dr. Dietrich from Stadtwerke Bad Urach now BGR. Helpful reviews by Dirk Schulze-Makuch and Jan Cramer and the help by Arne Bjørlykke are gratefully acknowledged.

References

Althaus, E. (1982) Geochemical problems in fluid-rock interaction. In: *The Urach Geothermal Project*, 123-133. Haenel, R. (editor) Schweizerbart'sche Verlagsbuchhandlung: Stuttgart.

Bartelsen, H., Lueschen, E., Krey, Th., Meissner, R., Schmoll, H., and Walter, Ch. (1982) The combined seismic reflection-refraction investigation of the Urach geothermal anomaly. In: *The Urach Geothermal Project*, 247-263. Haenel, R. (editor) Schweizerbart'sche Verlagsbuchhandlung: Stuttgart.

Bauer, F. (1987) Die Kristallinen Gesteine aus der Bohrlochvertiefung Urach 3 und ihre fluiden Einschlüsse: Eine Interpretation der hydrothermalen Überprägung anhand der Fluid-Daten aus Einschlußmessungen, *Dissertation at Universität (T.H.) Fridericiana Karlsruhe*, 118

Bertleff, B., Joachim, H., Koziorowski, G., Leiber, J., Ohmert, W., Prestel, R., Stober, I., Strayle, G., Villinger, E. and Werner, J. (1988) Ergebnisse der Hydrogeothermiebohrungen in Baden-Württemberg, *Jh. geol. Landesamt Baden-Württemberg*, 30, 27-116.

Carlslaw, H. and Jaeger, J.C. (1959) Conduction of heat in solids, *Clarendon Press, Oxford*, 510 p.

Chester, F.M. (1995) A rheologic model for wet crust applied to strike-slip faults, *Journal of geophysical Research*, 100, 13033-13044.

Dietrich, H.-G. (1982) Geological results of the Urach 3 Borehole and the Correlation with other Boreholes. In: *The Urach Geothermal Project.* Haenel, R. (editor) Schweizerbart'sche Verlagsbuchhandlung: Stuttgart.

Edmunds, W. M. and Savage, D. (1991) Geochemical characteristics of groundwater in granites and related crystalline rocks. In: *Applied Groundwater Hydrology, a British Perspective*, 199-216. Downing, R. A. and Wilkinson, W. B. (editors) Clarendon Press: Oxford/U.K.

Emmermann, R., Althaus, E., Giese, P. and Stöckhert, B. (1995) KTB Hauptbohrung Results of Geoscientific Investigation in the KTB Field Laborarory, Final Report: 0-9101 m, *KTB Report 95-2, Schweizerbart'sche Verl. Stuttgart.*

Emmermann, R. and Lauterjung, J. (1997) The German Continental Deep Drilling Program KTB, *Journal of geophysical Research,* 102, 18179-18201.

Frape, S. K. and Fritz, P. (1987) Geochemical trends for groundwaters from the Canadian shield. In: *Saline water and gases in crystalline rocks,* 19-38. Fritz, P. and Frape, S. K. (editors) The Runge Press Limited: Ottawa.

Frost, B.R. and Bucher, K. (1994) Is water responsible for geophysical anomalies in the deep continental crust? A petrological perspective, *Tectonophysics,* 231, 293-309.

Fuchs, K. (1986) Intraplate seismisity induced by stress concentration at crustal heterogeneities - the Hohenzollern Graben, a case history. In: *The nature of the lower continental crust,* 119-132. Dawson, J. B., Carlswell, D. A., Hall, J., and Wedepohl, K. H. (editors) Geological Society Special Publication.

Gascoyne, M. and Kamineni, D. C. (1993) The hydrogeochemistry of fractured plutonic rocks in the canadian shield. In: *Hydrogeology of Hard Rocks,* 440-449. Banks, S. B. and Banks, D. (editors) Geol. Survey of Norway: Trondheim.

Gringarten, A.C. and Ramey, H.J. (1974) Unsteady-state pressure distributions created by a well with a single horizontal fracture, partial penetration, or restricted entry, *Soc. Petrol. Engineers Journ.,* 413-426.

Gustavson, G. and Krasny, J. (1993) Crystalline rock aquifers: their occurrence, use and importance. In: *Hydrogeology of Hard Rocks,* 3-20. Banks, S. B. and Banks, D. (editors) Geological Survey of Norway: Trondheim.

Haak, V. and Hutton, R. (1986) Electrical resistivity in continental lower crust. In: *The nature of the lower continental crust,* 35-49. Dawson, J. B., Carswell, D. A., Hall, J., and Wedepohl, K. (editors) Geological Society Special Publication.

Jones, A. G. (1992) Electrical properties of the lower continental crust. In: *Continental Lower Crust.* Fountain, D. M., Arculus, R., and Kay, R. W. (editors) Elsevier: Amsterdam.

Jones, T. and Nur, A. (1982) Seismic velocity and anisotropy in mylonites and the reflectivity of deep crustal fault zones, *Geology,* 10, 260-263.

Jones, T. and Nur, A. (1984) The nature of scismic reflections from deep crustal fault zones, *Journal of geophysical Research,* 89b, 3153-3173.

Kozlovsky, Ye.A. (1984) The world's deepest well, *Scientific American,* 251, 106-112.

Kruseman, G.P. and De Ridder, N.A. (1991) Analysis and Evaluation of Pumping Test Data, *ILRI publication 47, 2nd ed. Wageningen / The Netherlands,* 377

Mair, J.A. and Green, A.G. (1981) High-resolution seismic reflection profiles reveal fracture zones within a 'homogeneous' granite batholith, *Nature,* 294, 439-442.

Markl, G. and Bucher, K. (1998) Composition of fluids in the lower crust inferred from metamorphic salt in lower crustal rocks, *Nature,* 391, 781-783.

Meissner, R. (1986) Twenty years of deep seismic reflection profiling in Germany - a contribution to our knowledge of the nature of the lower Variscan crust. In: *The nature of the lower continental crust*, 1-10. Dawson, J. B., Carlswell, D. A., Hall, J., and Wedepohl, K. H. (editors) Geological Society Special Publication.

Pauwels, H., Fouillac, C. and Fouillac, A.-M. (1993) Chemistry and isotopes of deep geothermal saline fluids in the Upper Rhine Graben: Origin of compounds and water-rock interactions, *Geochimica et Cosmochimica Acta,* 57, 2737-2749.

Schädel, K. and Stober, I. (1984a) Auswertung der Auffüllversuche in der Forschungsbohrung Urach 3, *Jh. geol. Landesamt Baden-Württemb.* 26, 27-34.

Schädel, K. and Stober, I. (1984b) Die Wärmeanomalie Urach aus geologischer Sicht, *Jh. geol. Landesamt Baden-Württemberg,* 26, 19-25.

Schädel, K. and Stober, I. (1984c) Gibt es thermische Stabilitätsgrenzen in der Erdkruste?, *Jh geol.Landesamtes Baden-Württemberg,* 26, 7-18.

Smithson, S. B., Brewer, J., Kaufman, J. S., Oliver, J. and Hurich, C. (1979) Structure of the Laramide Wind River Uplift, Wyoming, from COCORP deep reflection data and from gravity data, *Journal of geophysical Research,* 84, 5955-5972.

Stenger, R. (1982) Petrology and Geochemistry of the Basement Rocks of the Research Drilling Projekt Urach 3. In: *The Urach Geothermal Project,* 41-48. Haenel, R. (editor) Schweizerbart'sche Verlagsbuchhandlung: Stuttgart.

Stober, I. (1986) Strömungsverhalten in Festgesteinsaquiferen mit Hilfe von Pump- und Injektionsversuchen, *Geologisches Jahrbuch,* Reihe C, 204 p.

Stober, I. (1995) Die Wasserführung des kristallinen Grundgebirges, *Enke-Verlag, Stuttgart,* 191 p.

Stober, I. (1996) Researchers Study Conductivity of Crystalline Rock in Proposed Radioactive Waste Site, *EOS, Trans. American Geophysical Union,* 77, 93-94.

Stober, I. and Bucher, K. (1999a) Deep groundwater in the crystalline basement of the Black Forest region, *Applied Geochemistry,* 14, 237-254.

Stober, I. and Bucher, K. (1999b) On the origin of salinity of deep groundwater in crystalline rocks, *Journal of Conference Abstracts,* 586-587.

Warner, M. and McGeary, S. (1987) Seismic reflection coefficients from mantle fault zones, *Geophysical Journal of the Royal Astronomical Society,* 89, 223-230.

Wintsch, R. P., Christoffersen, R. and Kronenberg, A. K. (1995) Fluid-rock reaction weakening of fault zones, *Journal of geophysical Research,* 100, 13021-13032.

IN-SITU PETROHYDRAULIC PARAMETERS FROM TIDAL AND BAROMETRIC ANALYSIS OF FLUID LEVEL VARIATIONS IN DEEP WELLS: SOME RESULTS FROM KTB

KATJA C. SCHULZE AND HANS-JOACHIM KÜMPEL
Geological Institute, University of Bonn
Nußallee 8, 53115 Bonn
schulze@geo.uni-bonn.de

AND

ERNST HUENGES
GeoForschungsZentrum Potsdam
Postfach 600751, 14473 Potsdam
huenges@gfz-potsdam.de

Abstract. Natural fluid level fluctuations in sufficiently deep wells reflect pore pressure variations of the hydraulically connected formations, that are e. g. caused by tidal strain, changes in barometric pressure, or the passage of seismic wave fields. From comparison of the fluid level variations with the source signal in-situ petrohydraulic rock properties may be derived. Aiming to improve our knowledge about the petrohydraulic conditions and the volume strain spectrum in the intermediate crust, quasi continuous, high resolution fluid level registrations are carried out at the KTB (super-) deep boreholes in Bavaria, Germany, since summer 1996. The fluid level of the 4 km deep pilot hole shows a clear tidal signal with peak to peak amplitudes of 13 cm and a static confined barometric efficiency of about 60 to 65 %. Adopting poroelastic rheology, the areal strain sensitivity was found to be 0.16 hPa/nϵ (M2) and 0.19 hPa/nϵ (O1), which is consistent with a Skempton ratio B of roughly 0.40 to 0.45, respectively, when adopting a drained Poisson ratio ν of 0.24. The fluid level of the main hole (9.1 km deep) does not present any tidal or barometric signals, most likely because of some obstruction at depth, but yet shows a monotonous lowering.

Key words: deep wells, fluids in the crust, tides, poroelasticity.

I. Stober and K. Bucher (eds.), Hydrogeology of Crystalline Rocks, 79–104.

1. Introduction

Pore pressure in porous rocks is known to play an important role in transmitting crustal stresses. A fluid injection experiment in the KTB-main hole has shown that a pore pressure rise of 1 % of the ambient pressure is sufficient to trigger earthquakes (*Zoback & Harjes*, 1997). Natural pore pressure variaions are i.g. induced by tidal strain, barometric pressure or passing seismic wave fields. They can be monitored by observing fluid level changes in wells that are in hydraulic contact with confined aquifers. The phenomenon of tidal well level variations has been reported since at least as early as the last century (e.g. *Klonne*, 1880; see *Bredehoeft*, 1967). If the forcing functions are known, in-situ petrohydraulic rock properties can be derived. Such investigations are occasionally carried out in shallow boreholes. Likewise, deep boreholes offer the possibility to learn more about variable volume strains, fluid-rock interaction and hydraulic communication in petrohydraulic environments, believed to be similar to that in crustal seismogenic zones. The deep and super-deep KTB-boreholes allow to gain insights into the crystalline regime at 4 and 9 km depth. It is well known that petrohydraulic parameters resulting from laboratory measurements may differ by orders of magnitude from in-situ bulk values. In case of the KTB, *Brudy et al.* (1997) and *Huenges et al.* (1997) compiled the results from short term geohydraulic experiments, both from laboratory and from in-situ experiments. A broad band analysis of fluid level registrations obtained over several years reveals additional key rock parameters at natural strain and frequency conditions. At the same time, it allows to detect major changes in the stability of a borehole. In the following section, a brief outline of the method of analysis will be given. Section 3 summarises early experiences and results from ongoing studies at the KTB.

2. Method

2.1. FORCES AND MODELS

Fluid level fluctuations in wells reflect changes of pore pressure and thus changes in volume strain of the formations that are in hydraulic contact with the well (Fig. 1). The level variations may be driven by different forces which are entitled as forcing functions. Amplitudes and phases of the well level response depend on the geometries of the well and the pore space, the deformatonal behaviour of the rock to the specific forcing, and on the frequency spectra of the forcing functions. The forcing functions considered in this paper are mainly tidal straining of the earth (earth tides) and barometric loading. The former can be either measured using strainmeters or calculated for non-rigid earth model. The latter can be measured on site.

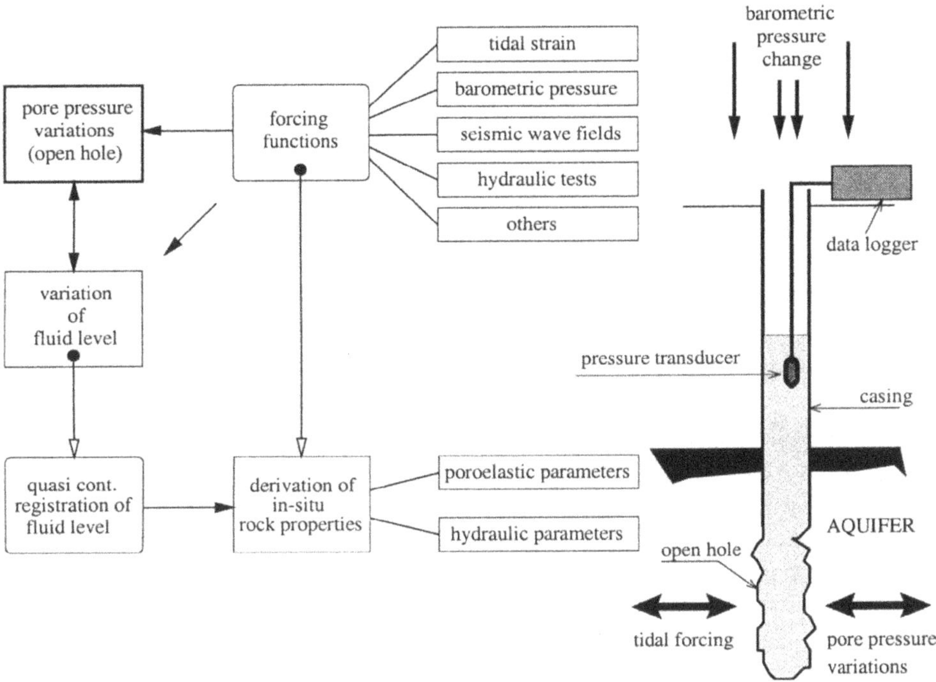

Figure 1. Fluids in boreholes are in hydrostatic equilibrium with the rock fluid pressure along the uncased section. From monitoring fluid level variations of a confined aquifer, valuable information about pore pressure changes and in-situ petrohydraulic properties can be obtained, in particular when the forcing functions are known. This is e.g. the case for tidal strain or barometric pressure fluctuations. The latter acts twofold: as a load on the fluid column in the borehole and as a load on the surface.

To be able to derive petrohydraulic parameters from the pore pressure signals, some assumptions about the forced medium are necessary. Porosity in crystalline rocks is in general dominated by fracture porosity. If fracture systems with few and large fractures prevail, those will determine the character of fluid flow and strain sensitivity. For a single, (dipped) penny shaped fracture or a single bi-wing fracture analytical solutions have been given by *Bower* (1983) and *Hanson* (1983), respectively. They allow to solve for fracture orientation, permeability and transmissibility. Yet, a useful analysis requires to know whether a single fracture exists, or e.g. a series with cracks of individual parameters each.

In another approach, the rock is seen as a macroscopically homogeneous poroelastic medium. For a review on the theory of poroelasticity see e.g. *Kümpel* (1991) or *Wang* (1993). Herein, the strain sensitivity of the formation is described by drained and undrained compressibilities and Poisson ratios as well as a measure of diffusivity. The method is also believed to be applicable if the pores are fractures of size small compared to the considered macroscopic volume. Interpretation becomes less constraining if different layers are intersected alongside the screened section of the borehole. Response of less strain sensitive layers will add to and thus attenuate the response of the more sensitive layers, both for deformation and induced fluid flow. This is analogous to the effect of water table drainage to the surface that attenuates the amplitude of fluid level variations in unconfined aquifers as compared to situations of fully undrained conditions (*Rojstaczer*, 1988). While some authors choose a dynamic formulation including frequency dependence of the well aquifer system and the forcing functions (e.g. *Hsieh et al.*, 1987) others restrict to the static confined description (e.g. *Van der Kamp & Gale*, 1983; *Rojstaczer & Agnew*, 1989, *Beavan et al.*, 1991). The latter is (almost) independent of the well's geometry.

The analysis of data presented here follows the homogeneous poroelastic and static confined description, but will be extended in future.

2.2. THE STATIC CONFINED APPROACH

Early analyses of the response of wells to earth tides and atmospheric loading and their relations to the formations' elastic properties date back to *Jacobs* (1940). More recently used formulations like those of *Van der Kamp & Gale* (1983) and *Rojstaczer & Agnew* (1989) will be used in this paper. Both assume a frequency independent and static confined response with no or little fluid flow (undrained conditions). The latter holds for a small radius of the well and a spatially extended, homogeneous strain field. A pore pressure change p will be hydrostatically reflected in a well level change w in the sense that $p = \rho_{fl} g w = w_p$, where ρ_{fl} is the fluid's density and g the gravitational acceleration. By w_p we denote, that the fluid level change is given in units of pressure of fluid column (see Tab. A1 for list of used symbols). A homogeneous, isotropic poroelastic half-space and a linear relation between strain and pore pressure are assumed. Using *Biot*'s (1941) fundamental relation between stress σ and deformation ϵ (both positive for extension) and introducing grain compressibility c_s, *Nur & Byerlee* (1971) and *Rice & Cleary* (1976) established

$$\epsilon_{ij} = c\left[\frac{(1+\nu)}{3(1-2\nu)}\sigma_{ij} + \frac{\nu}{(1-2\nu)}P_c\delta_{ij} + \frac{\alpha}{3}p\delta_{ij}\right] \quad \text{with} \quad i,j = 1,2,3 \quad (1)$$

where ν is the drained Poisson ratio, $P_c = -1/3(\sigma_{11} + \sigma_{22} + \sigma_{33})$ the confining pressure, c drained or matrix compressibility, and α the coefficient of effective stress, with

$$\alpha = 1 - \frac{c_s}{c}. \tag{2}$$

The volume dilatation is, accordingly,

$$\epsilon_v = \epsilon_{11} + \epsilon_{22} + \epsilon_{33} = c(\alpha p - P_c). \tag{3}$$

Considering the vertical stress component to be zero ($\sigma_{33} = 0$), which holds for a free surface, the areal strain $\epsilon_a = \epsilon_{11} + \epsilon_{22}$ is given by

$$\epsilon_a = c(\frac{2}{3}\alpha p - \Lambda P_c) \qquad \text{with} \qquad \Lambda = \frac{1-\nu}{1-2\nu}. \tag{4}$$

A notably meaningful parameter to describe a poroelastic medium is the Skempton ratio B (*Skempton*, 1954; *Rice & Cleary*, 1976). It quantifies the change in pore pressure per unit change in confining pressure for undrained conditions, i.e.

$$B \equiv \left(\frac{\partial p}{\partial P_c}\right)_{m=0} = \frac{(c - c_s)}{(c - c_s) + n(c_{fl} - c_n)}, \tag{5}$$

where $m = 0$ denotes that no fluid flow occurs, c_{fl} is fluid compressibility, and c_n pore compressibility and n porosity. Often, it is assumed that $c_n = c_s$ (compare *Kümpel*, 1991). Using eqs. (2) to (5), the ratio of areal strain to volume strain becomes

$$\frac{\epsilon_a}{\epsilon_v} = \frac{2\alpha B - 3\Lambda}{3(\alpha B - 1)} = \frac{1-\nu_u}{1-2\nu_u}, \qquad \text{where} \tag{6}$$

$$\nu_u = \frac{3\nu + B(1-2\nu)\alpha}{3 - B(1-2\nu)\alpha} \tag{7}$$

denotes the undrained Poisson ratio. If pore pressure phenomena are negligible, i.e. $c = c_s \Rightarrow B = \alpha = 0$, one finds $\epsilon_a/\epsilon_v = \Lambda$ as is valid for the surface of an isotropic body as given by e.g. *Hsieh et al.* (1987).

Strictly speaking, pore pressure changes and consequently well level changes respond to dilatation (volume strain). Accordingly, it is useful to define a volume strain sensitivity A_v. For the free surface condition, it is also reasonable to define the areal strain sensitivity A_a. The assumption of a free surface particularly holds for tidal forcing since tidal wavelengths are of global extension. If ϵ_v and ϵ_a denote tidal strains, the corresponding tidal sensitivities can be shown to be

$$A_v = -\frac{w_p}{\epsilon_v} = \frac{B}{c(1 - B\alpha)} = \frac{B}{c_u} \tag{8}$$

$$A_a = -\frac{w_p}{\epsilon_a} = \frac{3B(1-2\nu)}{c(3(1-\nu)-2\alpha(1-2\nu))} = \frac{1-2\nu_u}{1-\nu_u}\frac{B}{c_u}. \tag{9}$$

Herein, c_u is the undrained compressibility. Similarly, the loading efficiency $\gamma = -w_p/\sigma_{33}$ for a uniform and widespread load σ_{33} can be derived. *Rojstaczer & Agnew* (1989) extended the solution of *Van der Kamp & Gale* (1983), who only allowed for vertical strain ($\epsilon_a = 0$). They consider $\epsilon_a = H\epsilon_{33}$ where the dimensionless parameter H can take values between 0 and 1 and get

$$\gamma = -\frac{w_p}{\sigma_{33}} = \frac{B(1+H)(1+\nu_u)}{3(1-(1-H)\nu_u)}. \tag{10}$$

H=0 is equal to *Van der Kamp & Gale*'s assumption of zero areal strain, and H=1 is the solution for a laterally extended but finite load on a half space. Solving for B leads to

$$B = \frac{3\gamma(1-\nu)}{(1+\nu)+2\alpha\gamma(1-2\nu)} = \frac{3\gamma(1-\nu_u)}{1+\nu_u} \quad \text{for} \quad H=0 \tag{11}$$

$$B = \frac{3\gamma}{2(1+\nu)+\alpha\gamma(1-2\nu)} = \frac{3\gamma}{2(1+\nu_u)} \quad \text{for} \quad H=1 \tag{12}$$

Considering the barometric pressure as a uniform load on the earth surface, values for γ can be obtained by measuring air pressure and fluid level changes, simultaneously. Most wells are open to the air and, consequently, a barometric load P_b is an additional load on the fluid column itself which results in the definition of barometric efficiency as $\Gamma = (1-\gamma)$ (see also *Kümpel*, 1997).

The most simple expression for tidal sensitivity and barometric efficiency can be given in terms of the undrained parameters ν_u, c_u and B. Substituting eq. (12) into (9) yields

$$\frac{\gamma}{A_a} = \frac{2(1-\nu_u^2)}{3(1-2\nu_u)}c_u. \tag{13}$$

Fig. 2 shows the dependency of c_u from ν_u for various ratios γ/A_a. *Rojstaczer & Agnew* (1989) introduce an iterative method to find values for c and B, assuming A_a and γ (with $H=1$) to be known and c_s and ν to be adopted from other sources. Giving a start value for c, values for α and B (eqs. 2 and 12) can be calculated. After reordering eq. 9 and substituting values for α and B, c can be obtained. With this new value for c one iterative cycle is completed. The method works also for $H=0$ and converges after a few iterations.

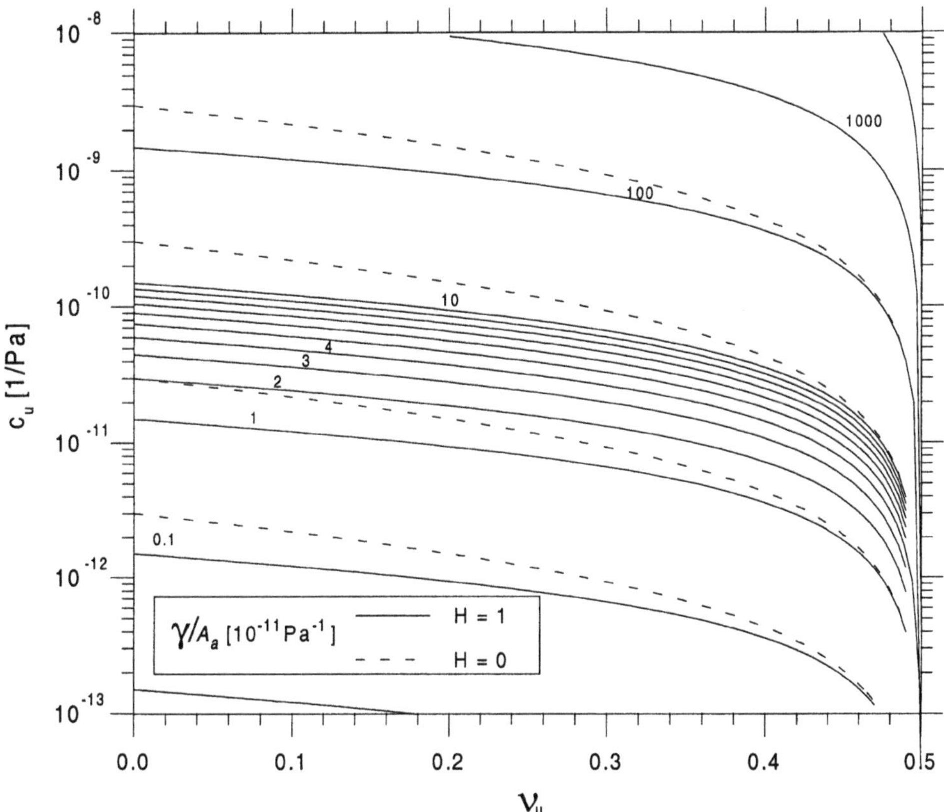

Figure 2. Undrained parameters ν_u and c_u for a set of ratios of loading efficiency γ over areal strain sensitivity A_a, when H=1 (solid lines) and H=0 (dashed lines). For H=0, γ/A_a also equals $(2\mu)^{-1}$, see eq. 14.

Having derived parameters for c, B, and α, other relevant parameters can be estimated. The porosity n can be found after reordering eq. 5. However, its value is strongly dependent on the adopted value for $(c_{fl} - c_n)$. *Van der Kamp & Gale* (1983) calculate the shear modulus μ from

$$p = -2\mu\gamma\epsilon_a \qquad \Leftrightarrow \qquad \mu = \frac{A_a}{2\gamma}, \tag{14}$$

assuming $H = 0$. They also give an expression for the specific storage coefficient if only vertical deformation occurs:

$$S_s = \rho_{fl}g\left[\alpha c\left(1 - \frac{\alpha(1-2\nu)}{3(1-\nu)}\right) + n(c_{fl} - c_n)\right]. \tag{15}$$

S_s is frequently used in hydrogeology. Assuming $H \neq 0$ and thus that loading also results in areal strain, *Rojstaczer & Agnew* (1989) define their three-dimensional storage coefficient for surface loading as

$$S_a = \rho_{fl} g \left[\alpha c \left(1 - \frac{\alpha(1-2\nu)}{3} \right) + n(c_{fl} - c_n) \right] \tag{16}$$

for $H = 1$.

2.3. DEEP BOREHOLES

In the following, boreholes will be considered as deep only if they reach some km of depth. Also, their open hole or screened section is assumed to be alongside the lowermost part. The results from the KTB-drilling project have shown that the intermediate crust is not dry but contains free fluids (e.g. *Kessels & Kück* 1995; *Huenges et al.*, 1997; *Möller et al.*, 1997; *Zoback & Harjes*, 1997; see *Emmermann & Lauterjung*, 1997 for a summary of KTB results). The fluid column in the pilot hole is stable at only a few meters below the surface. Since the open hole section extends from 3850 m to 4000 m this indicates nearly hydrostatic pore pressure which is also confirmed by other investigations (*Grawinkel & Stöckhert*, 1997; *Zoback & Harjes*, 1997). The great depth of the borehole's open hole section implies confined conditions, i.e. vertical flow to the free groundwater table can be excluded. The influence of meteorological effects will be dominated by air pressure loading and possibly small loading effects in case of heavy, laterally extended rain falls. Seepage induced pore pressure fluctuations in a deep environment are highly unlikely. In consequence, the major fluid level changes will be caused by volume strain variations in the formations that are in hydraulic contact with the water column in the borehole.

Roeloffs (1988) references studies reporting that tidally induced fluid level variations in deeper boreholes tend to have greater amplitudes. *Rojstaczer & Agnew* (1989) show that the strain sensitivity increases with decreasing porosity and matrix compressibility. Since it is more likely to find stiffer and less porous rocks in a deeper environment, the cited tendency can be linked to a physical reasoning. Yet, also shallow wells may have high strain sensitivities (e.g. *Kümpel et al.*, 1998). As the wavelengths of tidal forcing are of the order of thousands of km, even super-deep boreholes are shallow compared to tidal wavelengths. Accordingly, no difference to forcing of shallow well regimes are is expected from this point of view.

This is somewhat different for barometric loading effects in the sense that the derivation of loading efficiencies is only valid for (compared to a well's depth) laterally extended loads. Fig. 3 shows the dependency of σ_{33} from depth z beneath the centre of a uniform circular load P_0 of radius a

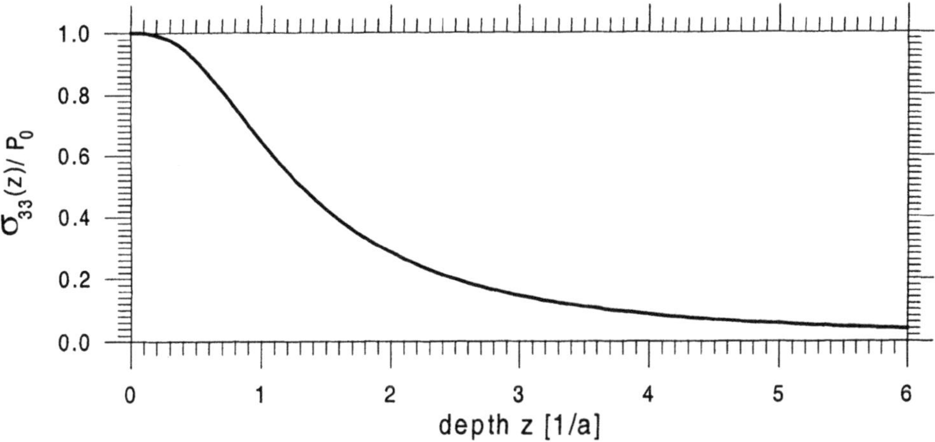

Figure 3. Normalised vertical stress σ_{33} below the centre of a uniform circular load P_0 with radius a at normalised depth z/a in a homogeneous elastic medium.

on the surface of a homogeneous elastic halfspace using

$$\sigma_{zz} = P_0 \left(1 - \frac{z^3}{(a^2 + z^2)^{3/2}}\right) \tag{17}$$

(*Davis & Selvadurai*, 1996). The graph shows that the effective vertical stress of a load with the radius of the well's depth ($a = z$) is only 65 % of the stress at the surface. *Liljequist & Cehak* (1984) consider a uniform spatial change of 1 hPa over 10 km to be large. Still, sometimes barometric phenomena may have higher pressure gradients. For example, a change of 10 hPa over a distance of 50 km was observed for a squall line over Southern Germany (*Haase-Straub et al.*, 1997). *Rabbel & Zschau* (1985) discuss the influence of (anti)cyclons, the spatial extensions of which are of some hundreds to thousands of km, on surface crustal deformation and gravity change. Even if barometric changes with longer periods tend to be related to laterally more extended barometric loads, there is no simple relation between temporal and spatial frequencies of air pressure variations. In conclusion, some attenuation of the barometric loading effect is expected (a) in very deep boreholes and (b) for spatially localised or high frequency barometric changes.

One may ask whether deformation due to tidal forcing of the borehole casing itself could result in major fluid level variations. The answer is no, even for super-deep boreholes. We may assume a well with casing radius r_1 and a fluid column of height h_1 which undergoes a deformation ϵ such as $r_2 = (1 - \epsilon)r_1$. Since the volume of water doesn't change ($v_1 = v_2$), a new fluid height $h_2 = h_1 + \Delta h$ will result. Then

$$\frac{r_1^2}{r_2^2} = \frac{h_2}{h_1} \qquad \Leftrightarrow \qquad \Delta h = \frac{1-(1-\epsilon)^2}{(1-\epsilon)^2} h_1. \tag{18}$$

For a typical tidal deformation of $5 \cdot 10^{-8}$ this yields roughly $\Delta h = 10^{-7} h_1$ which is equivalent to only 0.1mm/km water column.

3. Fluid level studies at KTB

3.1. ABOUT THE SITE

In summer 1996, we started long term high resolution monitoring of the fluid levels in the two KTB boreholes. The drill site is located in the Oberpfalz region, Southern Germany, close to the city of Windischeschenbach. Since January 1996, both boreholes serve as a deep crustal laboratory (*Kück et al.*, 1998), which allows to do long term observations. The lab is operated by the GeoForschungsZentrum Potsdam[1].

Table 1 summarises some of the borehole parameters. The pilot hole was drilled to a depth of 4000.1 m intersecting layers of metasediments, metabasites and alternations of both. The lower 150 m were left uncased and thus allow to study pore pressure changes within an amphibolite facies at that depth. The inner casing has a diameter of 14 cm except alongside the part below 3670 where it is 10.8 cm. The annulus of the casing is cemented between 3670 m and 3850 m. In 1990, a 4 month pump test was performed that produced 480 m^3 formation fluids (e.g. *Kessels & Kück*, 1995). Thereafter, the fluid in the borehole was exchanged with density 1.00 g/cm^3 fresh water (*Engeser*, 1995). In 1997, a density of 1.04 g/cm^3 was confirmed by a single fluid sample from depth 3960 m. Since the beginning of our study, the fluid level has been stable at about 17 m below surface which can be taken as a sign for stable borehole conditions in the open hole section.

The main hole reaches a depth of 9101 m, also intersecting layers of metasediments, metabasites and alternations of both. The lower 70 m which perforate these alternations are uncased, leaving an open diameter of 16.5 cm. In Januar 1995, the mud in the main hole was exchanged through a soda composition with a density of 1.17 g/cm^3. Due to borehole instability at final depth, the pipe for the circulation could not be installed deeper than 9080 m (*Engeser*, 1995) and consequently remainders of the drilling mud were left in the lower part of the borehole. In addition, 200 m^3 of a high density KBr/KCl brine had been injected in the open hole section

[1] see also http://icdp.gfz-potsdam.de/html/gfz_to.htm

during the 1994 induced seismicity experiment (*Zoback & Harjes*, 1997). Unlike the fluid level in the pilot hole it has not yet stabilised (see section 3.5).

TABLE 1. Some relevant parameters of the KTB pilot and main hole. The boreholes are cased except for the open hole section. The inner annuluses of pilot and main hole are cemented below 3670 m and 4350 m, respectively.

KTB	latitude [N]	longitude [E]	height a. s. l.	final depth	open hole section lower	open hole section diameter
pilot hole	49.816°	12.119°	513.4 m	4000.1 m	150 m	15.2 cm
main hole	49.816°	12.112°	513.8 m	9101 m	70 m	16.5 cm

3.2. DATA ACQUISITION

The monitoring in the main hole started end of June 1996, in the pilot hole late August 1996, and has been continuous since then, except for periods of other measuring activities in the boreholes. The longest gap-free periods lasted a few months. Our experience confirms the necessity of redundant registrations in the boreholes to minimise gaps due to instrumental failure and to filter out incorrect readings. Therefore, altogether four pressure transducers are in use, two of them measuring absolute pressure and the other two being air pressure compensated. The ranges of the transducers vary from $0.25 \cdot 10^5$ Pa to $3 \cdot 10^5$ Pa with resolutions of 0.01 % of the full scale deflections, respectively. The sampling period is set between 2 sec and 2 min, but is usually chosen to be 1 min to compromise between temporal resolution, logistical efforts and memory capacity of the loggers. Recently, a three months monitoring period with sampling frequencies up to 1 Hz was started, aimed to resolve teleseismic events. It will be possible to compare the results to readings from a borehole seismometer[2], which is installed at a depth of about 3800 m (*Schulz et al.*, 1998). Air pressure readings with a resolution of 0.1 hPa are taken on site with a sampling period of 5 min or shorter.

3.3. RESULTS FROM THE PILOT HOLE

Figures 4a and b show gap-free registrations of fluid level and air pressure fluctuations over a period of 3.5 months. The fluid level variations

[2]see also http://icdp.gfz-potsdam.de/html/ktbto/seisto/titelseiteDSL.htm

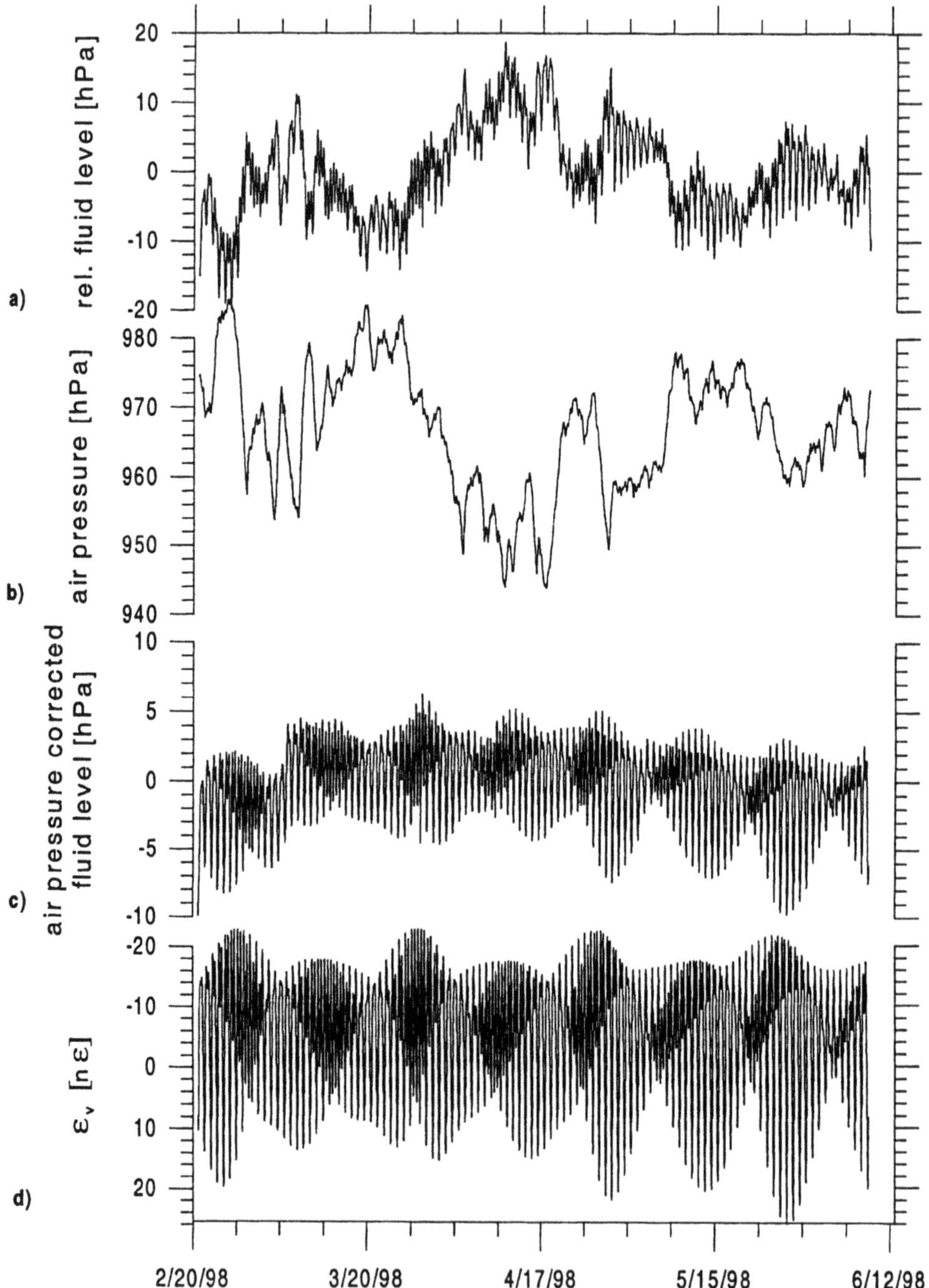

Figure 4. The fluid level of the pilot hole (a) shows a clear anticorrelation to the air pressure recording (b). After linear reduction of the air pressure effect tidal signals with peak to peak amplitudes of up to 13 hPa become the main fluctuations (c). The tidal volume strain ϵ_v at the KTB site for the same 3.5 months interval as calculated for an elastic earth model (d; *Wenzel*, 1996).

are clearly anticorrelated to air pressure changes and show diurnal and semidiurnal tidal signals. Obviously, fluid level variations of more than 30 hPa are generally due to air pressure forcing[3]. After reduction of the air pressure effect with a linear regression coefficient (-0.64) the tidal variations with amplitudes up to 13 hPa peak to peak can be resolved more clearly (Fig. 4c). Comparison with expected tidal volume strain (Fig. 4d) for the same time interval emphasises the high tidal sensitivity of the well. Figs. 5a and b show the amplitude spectra of the two original time series and Figs. 5c, d the results of a frequency dependent regression between the airpressure and fluid level registration plotted in Fig. 4a, b. Except for the diurnal and semidiurnal frequency band (= tidal frequencies) the correlation factor for frequencies between 0 and 10 cpd remains above values of 0.8 which confirms the strong influence of the airpressure variations on the fluid level. The regression factor varies between -0.67 and -0.55 indicating a barometric efficiency Γ in that range. A powerful tool to compare longer time series is to calculate their coherence and admittance in amplitude and phase (*Bevans et al.*, 1991).

A tidal analysis and the theoretical volume strain were computed with the earth tide program ETERNA3.30[4] (*Wenzel*, 1996, 1997). The program fits data to theoretical tides and additional (e.g. meteorological) parameters using the method of least squares adjustment. The strain tides are calculated for an elastic earth on the basis of the Wahr-Dehant model. Data of 3.5 months fluid level and barometric pressure recordings with sampling periods of 5 min were analysed and adjusted simultaneously. Prior to the analysis the data were high pass filtered (cutoff frequency 0.8 cycles per day). Fig. 6 shows the tidal components of the fluid level recording and the residual signal. The latter results from subtracting tidal and barometric effects from the original data. As demonstrated by the parallel recordings, the residua prove the existence of other than tidal or barometric pore pressure phenomena (Fig. 6b). Clearly, these variations are too small (few hPa) to be noticed without removal of tidal and barometric effects.

Comparison between the amplitude spectrum of the adjusted signal and the spectrum of the high pass filtered residuum demonstrates the high signal-to-noise ratio (Fig. 7a). The residuum still shows small diurnal variations. Most likely these reflect insufficient air pressure removal due to the linear regression technique. A frequency dependent adjustment of air pres-

[3]In the following the relative fluid level variations w are given in hPa meaning the resulting pressure change w_p above the pressure transducer as given by $w_p = \rho_{fl} g w$. It is 1 hPa=1 mbar$\simeq$1 cm H_2O.

[4]see also http://www-gik.bau-verm.uni-karlsruhe.de/~wenzel/eterna33.htm

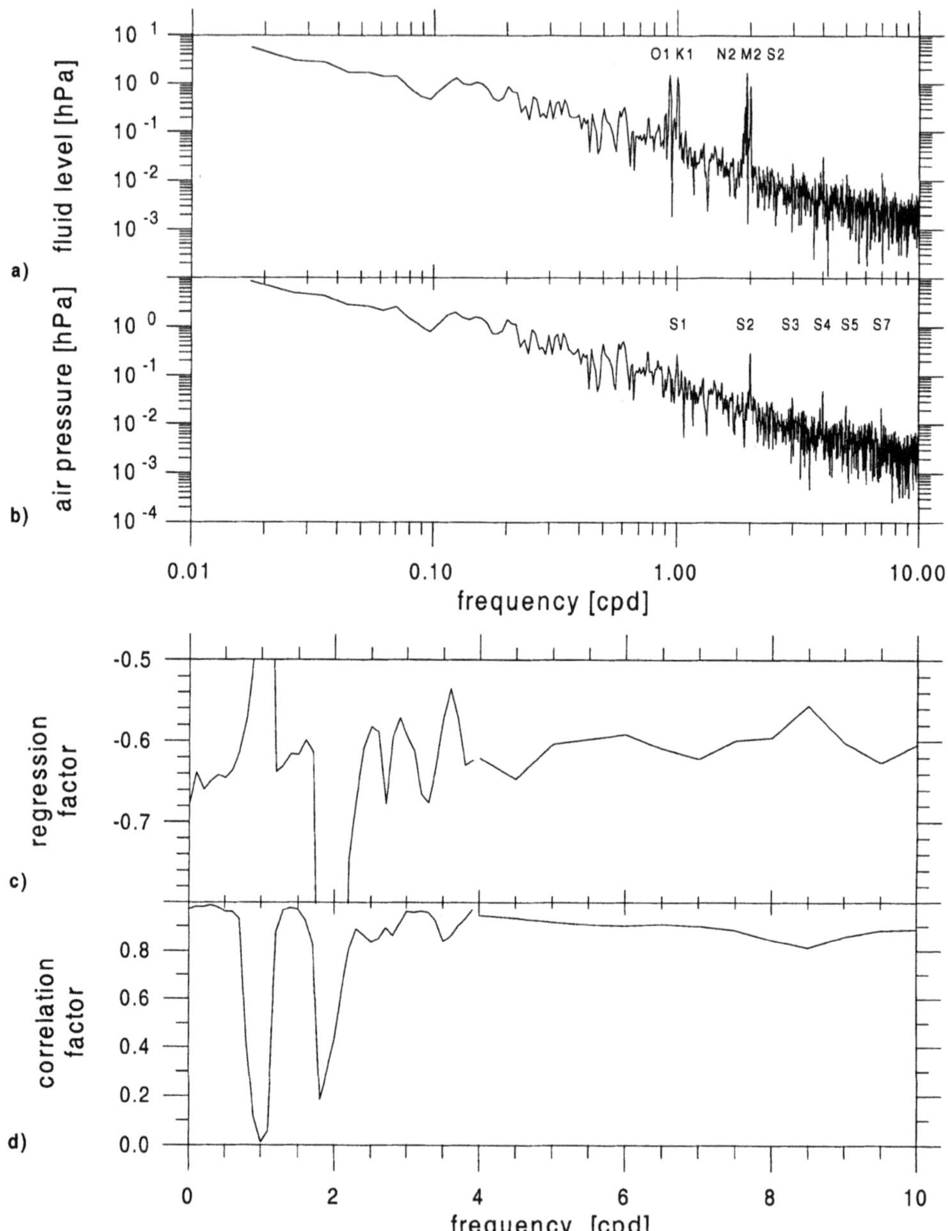

Figure 5. Fourier amplitude spectra of the 3.5 months recordings of (a) fluid level in KTB pilot hole and (b) air pressure as in Figs. 4a,b with mayor tidal constituents. (c) and (d) show the results of a frequency dependent regression of the original recordings. The regression factor Γ is also the frequency dependent barometric efficiency, and the correlation factor is a measure of the degree of crosscorrelation between the two series. For frequencies between 0 and 4 cpd a band width of 0.2 cpd and step width of 0.1 cpd, and for frequencies between 4 and 10 cpd a band width of 1 cpd and step width of 0.5 cpd were chosen.

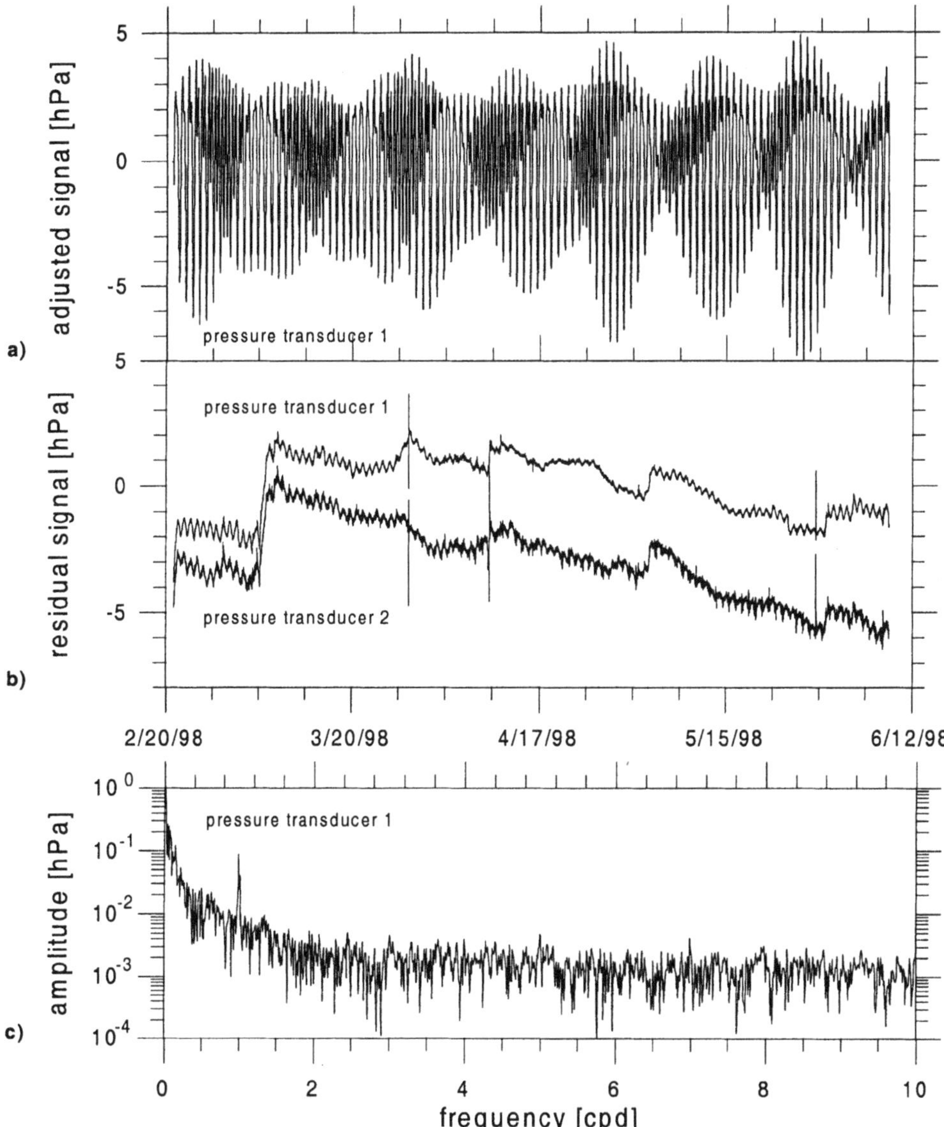

Figure 6. (a) Adjusted tidal signal (= tidal part of fluid level data) and (b) total residual signal of fluid level recording in KTB pilot hole (pressure transducer 1). The latter results from subtracting tidal effects and linear air pressure influence from the original data. The residuum of a second pressure transducer is plotted for comparison. The 3 spikes visuable in the residual data are caused by disturbances during installation/removal of a third pressure transducer. (c) Fourier amplitude spectra of residual signal (pressure transducer 1). See Fig. 7a for comparison with spectrum of adjusted signal.

sure influence might improve the result. Figures 7b and c summarise the results of the tidal analysis for the main tidal harmonics O1, P1S1K1, N2, M2, S2. The calculated areal strain sensitivities A_a range between 0.15 and 0.19 hPa/nϵ . Assuming $\nu = 0.25$ and a non-porous elastc rock ($\alpha = 0$) eqs. (6) to (9) yield $A_v = 1.50 A_a$. The large differences between the O1 and P1S1K1 amplitudes and phases are most likely due to a non sufficient reduction of air pressure influence in the diurnal band. This is also obvious in the amplitude spectra of the residual signal (Fig. 6c). The P1S1K1 wave is usually more affected because its frequency band (0.98-1.02 cpd) includes that of the diurnal atmospheric disturbance (1 cpd). While the noise level reaches 0.1 hPa for the diurnal periods it is more than one order of magnitude less for shorter periods including the semidiurnal band.

The phase shifts (Fig. 7c) of the main tidal waves suggest a slight frequency dependence of tidal sensitivities, since they are of opposite sign for the diurnal and the semidiurnal waves. Yet, the diurnal signal seems to lead the tidal forcing which would be surprising when assuming a homogeneous poroelastic medium. At least two more factors may explain this: (I) Even at locations far away from the ocean, loading effects of ocean tides can be of significant influence (*Beaumont & Berger*, 1975; *Berger & Beaumont*, 1976; *Jentzsch*, 1997). *Zaske* (1997), for example, analysed fluid level variations of a borehole at Soultz-sous-Forêts (about 500 km distance from the open sea) project and found a phase shift induced by ocean tide loading of +1.1 degrees for O1 and of −6.8 degrees for M2 (1 degree $\simeq$ 4 minutes). Accordingly, for a more detailed analysis, ocean loading effects need to be considered. (II) Phases and amplitudes of tidal strain as calculated for the earth model do not account for local geological heterogeneities. *Beaumont & Berger* (1975) and *Berger & Beaumont* (1976) showed for various wells in Northern America that the predicted tidal strain differs by up to a factor 2 from the observed, and that the calculated phases are not too reliable. Leading phases may also be explained assuming a single fracture with a certain orientation (*Bower*, 1983; *Hanson*, 1983).

The static confined barometric efficiency Γ for the presented data set is 0.64 (i.e. $\gamma = 0.36$). Note, that since the data was highpass filtered prior to the analysis this value holds only for frequencies higher than 0.8 cpd (periods shorter than 30 h). Again, more sophisticated regression analyses may be applied to resolve a frequency dependent barometric efficiency.

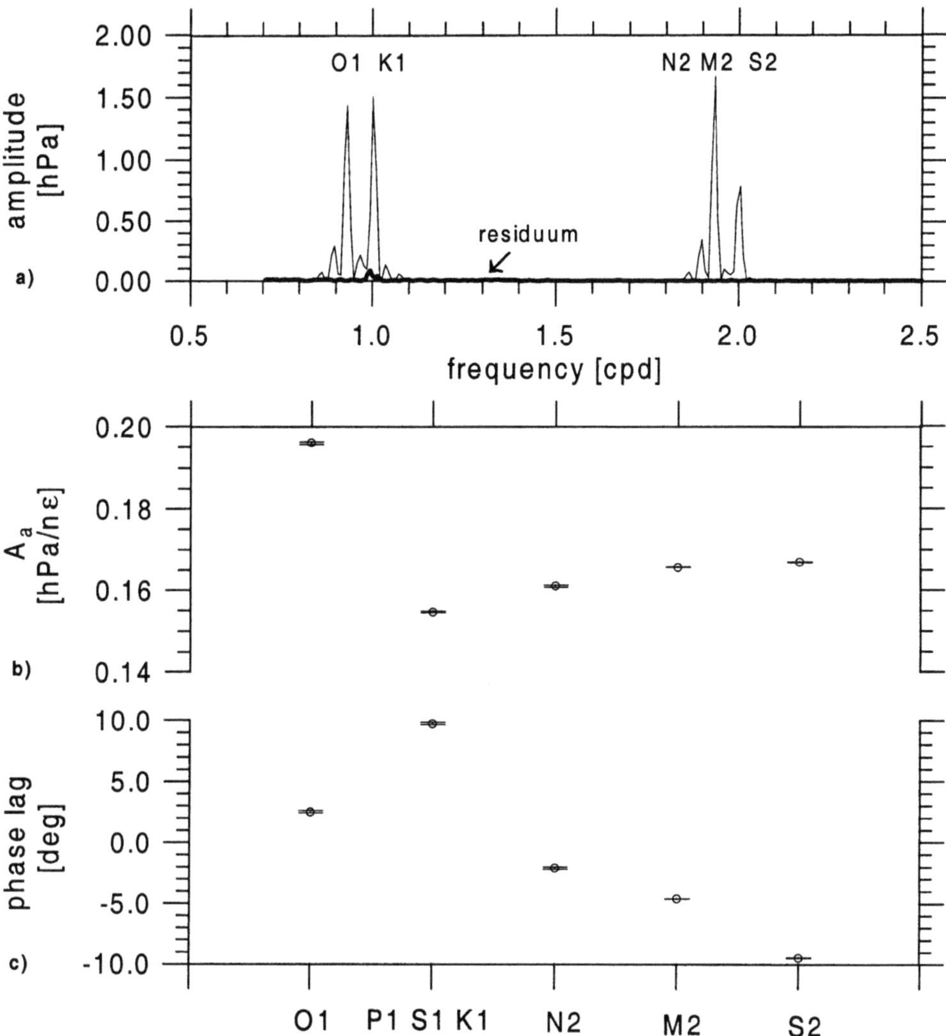

Figure 7. (a) Amplitude spectra of the adjusted fluid level signal in the KTB pilot hole and its high pass residuum. Note the high signal to noise ratio. (b) Areal strain sensitivity A_a and (c) phase shift for the main tidal constituents. Error bars symbolise standard deviations as given by ETERNA3.30. A negative phase lag indicates the signal response is delayed to the forcing.

3.4. CONSTRAINTS FOR ROCK PROPERTIES

To further constrain rock parameter values from the analysis of well level data some properties have to be adopted from other sources. At this point we will use the same ones as *Endom & Kümpel* (1994; see Tab. 2), who tentatively analysed fluid level measurements from a 10 months recording of 1992 in the KTB-pilot hole. They also used the single fracture approaches of *Bower* (1983) and *Hanson* (1983) to interpret the data. Yet, their analyses suffered from many data gaps, and partly corrupted data caused by malfunction of instruments after thunderstorms. *Endom & Kümpel* (1994) took values from density and seismic velocity studies on KTB rock published elsewhere (*Lippmann et al.* 1989; *Bücker et al.* 1990; *Rauen et al.* 1990; *Kern et al.* 1991) and tabulated rock parameters (*Landolt-Börnstein*, 1982; *Röhr et al.*, 1990.).

Usually, the areal strain sensitivities of O1 (period $T = 25.819$ h) and M2 ($T = 12.421$ h) are used to derive poroelastic parameters. They are less influenced by barometric effects because they are of lunar origin. Equations (2), (9) and (10) were applied for the iterative finding of B, α, and c, both for $H = 1$ and $H = 0$. Then, values for n, S_s, S_a, ν_u, and c_u were computed. Fig. 8 illustrates the values that c, B, and α take for different γ and A_a. Having in mind the often vague estimates for in-situ parameter values, this plot helps to validate such assessments. Table 3 summarises a set of consistent values. Fig. 2 and eq. (14) reveal already that c_u takes values between 1.2 and 1.9 $\cdot 10^{-11}\mathrm{Pa}^{-1}$ and the shear modulus μ between 20 and 27 GPa when $0.26 \leq \nu_u \leq 0.31$, γ ranges between 0.36 and 0.40, and A_a between 0.16 and 0.19 hPa/nϵ.

TABLE 2. Some of the rock parameters adopted by *Endom & Kümpel* (1994) for amphibolites at the open section of the pilot hole.

Density	ρ	$= 3000\ \mathrm{kg/m^3}$
Drained Poisson ratio	ν	$= 0.240$
Undrained Poisson ratio	ν_u	$= 0.249$
Shear modulus	μ	$= 40.8$ GPa
Matrix compressibility	c	$= 1.54 \cdot 10^{-11}\mathrm{Pa}^{-1}$
Grain compressibility	c_s	$= 1.26 \cdot 10^{-11}\mathrm{Pa}^{-1}$
Undrained compressibility	c_u	$= 1.47 \cdot 10^{-11}\mathrm{Pa}^{-1}$
Fluid compressibility	c_{fl}	$= 45 \cdot 10^{-11}\mathrm{Pa}^{-1}$
Fluid Density	ρ_{fl}	$= 980\ \mathrm{kg/m^3}$

Endom & Kümpel (1994) found $\gamma = 0.25$ and $A_a = 0.16$ hPa/nϵ (M2) or 0.22 hPa/nϵ (O1) from analysing a data series of only 50 days and thus, the values differ from those presented here. Also, they give results just for the case $H = 0$, and erroneously for S_a, which is valid only for $H = 1$.

As can be seen from Tab. 3 and Fig. 8, a high uncertainty in the calculated values in fact comes from the choice of H. For deep boreholes it

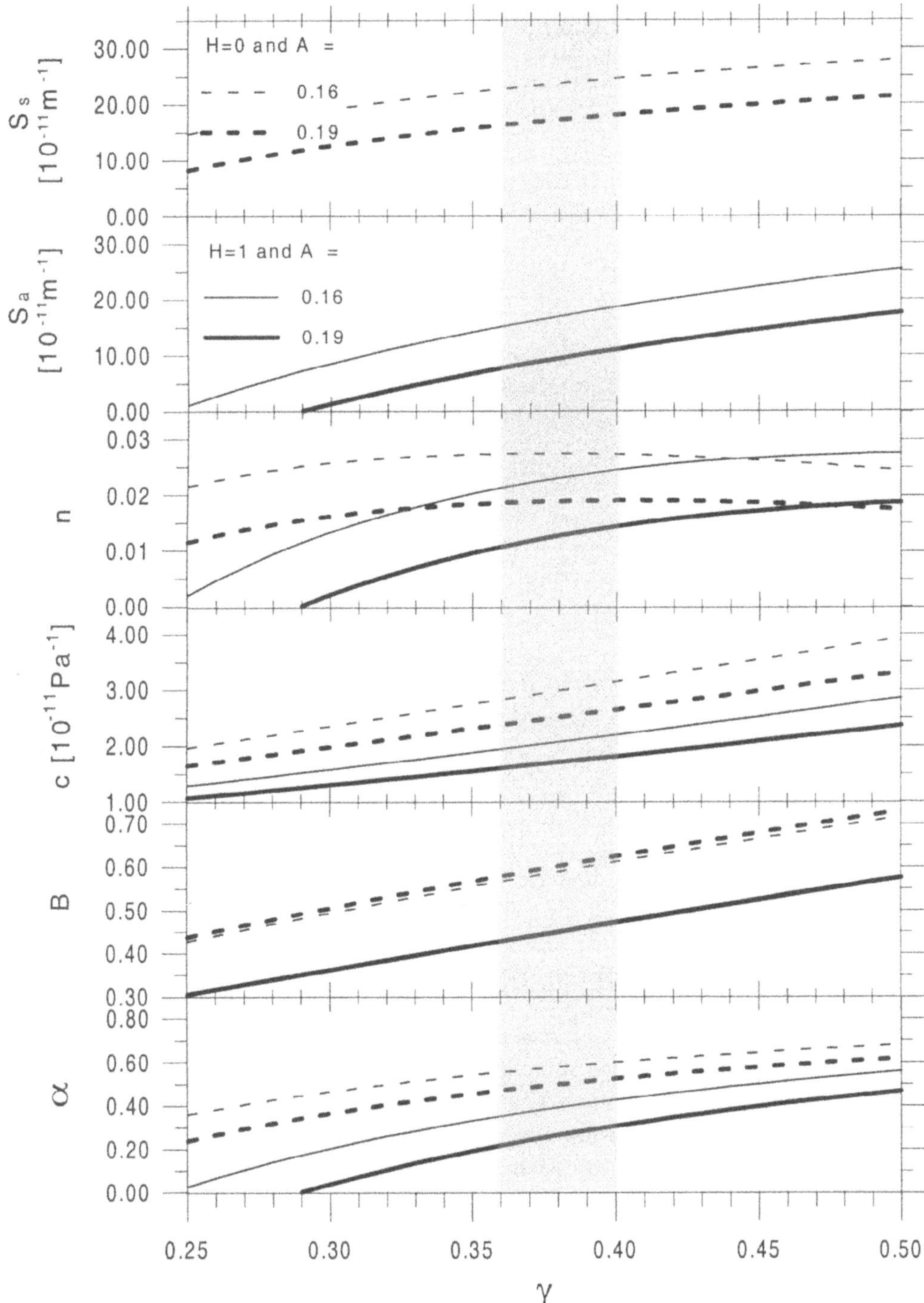

Figure 8. Poroelastic rock parameters by using the iterative method of *Rojstaczer & Agnew* (1989) for $H = 1$ (solid lines) and $H = 0$ (dashed lines) for a range of loading efficiencies γ and two different values of areal strain sensitivity A_a. The Poisson ratio for drained conditions was set to $\nu = 0.24$, grain compressibility $c_s = 1.26\ 10^{-11}\mathrm{Pa}^{-1}$, and

TABLE 3. Four sets of consistent parameters for a loading efficiency $\gamma = 0.36$ and the given areal strain sensitivity A_a for the pilot hole. $c_s = c_n = 1.26 \cdot 10^{-11}$ Pa^{-1} and ν =0.24 were adopted.

H	A_a hPa/nϵ	c [10^{-11} Pa^{-1}]	B	α	n	S_s or S_a $10^{-8}m^{-1}$	ν_u	c_u [10^{-11} Pa^{-1}]
0	(O1) 0.19	2.83	0.57	0.55	0.03	16.32	0.31	1.63
0	(M2) 0.16	2.38	0.58	0.47	0.02	22.80	0.30	1.73
1	(O1) 0.19	1.95	0.42	0.35	0.02	7.76	0.27	1.66
1	(M2) 0.16	1.61	0.43	0.22	0.01	15.15	0.26	1.45

seems to be more reasonable to assume that barometric loading leads to areal as well as vertical strain ($H > 0$). Nevertheless, as also mentioned by *Rojstaczer & Agnew* (1989), another uncertainty results from the difference between the expected (=theoretical) and the observed tidal strain. Assuming the observed tidal strain to be 2 times less than the theoretical and thus $A_a = 0.08$ hPa/nϵ (M2), the derived parameters would be $\alpha = 0.69$, $B = 0.41$, $c = 4.07 \cdot 10^{-11}$ Pa^{-1}, $n = 0.09$, and $S_a = 62.03 \cdot 10^{-8}m^{-1}$ (with ν, c_s as before and $H = 1$). If instead the observed strain tide was 2 times larger, i.e. $A_a = 0.32$, negative values for α and c would be the consequence. Changing c_s to $0.92 \cdot 10^{-11}Pa^{-1}$ (maximum value for positive α) yields $\alpha = 0.01$, $B = 0.44$, and $c = 0.93 \cdot 10^{-11}Pa^{-1}$ ($n = 0.01$, $S_a = 0.28 \cdot 10^{-8}m^{-1}$). This shows that after all the value of B is rather well constrained. Accordingly, roughly 40 to 45 % of any change in confining pressure acting on the rock matrix is directly transferred to the pore fluid. If this reflects a typical situation at a depth of several kilometers, this may have far reaching consequences for the understanding of stress transfer in the continental crust including earthquake mechanics.

3.5. RESULTS FROM THE MAIN HOLE

The main hole's fluid level curve shows a different, formerly unexpected signature (Fig. 9). Before we started our recording, the fluid column was filled up to the top of casing. Since then, the fluid level is lowering monotonously; in March 1997, the lowering speeded up for no obvious reason. At the same time, the fluid level in the pilot hole did not show any anomalous signal. In August 1998 the level was found almost 250 m below the surface which reveals a total fluid loss to the formation of approximately 19 m^3.

Despite the obvious existence of a hydraulic connection to rocks, no tidal or barometric signals can be resolved in the fluid level curve. A plausible

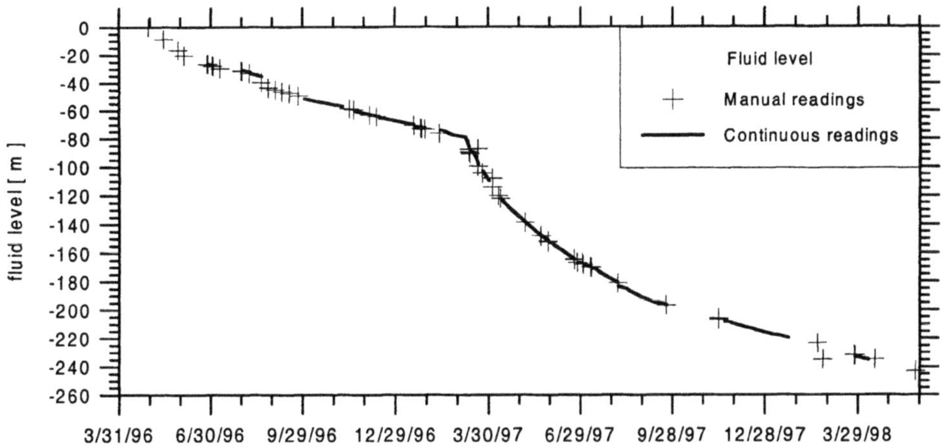

Figure 9. Monotonous decrease of the fluid level in the main hole as observed since April 1996. No reason for the speed up in fluid loss in March 1997 could be found so far.

reason could be that the pore space of the rocks in the open hole section is contaminated by highly viscous drilling mud. Indeed, the pilot hole started showing tidal signals only after a period of extensive pumping which seems to have cleaned the access to the pore space (*Endom & Kümpel*, 1994). Logging in the main hole gave evidence for some obstruction around 8.6 km depth, consisting of a highly viscous material (possibly a fall out product of a chemical reaction in the borehole fluid; *J. Kück, GFZ Potsdam*, pers. comm. 1996). The apparent contradiction between the high fluid loss on the one hand, and the absence of tidal signals on the other shows, that the static confined approach used in this paper cannot be applied in case of the main hole. However, if fluid flow is taken into account, a frequency dependence of the attenuation of pore pressure signals will result that yields higher amplitudes for lower frequencies (e.g. *Narasimhan et al.*, 1984; *Hsieh et al.*, 1987;). Similary, a higher viscosity of the pore fluid will reduce the amplitude of the fluid level changes. Thus, the surrounding rock of the open hole section as well as the obstruction most likely act as a low pass filter.

A leakage of the casing, that might also explain the higher fluid loss since March '97, can be ruled out alongside the upper 4350 m. This became evident after a 14 days lasting 50 bar pressurisation of the annulus and simultaneous control of fluid levels in the main and the pilot hole (*Schulze et al.*, 1998). The annulus reaches down to 6014 m and is cemented below 4350 m. Regarding the speed up in fluid level loss we belive that a change in the properties of the obstruction caused the more rapid fluid level fall. The situation of the main hole is presently subject of further investigations from different participating researchers of the deep crustal lab.

4. Conclusions and outlook

Fluid level registrations in deep and super-deep boreholes are a valuable data source to study the mobility of pore fluids and their mechanical interaction with rocks. As could be expected for deep wells, confined conditions and a high strain sensitivity were found in case of the KTB-pilot hole. The main fluid level variations are caused by barometric and tidal forcing, but additional small amplitude pore pressure variations are present. The former results in fluctuations up to 40 cm, tidal amplitudes reach up to 13 cm, peak to peak. Accordingly, from natural forcing, there is a steady exchange between rock fluids and the fluid column in the borehole of up to 0.4 l/h.

From analysis of a 3.5 months registration of the fluid level in the pilot hole, the tidal sensitivity (0.15-0.22 hPa/$n\epsilon$) and the static confined barometric efficiency (60-65 %) were derived. Both values can not be obtained by other methods. They were used to find constraints for in-situ rock parameters which represent static confined bulk properties on a scale from some tenth to few hundred meters. The Skempton ratio was found to be in the range 0.40 to 0.45 and is remarkably high for crystalline rocks at 4 km depth. This means that about 40 to 45 % of any change in confining pressure is reflected in the pore pressure and emphasises the important role of fluid pressure in transmitting crustal stresses.

Fluid level registrations also allow to control the stability of the open section of a borehole. In case of the KTB-pilot hole, the almost constant height of the fluid column in the borehole and constant response to tidal strain indicate stable petrohydraulic conditions. As for the KTB-main hole. although, so far, no relevant rock properties could be estimated, the continuous fluid level registrations gave important information on the situation here. The lack of tidal and barometric signals is probably caused by an obstruction below 8600 m in the borehole or by hindered access to the pore space of the surrounding rock. To obtain good quality data, a redundant registration has proven to be a must. A minimum registration period of one month is necessary to carry out first approach tidal and barometric analyses. Longer registration periods can be used to derive frequency dependent parameters. With moving window techniques possible changes of the relevant parameters in time can be studied. This, as well as a fracture approach, are planned to be applied to the data from the KTB-pilot hole next. Also, the effects of ocean loading will be included as well as the influence of barometric changes on different spatial scales. The former is especially important to derive correct phase shifts of tidal variations which are necessary for making use of a dynamic formulation of well-aquifer interactions. Finally, residual fluctuations in the level curves (Fig. 6) will be interpreted in view of other influential quantities (rainfall, anthropogenic, instrumental etc.).

Monitoring fluid level variations in the KTB boreholes will continue within the year 1999. There is some chance that a recently forced draw-down of the fluid level in the main hole down to 630 m (*Kück & Wohlgemuth*, GFZ-Potsdam, pers comm., 1998) will stimulate the supposed obstruction in the hole's deep part so that more details about fluid level variations will become visible. It is also planned to include other deep boreholes in this study, first of all the super-deep Kola-SG3, which from a 9-day registration has already proven to show some tidal and barometric signals. This will serve to compare the relevant petrohydraulic parameters from different locations and further help to improve our knowledge of fluid-rock interaction in the intermediate crust.

Acknowledgements

This research is funded by the German Research Association (DFG) as part of the ICDP/KTB program by grants Ku 583/10 and Hu 700/1. We thank J. Kück and K. Bohn of the GFZ-deep-crustal-lab for their steady support without which this research would not be possible. Some of the data handling and analysis were performed with the useful program T-Soft by P. Vauterin (Royal Observatory of Belgium). We appreciate the helpful comments on the original version by reviewers I. Stober and H. Wilhelm. Further thanks to G. Grecksch and G. Zimmermann for valuable comments on the manuscript.

References

Beaumont, C., & Berger J., 1975: An analysis of tidal strain observations from the United States of America: I. The laterally homogeneous tide. *Bull. Seismol. Soc. Am.*, 65, 1613-1629.

Beavans, J., Evans, K., Mousa, S., & Simpson, D., 1991: Estimating aquifer parameters from analysis of forced fluctuations in well level: An example from the Nubian formation near Aswan, Egypt. 2. Poroelastic Properties. *J. Geophys. Res.*, 96, 12139-12160.

Berger, J., & Beaumont C., 1976: An analysis of tidal strain observations from the United States of America: II. The inhomogeneous tide. *Bull. Seismol. Soc. Am.*, 66, 1821-1846.

Biot, M. A., 1941: General theory of three-dimensional consolidation. *J. Appl. Phys.*, 12, 155-164.

Bower, D. R., 1983: Bedrock fracture parameters from the interpretation of well tides. *J. Geophys. Res.*, 88, 5025-5035.

Bredehoeft, J.D., 1967: Response of well aquifer systems to earth tides. *J. Geophys. Res.*, 72, 3076-3087.

Brudy, M., Zoback, M. D., Fuchs, K., Rummel, F., & Baumgärtner, J., 1997: Estimation of the complete stress tensor to 8 km depth in the KTB scientific drill holes: Implications for crustal strength. *J. Geophys. Res.*, 102, 18453 - 18475.

Bücker, C., Huenges, E., Lippmann, E., Rauen, R., Streit, K.M., Wienand, J.A., & Soffel, H.C., 1990: KTB-pilot hole; Results obtained in the KTB Field Laboratory. *KTB-Report 90-8*, Hannover, D1-D29.

Davis, R. O., & Selvadurai, A. P. S., 1996: *Elasticity and Geomechanics*, Cambridge University Press, 201 p.

Emmermann, R., & Lauterjung, J., 1997: The German Continental Deep Drilling Program KTB: Overview and major results, *J. Geophys. Res.*, 102, 18179-18201.

Endom, J., & Kümpel, H.-J., 1994: Analysis of natural well level fluctuations in the KTB-Vorbohrung: parameters from poroelastic aquifer and single fracture models. *Scientific Drilling* , 4, 147-162.

Engeser, B., 1996: Das Kontinentale Tiefbohrprogramm der Bundesrepublik Deutschland, KTB, Bohrtechnische Dokumentation. *KTB-Report 95-3*, Hannover, 800 p.

Grawinkel, A., & Stöckhert, B., 1997: Hydrostatic pore fluid pressure to 9 km depth - Fluid inclusion evidence from the KTB deep drill hole. *Geophys. Res. Lett.*, 24, 3273-3276.

Haase-Straub, S. P., Hagen, M., Hauf, T., Heimann, D., Peristeri, M., & Smith, R. K., 1997: The squall line of 21 July 1992 in Southern Germany: An observational case study. *Beitr. Phys. Atmosph.*, 70, 147-165.

Hanson, J.M., 1983: Evaluation of surface fracture geometry using fluid pressure response to solid earth strain. *Terra Tek. Research Techn. Report 82-26*, Salt Lake City, Utah.

Hsieh, P.A., Bredehoeft, J.D., & Farr, J.M., 1987: Determination of aquifer transmissivity from earth tide analysis. *Water Res. Res.*, 23, 1824-1832.

Huenges, E., Engeser, B., Erzinger, J., Kessels, W., & Kück, J., 1997: The permeable crust: geohydraulic properties down to 9100m depth. *J. Geophys. Res.*, 102, 18255-18265.

Jacobs, C.E., 1940: On the flow of water in an elastic artesian aquifer. *EOS Trans. AGU*, 27, 574-586.

Jentzsch, G., 1997: Earth tides and ocean tidal loading. – In: *Tidal Phenomena*, Wilhelm, H., Wenzel, H.-G., & Zürn, W. (eds.), Lecture Notes in Earth Sciences, Springer, Berlin, 145-171.

Kern, H., Schmidt, R., & Popp, T., 1991: The velocity and density structure of the 4000m crustal segment at the KTB drilling site and their relationship to lithological and microstructural characteristics of the rock: an experimental approach. *Scientific Drilling*, 2, 130-145.

Kessels, W., & Kück J., 1995: Hydraulic communication in crystalline rock between the two boreholes of the Continental Deep Drilling Project in Germany. *Int. J. Rock Mech. Min . Sci. & Geomech. Abstr.*, 32, 37-47.

Klonne, F.W. 1880: Die periodischen Schwankungen des Wasserspiegels in den inundierten Kohlenschächten von Dux in der Periode von 8. April bis 15. September 1879. *Sitzber. Kais. Akad. Wiss.*, 81.

Kück, J., Lauterjung, J., & Wohlgemuth L. 1998: KTB Deep Crustal Lab - Exploring the deep on the long-term. *Annales Geophysicae*, 16, Supplement I, C86 (abstract), 23th General Assembly of the European Geophysical Society, April 20-24, 1998, Nice, France.

Kümpel, H.-J., 1991. Poroelasticity: parameters reviewed. *Geophys. J. Int.*, 105, 783-799.

Kümpel, H.-J., 1997: Tides in Water-Saturated Rock. – In: *Tidal Phenomena*, Wilhelm, H., Wenzel, H.-G., & Zürn, W. (eds), Lecture Notes in Earth Sciences, Springer, Berlin, 277-291.

Kümpel, H.-J., Gupta, H.K., Chadha, R.K., Radhakrishna, I., & Grecksch, G., 1998: Well tides in an area of high reservoir induced seismicity.–In *Proceedings of the Thirteenth*

International Symposium on Earth Tides, Ducarme, B., Pâquet, P. (eds.), Brussels, 125-132.

Landolt-Börnstein, 1982: *Zahlenwerte und Funktionen aus Naturwissenschaft und Technik*, Neue Serie (1982), Springer Verlag, Berlin.

Liljequist, G.H., & Cehak, K., 1984: *Allgemeine Meteorologie*, 3rd ed., Fried. Viehweg & Sohn, Braunschweig, 396.

Lippmann, E., Bücker, C., Huenges, E., Rauen, A., Wienand, J., & Wolter, K.E., 1989: Gesteinsphysik im KTB-Feldlabor: Messungen und Ergebnisse. *KTB-Report 89-3*, Hannover, 120-130.

Möller, P., Weise, S.M., Althaus, E., Bach, W., Behr, H.J., Borchard, R., Bräuter, K., Drescher, J., Erzinger, J., Faber, E., Hansen, B.T., Horn, E.E., Huenges, E., Kämpf, H., Kessels, W., Kirsten, T., Landwehr, D., Lodemann, M., Machon, L., Pekdeger, A., Pielow, H.-U., Reutel, C., Simon, K., Walther, J., Weinlich, F.H., & Zimmer, M.: Paleofluids and recent fluids in the upper continental crust: results from the German Continental Deep Drilling Program (KTB). *J. Geophys. Res.*, 102, 18233-18254.

Narasimhan, T.N., Kanehiro, B.,Y., & Witherspoon, P.A. 1984: Interpretation of three deep, confined aquifers. *J. Geophys. Res.*, 89, 1913-1924.

Nur, A., & Byerlee J. D., 1971: An exact effective stress law for elastic deformation on rocks with fluids. *J. Geophys. Res.*, 76, 6414-6419.

Rabbel, W., & Zschau, J. 1985: Static deformation and gravity changes at the earth surface due to atmospheric loading. *J. Geophys.*, 56, 81-99.

Rauen, A., Huenges, E., Bücker, C., Wolter, K.E., & Wienland, J., 1990: Tiefbohrung KTB-Oberpfalz VB. Ergebnisse der geowissenschaftlichen Bohrungsbearbeitung im KTB-Feldlabor (Windischeschenbach), Teufenbereich: 3500-4000,1m. *KTB-Report 90-2*, D1-D64.

Rice, J. R., & Cleary M. P., 1976: Some basic stress diffusion solutions for fluid-saturated elastic porous media with compressible constituents. *Rev. Geophys. Space Phys.*, 14, 227-241.

Röhr, C., Kohl, J., Hacker, W., Keyssner, S., Müller, H., Sigmund, J., Stroh, A., & Zulauf, G., 1990: German Continental Deep Drilling Programme (KTB) - Geological survey of the pilot hole "KTB Oberpfalz VB". *KTB-Report 90-8*, B1-B55.

Roeloffs, E., 1988. Hydrologic precursors to earthquakes: a review, *Pure Appl. Geophys.*, 126, 177-209.

Rojstaczer, S., 1988: Intermediate period response of water levels in wells to crustal strain: Sensitivity and noise level. *J. Geophys. Res.*, 93, 13619-13624.

Rojstaczer, S., & Agnew, D. C., 1989: The influence of formation properties on the response of water level in wells to earth tides and atmospheric loading. *J. Geophys. Res.*, 94, 12403-12411.

Schulz, Th., Borm, G., Scherbaum, F., & Weber, M., 1998: Seismologisches Tiefenobservatorium in der KTB/ICDP-Bohrung. Abstract for the ICDP/KTB-Kolloquium, 04./05. June 1998, Bochum (unpubl.).

Schulze, K.C., Kümpel, H.-J., Kück, J., & Huenges, E., 1998: Drucktest im Ringraum der KTB-HB und Analyse von Gezeitenvariationen im Fluidpegel der KTB-VB und KOLA-SG3. Abstract for the ICDP/KTB-Kolloquium, 04./05. June 1998, Bochum (unpubl.).

Skempton, A.W., 1954. The pore-pressure coefficients A and B. *Geotechnique*, 4, 143-147.

Van der Kamp, G., & Gale J.E., 1983: Theory of earth tide and barometric effects in porous formations with compressible grains. *Water Res. Res.*, 19, 538-544.

Wang, H.F., 1993: Quasi-static poroelastic parameters in rock and their geophysical applications. *Pure Appl. Geophys.*, 141, 269-286.

Wenzel, H.-G., 1996: The nanogal software: Earth tide data processing package ETERNA 3.30. *Bulletin d'Informations Mareés Terrestres*, 124, 9425-9439, Bruxelles.

Wenzel, H.-G., 1997: Analysis of earth tide observations. – In: *Tidal Phenomena*, Wilhelm, H., Wenzel, H.-G. & Zürn, W. (eds.), Lecture Notes in Earth Sciences , Springer, Berlin, 59-74.

Zaske, J., 1997: Einfluß eines Langzeitinjektionstests auf die Gezeitenantwort eines Bohrlochpegels am HDR-Standort Soultz-sous-Forêts. Diploma Thesis, Univ. Karlsruhe (unpubl.).

Zoback, M.D., & Harjes, H.-P., 1997: Injection induced earthquakes and crustal stress at 9 km depth in the KTB deep drilling site, Germany. *J. Geophys. Res.*, 102, 18477-1849.

A. List of used symbols

TABLE A1. Parameters

Parameter	Dimension	Notation
A_v, A_a	Pa/nϵ	volume, areal strain sensitivity
B	–	Skempton ratio
c, c_u	Pa^{-1}	drained, undrained compressibility
c_s, c_n	Pa^{-1}	grain, pore compressibility
c_{fl}	Pa^{-1}	fluid compressibility
g	$\mathrm{m/s^2}$	gravitational acceleration
H	–	parameter of Rojstaczer & Agnew
n	–	porosity
p	Pa	pore pressure (change)
P_c	Pa	confining pressure
w	m	well level change
$w_p = \rho_{fl} g w$	Pa	well level change in pressure units
S_s	m^{-1}	specific storage coefficient
S_a	m^{-1}	three-dimensional storage coefficient
α	$(0 < \alpha < 1)$	coefficient of effective stress
δ_{ij}	–	Kronecker-symbol
ϵ_{ij}	–	strain component
ϵ_v, ϵ_a	–	volume, areal strain
γ	$(0 < \gamma < 1)$	loading efficiency
$\Gamma = (1 - \gamma)$	–	barometric efficiency
$\Lambda = (1 - \nu)/(1 - 2\nu)$	–	–
μ	Pa	shear modulus
ν, ν_u	–	drained, undrained Poisson ratio
ρ_{fl}	$\mathrm{kg/m^3}$	fluid density
σ_{ij}	Pa	stress component

THE ROLE OF WATER-CONDUCTING FEATURES IN THE SWISS CONCEPT FOR THE DISPOSAL OF HIGH-LEVEL RADIOACTIVE WASTE

MARTIN MAZUREK[1], ANDREAS GAUTSCHI[2], PAUL A. SMITH[3] and PIET ZUIDEMA[3]

[1]*Rock/Water Interaction Group (GGWW), Institutes of Geology and of Mineralogy and Petrology, University of Bern, Switzerland (mazurek@mpi.unibe.ch)*

[2]*Nagra, Wettingen, Switzerland*

[3]*SAM Ltd., Hathersage, UK*

Abstract

Crystalline basement rocks are considered as potential host formations for the disposal of radioactive waste in several countries. In northern Switzerland, six boreholes were drilled into the sediment-covered basement, and water-conducting features intersected by the boreholes were identified by hydraulic testing methods. The study of the corresponding core materials resulted in the distinction of three geological types of water-conducting features: faults, fractured zones and fractured aplite/pegmatite dykes. Conceptual models describing the spatial arrangement of channels (in which advection occurs) and of wallrock domains (where matrix diffusion and sorption occur) within the water-conducting features were derived. The migration of radionuclides released from a repository through water-conducting features was modelled, taking into account advection/dispersion, matrix diffusion, sorption and radioactive decay/ingrowth. Due to the diversity of the types of water-conducting features and of the small-scale geometric parameters, six model cases were considered that spanned the range of geometric parameter uncertainty. Calculated radiological doses are relatively insensitive to variations, within the ranges of uncertainty, of pathlength within the geosphere, longitudinal dispersion and the depth of the diffusion-accessible wallrock matrix. However, the small-scale spatial arrangement of channels within water-conducting features was identified as affecting the barrier function of the geosphere to radionuclide transport more strongly. For a given large-scale permeability, highest radionuclide fluxes from the geosphere were obtained for the model case that minimizes the spatial density of channels.

The variable, sometimes high degree of channelling in the crystalline basement of northern Switzerland is a direct consequence of the intense hydrothermal water/rock interactions that recurrently affected the rocks. Some regions within fractures were sealed, whereas openings were created by wallrock dissolution in others, thus leading to the development of channels within the fractures. In the transport model, a high degree of channelling results in a reduction of flow-wetted surface, and this reduces the degree to which migrating radionuclides are retarded by matrix diffusion and sorption. On the other hand, alteration of the wallrocks produced highly sorbing minerals (*e.g.* clays) and increased the microporosity of the diffusion-accessible matrix, and these effects increase the retardation of radionuclides. Hydrothermal alteration is thus identified as a key geological process that affects the efficiency of the geosphere as a barrier to contaminant transport.

1. Introduction

Deep geological disposal of radioactive wastes is a field in which great efforts are currently expended. Several countries consider crystalline formations as potential host rocks, *e.g.* Sweden, Finland, Canada and Switzerland. Research in underground test

I. Stober and K. Bucher (eds.), Hydrogeology of Crystalline Rocks, 105–125.

facilities has substantially increased knowledge about the hydrogeology of crystalline rocks and relationships to geological and hydrogeochemical features. Rock laboratories are currently in operation at Äspö (Sweden), Grimsel (Switzerland) and Whiteshell (Canada).

In most geological repository concepts for radioactive waste, the most likely route for radionuclides from the repository to the biosphere[1] is via groundwater transport. Predictive modelling of solute migration through the geosphere[2] is based on a large number of geological, hydrogeological and hydrochemical input data derived from field and laboratory studies. This paper presents some of the techniques and procedures applied to derive such a database, together with its use in the calculation of radiological doses, using the crystalline basement of northern Switzerland as an example. This region is a potential target area for a high-level radioactive waste repository. Whereas the data themselves are site-specific, the methodology is applicable to most other waste disposal programmes. Emphasis is placed on methodological aspects (*e.g.* how to derive relevant concepts and data on the basis of field investigations), on the recognition of relevant parameters by means of sensitivity analysis and possible consequences for future investigation programmes.

Crystalline rocks (more specifically, intrusive rocks and gneisses) are considered for deep disposal mainly for the following reasons:

- They are wide-spread in their geographic location and depth below surface;
- In general, drilling and tunnelling in crystalline rocks does not present major engineering difficulties (in contrast to some sedimentary formations);
- Generic knowledge of fault architecture and hydrogeology in crystalline rocks is substantial (overview in Mazurek 1999);
- The number of processes that affect the flow and transport properties is relatively small. For example, effects of osmosis, swelling, or contaminant complexation with naturally occurring organic ligands or colloids do not generally play a significant role in crystalline rocks;
- Hydrothermal or low-temperature alterations have affected crystalline rocks at many potential sites and have produced highly sorbing minerals in fractures, *e.g.* oxy-hydroxides or clay minerals;
- Hydrochemical investigations in tectonically inactive crystalline-rock environments indicate the presence of very old, stagnant groundwaters, at least in specific domains of the rocks, indicative of negligible flow over geologic timescales.

Less favourable properties include:

- Crystalline rocks are frequently highly heterogeneous, with discrete water-conducting features; transmissivities of water-conducting features may be relatively high, even at great depth;
- In the absence of good outcrop conditions, the large-scale structure of crystalline rock formations is difficult to characterize on the basis of surface-based investigation methods, such as vertical drilling or geophysics. Therefore, the explorability of rock volumes suitable for disposal is non-trivial as long as site-specific information from inclined boreholes, shafts and tunnels is not available.

---

1 Biosphere: The portion of the Earth's environment that is inhabited by living organisms. It comprises parts of the atmosphere, the hydrosphere and the lithosphere and includes the human environment.

2 Geosphere: The repository host rock and any surrounding or overlying strata which form a natural safety barrier for waste disposal.

Safety barrier system for high-level waste

Glass matrix (in steel mould)

- Low corrosion rate of glass
- High resistance to radiation damage
- Homogeneous radionuclide distribution
- Corrosion products take up radionuclides

Steel canister

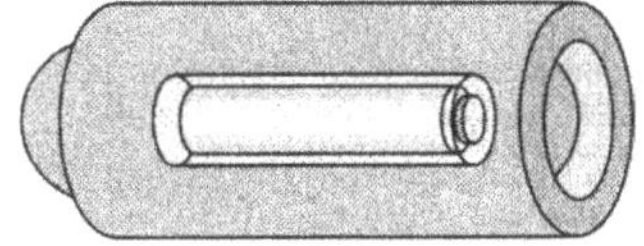

- Completely isolates waste for > 1000 years
- Corrosion products act as a chemical buffer
- Corrosion products take up radionuclides

Bentonite backfill

- Long resaturation time
- Low solute transport rates (diffusion)
- Retardation of radionuclide transport (sorption)
- Chemical buffer
- Low radionuclide solubility in pore water
- Colloid filter
- Plasticity (self-healing following physical disturbance)

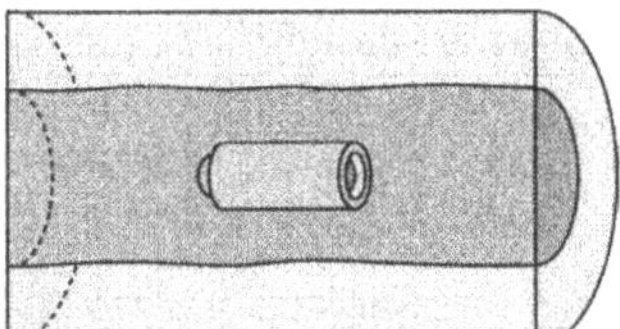

Geological barriers

Repository zone:

- Low groundwater flux
- Favourable hydrochemistry
- Mechanical stability

Geosphere:

- Retardation of radionuclides (sorption, matrix diff.)
- Reduction of radionuclide concentration (dilution, radioactive decay)
- Physical protection of the engineered barriers (e.g. from glacial erosion)

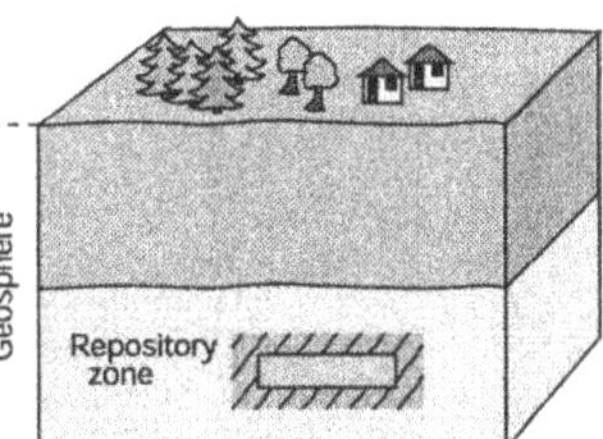

Figure 1. The system of safety barriers for disposal of high-level radioactive waste in the crystalline basement of northern Switzerland. Taken from Nagra (1994).

2. The Swiss disposal system

According to the Swiss concept, high-level radioactive waste will be isolated in a deep mined repository in which the waste packages (massive steel canisters containing vitrified waste from reprocessing) are horizontally emplaced in tunnels, backfilled with highly compacted bentonite (Figure 1). For the expected evolution of the repository system, the steel canisters will slowly degrade and, after a period of time (Swiss reference-case value: 1000 a), water will contact the glass matrix, which will begin to dissolve, and radionuclides will be released into deep groundwater following diffusion through the bentonite backfill. Quantitative evaluation of the extent and consequences of such releases utilizes a chain of models which calculate the flux of water through the repository, the corrosion/alteration/erosion of the engineered barrier system (glass matrix, steel canister and bentonite backfill), the release of radionuclides from the engineered barrier system, their transport through the geosphere to the biosphere and the resulting radiological doses to man. All of these models require site-specific geological and other input (Nagra 1994).

The role of the geosphere in such a concept of multiple barriers to radionuclide migration is twofold:

- It provides protection for the engineered barrier system and conditions that favour its longevity and performance, including mainly mechanical stability, limited water fluxes through the repository zone and stable chemical boundary conditions (*e.g.* reducing environment).
- Radionuclides released from the engineered barrier system are transported to water-conducting features in the geosphere and interact physically and chemically with the rocks that comprise these features. The interactions retard the release of nuclides to the biosphere and decrease their concentrations.

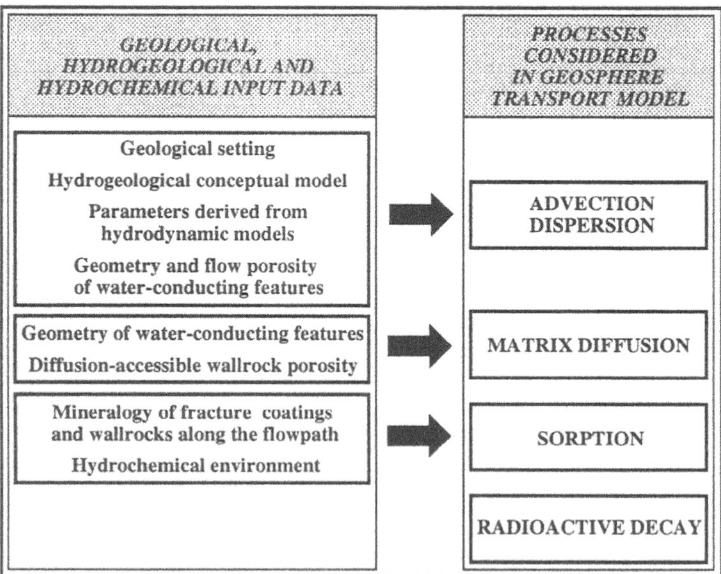

Figure 2. Use of geological, hydrogeological and hydrochemical data as input to radionuclide transport modelling - overview of input parameters and processes.

3. Processes that govern radionuclide transport through the geosphere

A computational model coupling all relevant processes that affect the transport of radionuclides through water-conducting features is used to quantify the efficiency of the geosphere as a barrier to the release of contaminants to the biosphere. Figure 2 gives an overview of the processes that are taken into account and also indicates what kind of geological, hydrogeological and hydrochemical data are needed for such a model. The field-derived database contains data from geology (geometry of water-conducting features, mineralogy, porosity), hydrogeology (transmissivities of water-conducting features, flow rates, flow vectors) and geochemistry (reference hydrochemical compositions and redox states, sorption properties). Many of the relevant system properties are spatially variable, but, in certain cases, the averaging of natural variability can be justified. For example, small-scale variability of the mineralogies of fracture infills averages out because transport paths are very long with respect to such heterogeneity. In other cases, ranges of parameter values need to be specified, considering also the possibility of correlations between parameters.

3.1 ADVECTION

Crystalline rocks are fractured media, and water-conducting features may be defined as roughly two-dimensional zones, with enhanced transmissivities, that correspond to structures generated by different types of faulting and fracturing. Advective transport of any radionuclides that are released from a repository would be expected to occur in a network of open fractures or channels within the water-conducting features. Flow models are used to calculate a range of parameters, including Darcy velocities, flow directions and the length of the transport paths between the repository and exfiltration areas. Flow rates through fractures or channels within water-conducting features are calculated from Darcy velocities by including information on the small-scale geometry of fractures or channels. Thus in addition to the hydraulic input data (such as transmissivity and head distribution, initial and boundary conditions), information is required on the size and spatial distribution of open fractures or channels (*i.e.* the flow porosity[3]) within water-conducting features.

3.2 MATRIX DIFFUSION

In addition to the open channels in which flow occurs, crystalline rocks have a micro-scale porosity in the surrounding rock matrix (*e.g.* along grain boundaries or crystallographic cleavage planes within minerals), with typical apertures in the range of nanometers to micrometers. Whereas flow through such small structures is negligible, even over geologic timescales, they are accessible for diffusion, provided the pore network is interconnected (see, for example, Hellmuth *et al.* 1995, Rasilainen *et al.* 1996, Siitari-Kauppi *et al.* 1998).

The microporosity of the unfractured rock matrix in northern Switzerland is ca. 0.25 vol% for unaltered granites and 1 vol% for gneisses (Mazurek 1998). Because water-conducting features were affected by recurrent stages of hydrothermal activity,

---

3 Flow porosity: Connected porespace in a rock through which flow occurs (also called effective porosity). In the case of fractured media, flow porosity occurs in fractures or channels within fractures.

alteration rims occur adjacent to fractures. Alteration resulted in a substantial increase of matrix porosity to 1.5 - 5 vol%. Microporosity of the rock matrix (whether fresh or altered) is higher than the porosity associated with fractures and channels (*i.e.* flow porosity) and thus represents the dominant fluid reservoir in the rocks. Unless the rock matrix is completely disconnected from flow porosity (*e.g.* by localized hydrothermal cementation), radionuclides derived from the repository may migrate from the channels or fractures into the rock matrix via diffusion (Neretnieks 1980). Even for very long-lived and poorly sorbing radionuclides, matrix diffusion may contribute to the barrier function of the geosphere because it dilutes the radionuclide concentrations and thus attenuates the peaks of the breakthrough curves in the exfiltration areas.

3.3 SORPTION

Along the flowpath, migrating radionuclides may sorb onto mineral surfaces on the channel or fracture walls (the *flow-wetted surface*[4]). In addition, the walls of the diffusion-accessible wallrock porosity provide a large mineral-surface area on which this process can occur. Element-specific sorption distribution coefficients (K_d) relate the nuclide concentrations on minerals and in solution and are a function of mineralogy, groundwater composition and of redox conditions. Several radioelements sorb effectively on mineral surfaces (*e.g.* Am, Cm, Pu), whereas others are weakly to non-sorbing (*e.g.* Se; see Stenhouse 1994). The retardation[5] produced by the combined effects of matrix diffusion and sorption can give rise to transport times through the geosphere that exceed the half lives of many safety-relevant radionuclides. The releases of these radionuclides are therefore substantially attenuated by the geosphere transport barrier. Table 1 lists the key nuclides (from a larger radionuclide inventory) in the Swiss high-level reprocessing waste and some of their relevant physical and chemical characteristics.

4. Water-conducting features in the crystalline basement of northern Switzerland

4.1 OVERVIEW OF AVAILABLE FIELD DATA

The crystalline basement of northern Switzerland is covered by sedimentary rocks of several hundred meters thickness. Most field data have, therefore, been collected from 6 deep boreholes, which yielded almost 6 km of core profile length in the crystalline basement to a maximum depth of 2500 m below the surface (Figure 3). All boreholes have been subjected to extensive hydraulic testing (hydraulic packer tests, fluid logging, long-term monitoring; Voborny *et al.* 1994), and numerous groundwater samples have been taken for hydrochemical studies (Schmassmann *et al.* 1992, Pearson *et al.* 1991, Michard *et al.* 1996). Geological, geochemical and petrophysical data were derived from the cores, and standard geophysical logging has been performed in all boreholes (Thury *et al.* 1994).

---

4 Flow-wetted surface: Surface area of the fracture or channel walls that is in direct contact with flowing water. In the case of flow through a homogeneous fracture with constant aperture, flow-wetted surface is equal to 2 times the fracture-surface area.

5 Retardation: Reduction in the rate of radionuclide migration through the environment due to interaction between dissolved radionuclides and mineral surfaces. Sorption and matrix diffusion are examples of retardation mechanisms.

TABLE 1. Properties of a selection of safety-relevant radionuclides in reprocessed Swiss high-level waste (data from Nagra 1994). "High" solubility means that no solubility limit was considered in the transport calculations

nuclide	half life, a	solubility, M	K_d, m^3/kg
^{79}Se	6.50×10^4	10^{-8}	0.01
^{99}Tc	2.13×10^5	10^{-7}	0.5
^{135}Cs	2.30×10^6	"high"	0.042
^{235}U	7.04×10^8	10^{-7}	1
^{237}Np	2.14×10^6	10^{-10}	1
^{239}Pu	2.41×10^4	10^{-8}	5
^{242}Pu	3.76×10^5	10^{-8}	5

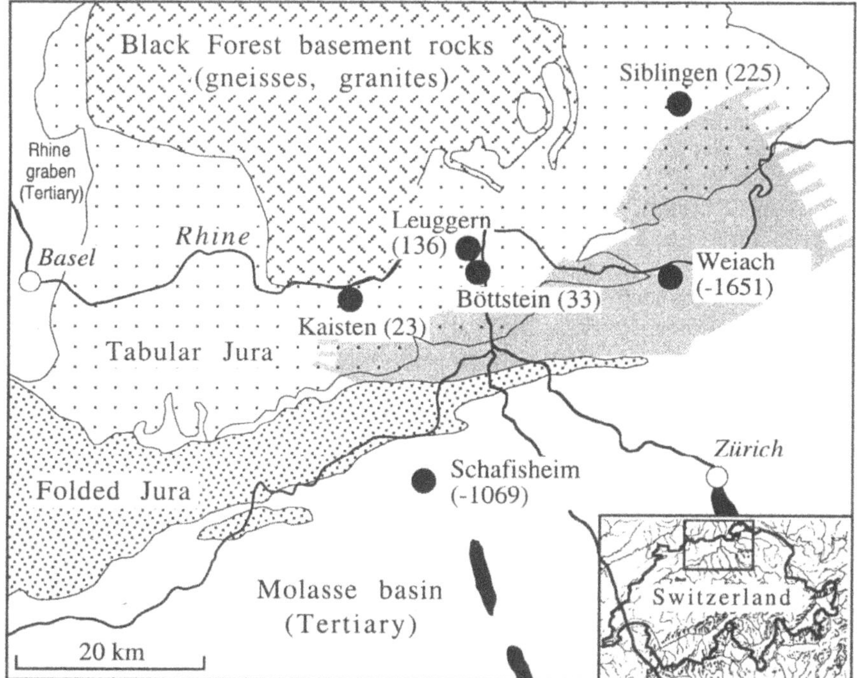

Figure 3. Simplified geological map of northern Switzerland, with positions of Nagra boreholes. Numbers in brackets indicate the top of the crystalline basement in m above sea level. Grey area indicates the position of the deep part of the Permo-Carboniferous trough of northern Switzerland that underlies the Mesozoic sediments (derived from seismic surveys).

Although the Nagra borehole studies provided a detailed geological characterization of water-conducting features on a small-scale (Mazurek 1998), information on regional geological structure and its effect on groundwater fluxes, gradients and flow vectors had to be supplemented by data from other boreholes, seismic surveys and field mapping in the nearby Black Forest area, where the basement rocks crop out or are accessed by mines and hydropower tunnels (Daneck 1994, Biehler 1995).

4.2 GEOLOGIC EVOLUTION

The regional framework of the study area is described in Diebold *et al.* (1991), Laubscher (1986a,b, 1987) and Mazurek (1998). The Nagra boreholes in northern Switzerland penetrated Pre-Variscan high-grade metamorphic gneisses and migmatites (mostly metapelites and meta-graywackes) that were intruded by Variscan plutonites (mainly S-type granites) and dykes (aplites, pegmatites, granite/rhyolite porphyries, lamprophyres). The late and post-Variscan evolution is characterized by a series of tectono-hydrothermal events, *i.e.* brittle deformation phases (faulting, fracturing) associated with hydrothermal alteration due to interaction with fluids circulating in the brittle structures (Peters 1987, Meyer 1987, Mazurek 1999). The Late Carboniferous high-temperature phase resulted in greenschist-facies alteration of the wallrocks (mainly albitization/sericitization of plagioclase and chloritization of biotite), whereas the Early Permian low-temperature phases were dominated by argillic alteration (illite, illite/smectite). Younger events include a kaolinitic alteration, which is accompanied only by subordinate brittle deformation, the formation of vugs (open channels) in pre-existing discontinuities by dissolution of pre-existing fracture infills, a calcite precipitation in fractures and the formation of ore and mineral veins. Hydrothermal alteration may penetrate up to several meters into the wallrock away from fractures, which are quite often healed by hydrothermal infills or cataclastic matrices.

Stages of brittle deformation include both cataclasis and purely tensile jointing. The correlation of brittle structures with the greenschist- or low-temperature alteration phases is based on the contrasting mineralogies of fracture infills and altered-wallrock rims, whereas no geometric distinction (anatomy, orientation) could be made between different stages of deformation on the basis of borecore data. In a few cases, systematic variation of structure orientations between different phases could be identified, but they do not result in a consistent regional pattern. Younger deformation events quite often reactivate pre-existing structures, resulting in complex interference patterns.

4.3 IDENTIFICATION OF INFLOW POINTS AND CORRELATION WITH GEOLOGICAL STRUCTURES

Over a total profile length of 5800 m drilled in the crystalline basement, 138 discrete inflow points of formation water into the boreholes were identified by hydraulic packer tests and fluid logging techniques (Thury *et al.* 1994). All of these inflow points were related to core sections in which the rock was affected by significant brittle deformation postdating the emplacement of the rock types. At most inflow points, a complex, interconnected system of several fracture planes has been identified rather than a single fracture, reflecting the long tectonic history of the crystalline basement. In general, the intensity of brittle deformation does not vary systematically between rock types, and the frequency and transmissivity of inflow points in granites and gneisses cannot be distinguished. The only important exception are aplite and pegmatite dykes, where brittle deformation and therefore inflow points are concentrated (see Mazurek 1999 for details). Faults and fractures that carry water today have been hydraulically active throughout geological history, at least episodically. The passage of hydrothermal paleofluids is recorded by a variety of fracture mineralizations and alteration features in the adjacent rock matrix (see also Mazurek 1998, 1999). The effects of the greenschist-grade alteration, subsequent argillic alteration and younger stages of water/rock interaction are often recorded in the same structure, indicative of recurrent activity.

4.4 CLASSIFICATION OF WATER-CONDUCTING FEATURES

The most suitable criteria for a classification of water-conducting features with respect to their flow and transport properties include the brittle deformation mechanism and mineralogy/lithology (Mazurek 1998). Due to the complexity of the regional hydrothermal evolution, the type of alteration was not used as a classification criterion, as the effects of several alteration stages are superposed within single water-conducting features. Following this scheme, a systematic pattern could be derived that is valid on a regional scale. Flow occurs in the following types of water-conducting features:

1. Faults (cataclastic zones; 43 % of all inflow points)
2. Fractured zones (32 %)
3. Fractured aplite and pegmatite dykes (23 %).

Three additional inflow points in the Leuggern borehole are associated with Paleozoic mineral veins[6] that do not fit the regional classification scheme. Geometric, hydrogeological and hydrochemical information shows that all three types of water-conducting features are interconnected and form a 3-dimensional network of potential transport paths. The classification scheme derived from the borehole data compares well with findings from the Black Forest (borehole, surface and tunnel observations; Daneck 1994, Stober 1995, Biehler 1995). One major difference is the clustered occurrence of highly conductive ore-vein systems in the Black Forest, which were not found to the same extent in northern Switzerland, either because such systems do not exist in this region or because they were not identified due to the limitations of the investigation programme.

The first two types are distinguished by the nature of the deformation process, which is cataclasis for type 1 (often with cataclasite and/or fault gouge/breccia infills) and fracturing without clear indication of brittle shear deformation for type 2. By definition, cataclastic zones show evidence of shear deformation and generally consist of networks of zones of movement and accompanying fracturing. At least a part of the cataclastic zones can be expected to extend over large distances (tens of meters and more), which is not necessarily the case for fractured zones. If fractured zones consist of tensional joints, the size of the individual structures is expected to be in the range of meters to a few tens of meters at most.

Hololeucocratic rocks of largely two-dimensional shape, namely aplite/pegmatite dykes and aplitic gneisses[7], are classified in a separate group because, irrespective of the brittle deformation process, both their geometric and mineralogic characteristics are systematically different from those of all other rock types. Fracturing and the occurrence of vugs (open channels) are much more prominent than in the adjacent country rocks of the dykes, and it is therefore likely that the dykes may contain interconnected fracture

---

[6] K/Ar dates of illite from these veins yield 275 - 278 Ma, and a Rb/Sr isochron of fresh and altered whole-rocks and illites from veins results in an age of 279 ± 5 Ma. These data are documented in Mazurek (1998).

[7] Aplitic gneisses form concordant, meter-thick intercalations within the paragneiss series at Kaisten and Leuggern and represent the metamorphosed equivalents of rhyolitic sheets. The term *aplitic* refers to the fine-grained texture and quartzo-feldspathic composition of this rock type, while no genetic relation with aplite dykes is made. The large-scale geometry of the aplitic gneisses is also similar to that of the dykes in that both are largely two-dimensional rock bodies (slabs) with thicknesses measured in decimeters to meters and extents that could well exceed 100 m. Intense fracturing and the scarcity of hydrothermal neoformations in fractures are common to both. The effects of these two rock types on present-day groundwater flow and solute transport are very similar, and so they are grouped together in one type of water-conducting feature.

networks whose large-scale geometry is entirely determined by the size and shape of the dykes.

Figure 4 shows the distribution of hydraulic conductivities (derived from hydraulic packer testing) and types of inflow points in the boreholes at Böttstein, Kaisten and Leuggern. The geological types of water-conducting features cannot be discriminated with respect to hydraulic conductivity or depth below surface. In the uppermost 350 - 650 m of the crystalline basement, conductivities are relatively high in all boreholes. In the deeper parts, they are substantially lower in some boreholes (*e.g.* Böttstein) or similar to those of the shallower levels (*e.g.* Kaisten).

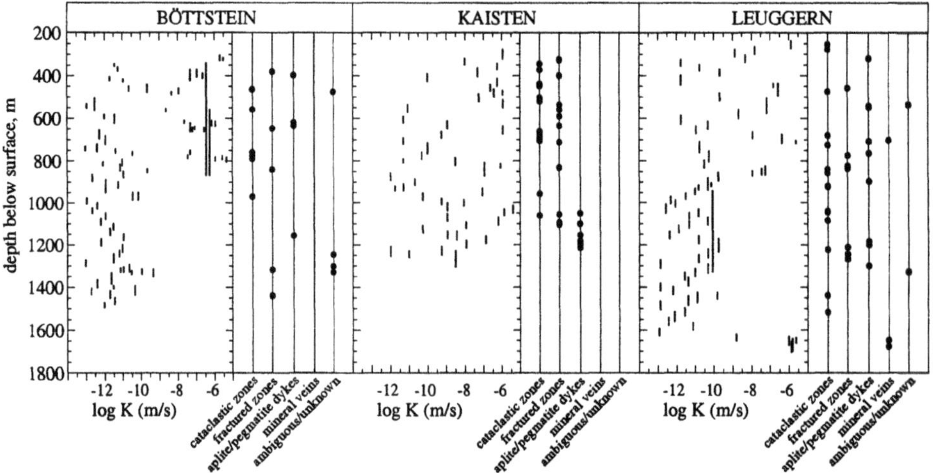

Figure 4. Hydraulic conductivities derived from packer testing (from Küpfer *et al.* 1989) and positions of inflow points (from Mazurek 1998) in the the Böttstein, Kaisten and Leuggern boreholes.

5. Small-scale conceptual models of water-conducting features

5.1 REPRESENTATION OF WATER-CONDUCTING FEATURES IN TRANSPORT MODELS

Many computational models that quantify contaminant transport through the geosphere are based on simple concepts (*e.g.* water flow through one or several representative channels with constant water chemistry, flow rate, wallrock mineralogy and porosity along the whole transport path). Natural complexity, therefore, needs to be reduced, and some degree of averaging is necessary. In order to provide input for the transport calculations, the geometry of water-conducting features, including the spatial arrangement of flow porosity (*e.g.* open fractures, channels), of fracture infills (*e.g.* cataclastic matrices, vein mineralizations) and of wallrock lithologies (*e.g.* altered-wallrock rims around fractures, fresh rocks) is simplified in conceptual geological models.

In the reference concept for high-level waste disposal in northern Switzerland, a repository will be overlain by at least 500 m of crystalline basement rocks. Because of the good interconnection of all types of water-conducting features, it seems unlikely that any radionuclides released from the repository would remain in the same type of water-conducting feature during its transport through the crystalline basement. The large-scale spatial arrangement of water-conducting features is, however, not well known at present, and so the performance of the geosphere with respect to radionuclide transport is calculated separately for each type. Because the type that results in the most limited retardation of migrating radionuclides is adopted as the reference case, this approach may lead to an underestimation of the barrier function of the geosphere, but the advantage is that it does not require detailed information on the spatial arrangement (orientations, sizes) of water-conducting features. In the following, one particular type of water-conducting feature, namely faults, will be discussed in more detail.

5.2 CONCEPTUAL MODEL FOR FAULTS (CATACLASTIC ZONES)

Faults in the basement of northern Switzerland occur on a wide range of scales. Their thickness may reach tens of meters in infrequent regional lineaments, whereas minor structures are only visible in thin-section. The typical thickness of faults associated with inflow points in the boreholes is in the order of 50 - 100 cm, and such faults consist of complex fracture networks. Even thinner structures with only 1-2 fracture planes exist, and these represent either simple segments within otherwise geometrically more complex structures or, alternatively, minor features of limited size. The lateral extent of the faults cannot be determined directly from core studies. By analogy with basement outcrops in the Black Forest, the length of such structures can be constrained to the range 100 - 1000 m. The considerable uncertainty related to the size distribution of faults does not critically affect the results of the transport calculations (see section 5.1).

On a smaller scale, a typical fault structure consists of an interconnected network of individual fractures (Figure 5, upper left). This pattern is represented in the model abstraction (Figure 5, upper right) by a set of parallel plates. Each individual fracture is observed to be partially filled by fault gouge, cataclastic matrix or hydrothermal infill material, but typically contains portions with open channels where advective transport predominantly occurs (Figure 5, bottom left). These relationships are represented in the conceptual model (Figure 5, bottom right) by a planar fracture infill of constant thickness (typically 1 - 2 mm), which is interspersed by the channels. The latter are represented by conduits of rectangular cross-sections and form a rectangular mesh within the fracture plane. Both the size of the channels and their spacing within the fracture plane vary widely in natural occurrences. These variations are expressed by the ranges of 1-10 cm for the width and 10 - 500 cm for the spacing of the channels in Figure 5. For transport modelling (see section 6), two alternative arrangements of channels will be considered, namely widely-spaced, broad channels (width: 10 cm; spacing: 500 cm) and closely-spaced, narrow channels (1 cm; 10 cm).

The wallrock of each cataclastic fracture is altered to some degree, and this has changed the primary mineralogy considerably. The thickness of the altered rims is variable; typical values are in the order of 10 cm, and never drop below 1 cm. While the mineralogy of the altered rock strongly depends on the type of alteration, the increased open porosity in the altered zones is common to all alteration schemes.

In summary, the anatomy of faults is represented conceptually by the following domains in Figure 5:

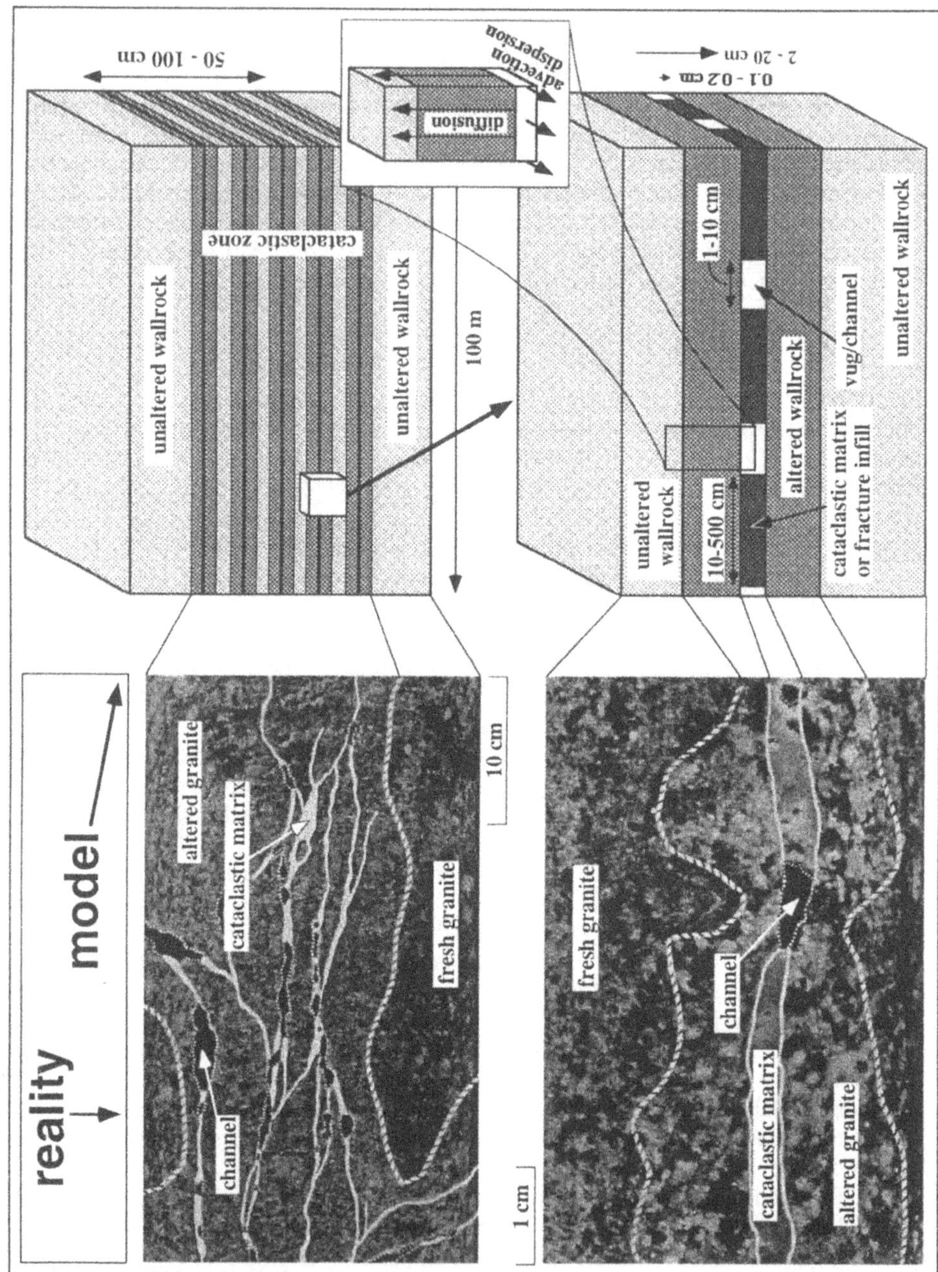

Figure 5. Small-scale conceptual model of faults in the crystalline basement of northern Switzerland, including geometric, mineralogical and porosimetric data.

- Cataclastic matrix or fault gouge (fracture infill) in the fracture planes, containing water-conducting channels (representing flow porosity);
- Altered wallrock around each fracture;
- Unaltered wallrock outside the cataclastic zone.

Based on the numerous analyses performed on rock samples, data for matrix porosity and mineralogy can be assigned to each of the domains in Figure 5 (not shown).

The possible existence of a fault damage zone embedding the fault core (see *e.g.* Caine *et al.* 1996, Mazurek 1999) has not been considered in the conceptual model. This is mainly due to the difficulty in distinguishing fractures that were created or reopened by the drilling process from naturally open fractures. This omission is conservative in the sense that the flow-wetted surface is likely to be underestimated, and thus so is radionuclide retardation due to sorption and matrix diffusion.

6. Transport modelling, model parameters and sensitivity analysis

6.1 THE TRANSPORT MODEL AND INPUT PARAMETERS

The presence of channels and an interconnected wallrock porosity in all types of water-conducting features indicates that the processes described in section 3, namely advection and retardation due to matrix diffusion and sorption, should be included in a transport model. Such a dual-porosity model must take account of:

1) water-conducting channels, equivalent to the flow porosity (Figure 5), where advective transport and dispersion (due to the variability in travel times through a number of channels) occur and
2) adjacent wallrock, where solute diffusion and sorption take place within the open connected microporosity.

Sorption of migrating species can occur on the surfaces of the channels as well as in the open wallrock pores. This dual-porosity model, the governing equations and parameter values for which are presented in Nagra (1994), was solved using the code RANCHMD (**RA**dio**N**uclide **CH**ain transport with **M**atrix **D**iffusion; Hadermann and Rösel 1985). RANCHMD couples one-dimensional advection/dispersion in channels with diffusion into the wallrocks, sorption on the walls of the accessible matrix porosity, and radioactive decay and ingrowth of daughter nuclides.

Based on the mineralogical compositions and accounting for groundwater chemistry and redox state, specific distribution coefficients for sorption of solute contaminants can be assessed for all rock domains. Structural and hydrogeological input data to the transport model comprise the flow rates through the water-conducting features and their small-scale geometric characteristics (size and spatial density of channels, flow-wetted surface, matrix porosity), and dispersion length. Modelling studies to date (*e.g.* Nagra 1994) have aimed at the identification of critical parameters, in order to focus further field and laboratory investigations and reduce uncertainties. Table 2 lists the geoscientific parameters for the transport model, together with the reference-case values that were used for the "Area West", which corresponds to the region defined by the boreholes Böttstein, Kaisten and Leuggern (Figure 3).

TABLE 2. Selected geological input parameters used in modelling radionuclide transport through the crystalline basement of northern Switzerland. Parameter values are taken from Nagra (1994) and refer to the reference case of area West (i.e. broad, widely-spaced channels in faults in the region Böttstein-Leuggern-Kaisten)

parameter	derivation method	information used for the derivation	use of parameter	parameter value (reference case)
q Darcy velocity	flow modelling	transmissivity and head distributions	advective/ dispersive transport	2.4×10^{-5} m/a
a_L dispersion length	Peclet number = 10 from literature (Gelhar *et al.* 1992)	length of flowpath = 200 m	advective/ dispersive transport	20 m
2 H flow-wetted surface	small-scale geological characterization of water-conducting features	-	matrix diffusion	0.0048 m^2/m^3
n connected wallrock microporosity	laboratory or in-situ measurement techniques	arrangement of wallrock domains along channels (e.g. alteration rims)	matrix diffusion	altered rock: 1.5 - 5 vol%; fresh rock: 0.25 - 1 vol%
D_p pore diffusivity of nuclides	$D_p = D_W * G$, where D_W = diffusivity in free water G = geometry factor data based on laboratory experiments (summarized in Frick 1993)		matrix diffusion	0.001 m^2/a
K_d distribution coefficients for sorption	laboratory experiments, chemical analogy	mineralogy of wallrock domains	sorption	element-specific, see Table 1

6.2 RESULTS

Given the existence of different types of water-conducting features and the variability of their small-scale geometry (channels, types of wallrock), six parameter sets that cover the likely spectrum were used for the calculations:

- Faults with widely-spaced, broad channels
- Faults with closely-spaced, narrow channels
- Aplite/pegmatite dykes,

where, for each case, matrix diffusion was either assumed to be unlimited or restricted to the altered-wallrock rims. In the case of aplite/pegmatite dykes, it was not necessary to distinguish different channel geometries because the natural variability is more limited than in faults (Mazurek 1998). Fractured zones were not treated separately because, at the level of conceptual simplification shown in Figure 5, they are not significantly different from faults.

Results of near-field/geosphere/biosphere chain calculations are given in Table 3 in the form of maximum annual individual doses for selected radionuclides. Such calculations take into account the radionuclide releases from the repository system to the geosphere and then to the surface environment, where radionuclides may enter the foodchain (Klos and van Dorp 1994). Calculated radionuclide uptakes by man are

TABLE 3. Results of calculations of radionuclide transport through the geosphere using the code RANCHMD. Numbers refer to the logarithm of maximum annual individual doses of a selection of key nuclides in units of mSv/a. Such peak values are attained at different times after repository closure (*e.g.* ^{135}Cs: 10^5 - 10^7 a). The reference case around which parameter variations are made is indicated in bold. Parameter variations are shown in the lower half of the Table. All parameters except the one that is subjected to variation are kept constant at reference-case values. All results from Nagra (1994)

		^{79}Se	^{99}Tc	^{135}Cs	^{237}Np
transport in geosphere neglected (repository near-field coupled directly to the biosphere) - pessimistic, hypothetical case		-4.6	-4.3	-3.4	-4.3
case 1	**faults: broad, widely-spaced channels, matrix diffusion in altered wallrock only (reference case)**	**-4.9**	**-6.3**	**-3.6**	**-5.4**
case 2	faults: broad, widely-spaced channels, unlimited matrix diffusion	-5.2	-5.4	-4.0	-5.4
case 3	faults: narrow, closely-spaced channels, matrix diffusion in altered wallrock only	≤-15	≤-15	-7.1	≤-15
case 4	faults: narrow, closely-spaced channels, unlimited matrix diffusion	≤-15	≤-15	≤-15	≤-15
case 5	aplite/pegmatite dykes: matrix diffusion in altered wallrock only	-6.9	≤-15	-4.6	-8.9
case 6	aplite/pegmatite dykes: unlimited matrix diffusion	-11.4	≤-15	-6.5	≤-15
parameter variations	0.1 x base case Darcy velocity	-7.8	≤-15	-5.2	-9.6
	10 x base case Darcy velocity	-3.6	-3.6	-3.0	-3.5
	100 x base case Darcy velocity	-3.1	-2.9	-2.8	-2.2
	2.5 x base case path length = 500 m	-5.4	-8.3	-3.9	-6.1
	0.5 x base case path length = 100 m	-4.8	-5.4	-3.5	-5.1
	0.2 x base case longitudinal dispersion	-4.9	-7.1	-3.6	-5.5
	5 x base case longitudinal dispersion	-4.9	-5.8	-3.6	-5.1
	accessible matrix depth reduced to 1 cm	-4.7	-4.8	-3.5	-4.9
regulatory dose limit		-1			
average dose from natural radiation in Switzerland		+0.5			

converted to equivalent doses (in units of mSv/a) according to their radio-toxicity. For all radionuclides listed in Table 3, faults with broad, widely-spaced channels give rise to the highest doses, whereas aplite/pegmatite dykes have a stronger retardation effect and consequently give rise to lower doses. In faults with closely-spaced, narrow channels, all radionuclides with the exception of ^{135}Cs largely decay within the geosphere. Whereas the case of broad, widely-spaced channels in faults is much less likely to occur in nature than the case of narrow, closely-spaced channels, it was taken as the reference case for reasons of conservatism, and parameter sensitivity studies were performed around this case (see below). It is evident that even this least favourable case provides a reduction of the peak doses when compared to the hypothetical case of direct release from the engineered barrier system in the repository to the biosphere (factor of 1.6 for ^{135}Cs, 100 for ^{99}Tc). Computational cases that assume unlimited matrix diffusion or different geometric arrangements all provide higher retardation, and in several cases, the nuclides almost completely decay within the geosphere (*i.e.* the doses are $\leq 10^{-15}$ mSv/a in Table 3).

6.3 DISCUSSION

Ideally, the objective of transport modelling would be to quantify mass fluxes that, according to the best current state of system understanding, represent the behaviour of the actual system as closely as possible. In reality, however, some input parameters show variability in space and time that cannot be incorporated fully into existing transport models (even if adequately known), whereas other parameters are not well characterized due to the limitations of the currently applied characterization techniques. Thus, a range of uncertainty is inherent in the results of such models. The approach adopted for a safety assessment is, where there exists uncertainty, to adopt model assumptions and parameter values that err on the side of pessimism, *i.e.* tend to overestimate the resulting radiological doses. Whereas such a cumulation of pessimistic model assumptions and parameters may have little to do with reality, it has the advantage of being easier to defend ("it cannot be worse than ..."). In the most recent Swiss safety assessment for crystalline rocks (Nagra 1994), a number of pessimistic assumptions regarding geosphere transport have been made, for example:

- A very high degree of small-scale flow channeling is assumed, which minimizes retardation by matrix diffusion and sorption. In faults and in fractured zones, only 2 - 10 % of the existing fracture surfaces are accessible for flow (Figure 5), and in fractured aplite/pegmatite dykes this value is 50 % due to the scarcity of hydrothermal fracture infills (see section 6.5).
- Of the six representations of water-conducting features considered, the one that results in the highest doses is taken as the reference case, even though there are no geological criteria for such a choice.
- None of the small-scale geometric conceptual models of water-conducting features accounts for the likely presence of a fault-damage zone that provides additional surfaces for sorption and diffusion.
- Sorption and matrix diffusion in fracture infills (which are often clay-rich) are neglected.
- Sorption on mineral surfaces is assumed to be fully reversible. Thus, when radionuclide concentrations released from the repository system decline at times post-dating the peak releases, desorption is accounted for in the model.
- Transport through the host rock is modelled with a reference flowpath length of 200 m, whereas subsequent transport through major faults, higher-permeability crystalline rocks or through the sedimentary cover rocks is assumed to be instantaneous.
- Transmissivity and hydraulic gradient in water-conducting features are considered to be independent parameters, whereas, in nature, highly transmissive features are expected to have the relatively lowest gradients.
- Transverse dispersion is neglected.

The combination of pessimistic assumptions may in some cases lead to the use of parameter values that are in contradiction to independent observations and measurements at the site. For example, the parameters used for the calculations (Table 2) would imply that groundwater at repository level would have to recharge within hundreds to a few thousands of years[8]. This is in conflict with evidence based on environmental tracers in the deep groundwaters (Pearson *et al.* 1991, Thury *et al.* 1994), which suggest underground residence times >70 ka for the groundwaters sampled in the "Area West". It is clear that calculations whose results are given in Table 3 substantially overestimate radiological doses. However, even these values are all below the Swiss regulatory limit of 0.1 mSv/a (HSK and KSA 1993). For comparison, natural radiation in Switzerland accounts for an average individual dose of 3.4 mSv/a.

6.4 SENSITIVITY ANALYSIS AND KEY PARAMETERS

Of the field-derived, site-specific parameters that served as input to the transport calculations, variations in those that describe the small-scale geometry of flow porosity influence the model results most strongly, as shown by the contrasting doses evaluated by the six calculation runs (Table 3, upper part). In particular, the size and spacing of channels within single water-conducting features are very sensitive (*cf.* Rasmuson and Neretnieks 1986, Tsang *et al.* 1988), and the efficiency of the geosphere barrier decreases with increasing channel spacing. This is because

- access to matrix porosity occurs through a decreasing proportion of the fracture/wall interface, and thus retardation due to matrix diffusion and sorption decreases, and
- the flow rate through individual channels increases with lateral channel spacing for a given Darcy velocity in the geosphere.

The results also showed sensitivity to varying the Darcy velocity by factors 0.1, 10 and 100 (Table 3). However, the factors 10 and 100 correspond to flow rates through channels that are even less realistic from the viewpoint of the information provided by environmental tracers in the groundwaters (see above). With the exception of ^{99}Tc, variations in pathlength, longitudinal dispersion and depth of the diffusion-accessible wallrock matrix are insensitive model parameters (Table 3).

6.5 HYDROTHERMAL ALTERATION - THE KEY GEOLOGICAL PROCESS

The sensitivity analysis of the geological input parameters indicates that the small-scale properties of water-conducting features are important for the quantification of radionuclide transport through the geosphere. Many of these properties are genetically linked to the hydrothermal activity that very strongly affected the crystalline basement of northern Switzerland. From the perspective of radionuclide transport, the following effects of hydrothermal activity are relevant:

a) Alteration of wallrock domains along fractures substantially affects mineralogy. Hydrothermal products, such as sheet silicates (in particular clay minerals) or oxides (such as hematite) have much higher sorption K_d coefficients when compared to primary (magmatic or metamorphic) minerals (Stenhouse 1994).

---

8 The parameters given in Table 2, together with an assumed channel aperture (a) of 1 mm, imply an advection velocity of approx. 10 m a^{-1}.

b) Alteration, at least under the hydrothermal conditions that affected the crystalline rocks in northern Switzerland, enhanced the microporosity of the wallrocks and so increased the volume of the porewater reservoir that is accessible to matrix diffusion.

c) Within the fracture itself, hydrothermal activity resulted in enhanced small-scale heterogeneity (see also Mazurek 1999). Parts of the surface area of a fracture were sealed by hydrothermal minerals, whereas dissolution of the fracture walls enhanced channel apertures and widths in other areas. The consequence of such a redistribution of flow porosity within individual fractures is a channeling effect, *i.e.* flow occurs only within a fraction of the total fracture-surface area (Figure 5). This results in a reduction of flow-wetted surface and thus in a reduced ability of the wallrock to retard radionuclide transport by matrix diffusion and sorption. Moreover, the degree to which the fracture-surface area is affected by hydrothermal cementation or dissolution is spatially variable, which results in a substantial widening of the parameter ranges that describe the small-scale channel geometry (as indicated in Figure 5).

Effects a) and b) improve the barrier function of the geosphere for radionuclide transport, whereas effect c) reduces the geosphere efficiency. However, the impact of effect c) on the model results is quantitatively more important than that of effects a) and b). For example, the ^{79}Se doses are reduced by more than 10 orders of magnitude relative to the reference case when channels are assumed to be closely spaced (compare cases 1 and 3 in Table 3). On the other hand, keeping the channel geometry constant, the dose reduction due to unlimited matrix diffusion, when compared to diffusion in the altered wallrock only, is relatively small (compare cases 1 and 2 in Table 3). It follows that, from a safety assessment point of view, the net impact of hydrothermal alteration is to reduce the barrier function of the geosphere. This result is surprising and at least partially the result of the conservatism that underlies the calculations. Because the effects of hydrothermal alteration are much more pronounced in gneisses/granites when compared to aplite/pegmatite dykes, the geometric input parameters have broader ranges. The case of closely-spaced channels in faults provides a geosphere efficiency that is better than that of the aplite/pegmatite dykes, as would be expected intuitively. However, assuming that groundwater flow in faults is highly channelled (one opening every 5 m, only 2 % of the fracture surface accessible for flow) results in a retardation that is more limited than that of the aplite/pegmatite dykes.

7. Summary and conclusions

Groundwater flow in the crystalline basement of northern Switzerland occurs through discrete water-conducting features. Based on a hydraulic identification of inflow points in deep boreholes, water-conducting features have been classified into three types: faults, fractured zones and fractured aplite/pegmatite dykes. The first two types are distinguished by the brittle deformation mechanism, whereas the last type is defined by primary mineralogy. Simplified conceptual models of the water-conducting features describing the small-scale arrangement of channels (where advection occurs) and embedding wallrock types (accessible by diffusion and providing a large surface area for sorption) were used as a basis for the modelling of radionuclide transport.

The ranges given for the geometric parameters describing the small-scale structure of water-conducting features are large and represent both *uncertainty* (*i.e.* the limitations in the methods of characterization) and *natural variability*. In order to make a safety case, a pessimistic combination of input parameters is chosen as a reference. Thus the objective of a safety assessment is not to achieve a good representation of "reality", but rather to produce a defensible model calculation that, even when using extreme parameter combinations, meets regulatory requirements.

For the model calculations, sets of parameter combinations are used that represent the bounding cases, and the extent to which radionuclides are retarded in the water-conducting features is explored separately for each case. Faults and fractured zones with closely-spaced, narrow channels provide the highest degree of retardation. Fractured aplite/pegmatite dykes take an intermediate position. On one hand, their mineralogic and porosimetric parameters are less favourable than those of the faults, but the lesser degree of hydrothermal cementation provides a better connection between flow porosity and the matrix microporosity. The lowest retardation is provided by faults and fractured zones with widely-spaced, broad channels because the extreme assumption of a channel spacing within fractures of 5 m minimizes the retarding effects of matrix diffusion and sorption.

Sensitivity analysis of radiological doses to geological parameters indicates that flow rate through channels strongly affects the model results. This parameter can, however, be constrained to some extent by hydrochemical investigations, and, in particular, by studying environmental tracers that indicate the age structure of the groundwaters. Other parameters, such as path length, dispersion and depth of diffusion-accessible wallrock matrix are relatively insensitive parameters.

Hydrothermal alteration (including precipitation/dissolution reactions in fractures) is identified as a key geological process that has the highest impact on the results of geosphere transport calculations. Wallrock alteration creates additional porespace in the rock matrix and, together with the precipitation of highly sorbing minerals at the expense of primary phases, increases the retardation provided by the matrix adjacent to fractures. On the other hand, the partial cementation of the fractures themselves leads to channelled flow that reduces the barrier performance. The spatial arrangement of the channels is one of the most sensitive geological parameters, and, in calculations that assume a high degree of channelling, the efficiency with which the geosphere retards migrating radionuclides is minimized. Calculations assuming a lower degree of channelling result in almost complete decay of safety-relevant radionuclides within the geosphere (e.g. cases 3 and 4 in Table 3). The results of the model calculations are more sensitive to the arrangement of channels than to the type of water-conducting feature. Future field investigations should therefore address the small-scale anatomy of water-conducting features, in order to constrain the possible spectrum of geometric parameter values.

Acknowledgements

The authors wish to acknowledge reviews provided by E. Frank (HSK, Würenlingen) and A. Matter (Uni. Bern). Discussions with Tj. Peters, H. N. Waber (both Uni. Bern) and S. Vomvoris (Nagra, Wettingen) are also contributed to the paper.

References

Biehler, D. (1995) Kluftgrundwässer im kristallinen Grundgebirge des Schwarzwaldes - Ergebnisse von Untersuchungen in Stollen, Tübinger Geowiss. Arb. C22.

Caine, J. S., Evans, J. P. and Forster, C. B. (1996) Fault zone architecture and permeability structure, *Geology* **24**, 1025-1028.

Daneck, T. (1994) Platznahme und mechanisches Verhalten von Ganggesteinen im Grundgebirge des Südschwarzwaldes, Mitt. ETH Zürich, NF 296.

Diebold, P., Naef, H. and Ammann, M. (1991) Zur Tektonik der zentralen Nordschweiz, Nagra Technischer Bericht NTB 90-04, Nagra, Wettingen, Switzerland.

Frick, U. (1993) An evaluation of diffusion in the groundwater of crystalline rocks, unpubl. Nagra Report, Nagra, Wettingen, Switzerland.

Gelhar, L. W., Welty, C. and Rehfeldt, K. R. (1992) A critical review of data on field-scale dispersion in aquifers, *Water Resources Res.* **28**, 1955-1974.

Hadermann, J. and Rösel, F. (1985) Radionuclide chain transport in inhomogeneous crystalline rocks: limited matrix diffusion and effective surface sorption, Nagra Technical Report NTB 85-40, Nagra, Wettingen, Switzerland.

Hellmuth, K. H., Klobes, P., Meyer, K., Röhl-Kuhn, B., Siitari-Kauppi, M., Hartikainen, J. and Timonen, J. (1995) Matrix retardation studies: size and structure of the accessible pore space in fresh and altered crystalline rock, *Z. geol. Wiss.* **23**, 691-706.

HSK and KSA (1993) Guideline for Swiss nuclear installations: Protection objectives for the disposal of radioactive wastes, Report HSK-R-21/e, Swiss Federal Nuclear Safety Inspectorate (HSK) and Federal Commission for the Safety of Nuclear Installations (KSA), Villigen, Switzerland.

Klos, R. A. and van Dorp, F. (1994) Biosphere datasets for the Kristallin-I assessment, Nagra Technical Report NTB 93-11, Nagra, Wettingen, Switzerland.

Küpfer, T., Hufschmied, P. and Passquier, F. (1989) Hydraulische Tests in Tiefbohrungen der Nagra, *Nagra informiert* **11**, 7-23.

Laubscher, H. P. (1986a) Struktur des Grundgebirges und des Paläozoikums in der Nordschweiz, Expertenbericht zum Projekt Gewähr, Hauptabteilung für die Sicherheit von Kernanlagen (HSK), Würenlingen, Switzerland.

Laubscher, H. P. (1986b) The eastern Jura: Relations between thin-skinned and basement tectonics, local and regional, *Geol. Rdschau* **75**, 535-553.

Laubscher, H. P. (1987) Die tektonische Entwicklung der Nordschweiz, *Eclogae Geol. Helv.* **80**, 287-303.

Mazurek, M. (1998) Geology of the crystalline basement of northern Switzerland and derivation of geological input data for safety assessment models, Nagra Technical Report NTB 93-12, Nagra, Wettingen, Switzerland.

Mazurek, M. (1999) Geological and hydraulic properties of water-conducting features in crystalline rocks, in I. Stober and K. Bucher (eds.) *Hydrogeology in crystalline rocks,* Kluwer, this volume.

Meyer, J. (1987) Die Kataklase im kristallinen Untergrund der Nordschweiz, *Eclogae geol. Helv.* **80**, 323-334.

Michard, G., Pearson, F. J. and Gautschi, A. (1996) Chemical evolution of waters during long term interaction with granitic rocks in northern Switzerland, *Appl. Geochem.* **11**, 757-774.

Nagra (1994) Kristallin I - Safety assessment report, Nagra Technical Report NTB 93-22, Nagra, Wettingen, Switzerland.

Neretnieks, I. (1980) Diffusion in the rock matrix: An important factor in radionuclide retardation ? *J. Geophys. Res.* **85**, 4379-4397.

Pearson, F. J., Balderer, W., Loosli, H. H., Lehmann, B. E., Matter, A., Peters, T., Schmassmann, H. and Gautschi, A. (1991) Applied isotope hydrogeology - A case study in northern Switzerland, Elsevier, Amsterdam and Nagra Technical Report NTB 88-01, Nagra, Wettingen, Switzerland.

Peters, Tj. (1987) Hydrothermal alteration of a Variscan granite, magmatic autometasomatism and fault related vein metasomatism, in H. C. Helgeson (ed.) *Chemical transport in metasomatic processes,* Reidel Publ., pp. 577-590.

Rasilainen, K., Hellmuth, K. H., Kivekäs, L., Melamed, A., Ruskeeniemi, T., Siitari-Kauppi, M., Timonen, J. and Valkiainen, M. (1996) An interlaboratory comparison of methods for measuring rock matrix porosity, VTT Research Notes 1776, Technical Research Centre of Finland, Espoo, Finland.

Rasmuson, A. and Neretnieks, I. (1986) Radionuclide transport in fast channels in crystallline rock,*Water Resources Res.* **22**, 1247-1256.

Schmassmann, H., Kullin, M. and Schneemann, K. (1992) Hydrochemische Synthese Nordschweiz. Buntsandstein-, Perm- und Kristallin-Aquifere, Nagra Technical Report NTB 91-30, Nagra, Wettingen, Switzerland.

Siitari-Kauppi, M., Flitsiyan, E. S., Klobes, P., Meyer, K. and Hellmuth, K. H. (1998) Progress in physical rock matrix characterization: Structure of the pore space, *Mat. Res. Soc. Symp. Proc.* **506**, 671-678.

Stenhouse, M. J. (1994) Compilation and review of experimental sorption data with recommended K_d values, Nagra Technical Report NTB 93-06, Nagra, Wettingen, Switzerland.

Stober, I. (1995) Die Wasserführung des kristallinen Grundgebirges, Enke, Stuttgart.

Thury, M., Gautschi, A., Mazurek, M., Naef, H., Voborny, O., Vomvoris, S. and Wilson, W.E. (1994) Geology and hydrogeology of the crystalline basement of northern Switzerland, Nagra Technical Report NTB 93-01, Nagra, Wettingen, Switzerland.

Tsang, Y. W., Tsang, C. F., Neretnieks, I. and Moreno, L. (1988) Flow and tracer transport in fractured media: a variable aperture channel model and its properties, *Water Resources Res.* **24**, 2049-2060.

Voborny, O., Resele, G., Hürlimann, W., Lanyon, W., Vomvoris, S. and Wilson, W. (1994) Hydrodynamic modelling of crystalline rocks, northern Switzerland, Nagra Technical Report NTB 92-04, Nagra, Wettingen, Switzerland.

THE SCALING OF HYDRAULIC PROPERTIES IN GRANITIC ROCKS

DIRK SCHULZE-MAKUCH
Department of Geological Sciences, University of Texas at El Paso
El Paso, TX 79968-0555, email: dirksm@geo.utep.edu

PETER MALIK
Geological Survey of Slovak Republic
Mlynska dolina 1, 817 04 Bratislava, Slovakia, email: malik@gssr.sk

Abstract

Crystalline carbonates from 25 sites display an exponential increase of hydraulic conductivity with scale of measurement of 0.5 for porous-flow media, between 0.5 and 1.0 for dual-porosity media and about 1.0 for fracture- and conduit-flow media. Granitic rocks from 3 sites were analyzed for variations of hydraulic properties (hydraulic conductivity and transmissivity) to determine whether they follow the same trend as fractured crystalline carbonates do. Granitic rocks from two of the sites show a definite increase of hydraulic properties with scale of measurement while data are lacking to obtain a significant relationship for the third site. When establishing a scale relationship, transmissivity appears to be a more suitable parameter for granitic rocks than does hydraulic conductivity. This is due to (1) a correlation of hydraulic conductivity to depth which was observed at two sites, and (2) the presence of transmissive fractures outside the well screen that allow additional discharge to the borehole. More data are needed to verify whether the relationship found for fractured crystalline carbonates is also valid for granitic rocks because of the larger local variations in hydraulic conductivity and transmissivity for granitic rocks.

Introduction and Previous Work

The investigation of scaling properties of fractured rock is more challenging than the investigation of scaling properties of porous media. In porous media, hydraulic conductivity (K) and transmissivity (T) are determined from well-accepted standard methods, such as the one by Theis (1935). This is not the case for fracture-flow dominated rocks. Hydraulic properties in fracture-flow controlled media are primarily dependent on fracture aperture, fracture distribution and fracture interconnectivity. Thus, K and T measurements will exhibit large variations depending on whether a measurement was taken near a transmissive fracture or in the poorly-fractured or non-fractured matrix. In addition, fractured rocks are sensitive to overburden and internal fluid-flow-pressure, either of which tends to significantly affect fracture openings and the hydraulic properties of the rock with depth. The jointed part of the rock is much more sensitive to pressure and will react in a different way than unjointed parts of the rock body (Kranz et al., 1979).

I. Stober and K. Bucher (eds.), Hydrogeology of Crystalline Rocks, 127–138.

Probably the most comprehensive study of the scale behavior of hydraulic properties in granitic rocks is that by Clauser (1992). His study indicates an increase of permeability with scale of measurement from the laboratory scale to the field scale. However, beyond the field scale, permeability appears to remain constant with scale of measurement. The study also demonstrates the large variations of permeability within individual sampling locations, which is the main problem in defining the scaling relationship for granitic rocks. Scaling behavior of hydraulic conductivity in granitic rocks was observed by Guimera et al. (1995). However, Guimera et al. (1995) argued that the observed increase of median Ks with measurement scale in their data set was based on sampling bias. Schulze-Makuch et al. (in review) collected data from 39 different geological media and proposed that the relationship of K is a function of the type of flow and degree of heterogeneity present in a geological medium. Based on their data set they proposed that the K increase with scale of measurement can be described with the empirical equation

$$K = c\,(V)^m, \tag{1}$$

where K is hydraulic conductivity (L/T)
c is a medium-characteristic parameter ($L^{1-3m}T^{-1}$),
V is the volume of material tested (L^3), and
m is the scaling exponent of the relationship.

The scaling exponent was 1.0 for fracture-flow and conduit-flow media, 0.5 for porous-flow media and between 0.5 and 1.0 for dual-porosity media. The investigation was based on sedimentary and metamorphic rocks and did not include plutonic rocks. However, their data set also included 23 types of crystalline carbonates that may behave comparably to granitic rocks in regard to fluid flow and scaling behavior. Results for the 23 different carbonate rocks are summarized in Table 1. The carbonate rocks follow the same scaling rule as equation (1). Because flow in a fractured crystalline carbonate is in many ways comparable to flow in a granitic rock, the hypothesis of the study is that they exhibit the same scaling behavior. A study of 3 granitic sites was initiated to investigate whether (1) an increase of hydraulic conductivity with scale of measurement can be confirmed for granitic rocks and (2) any scale variations of K are consistent with those found for fractured crystalline carbonate rocks.

Study Area

Two granitic sites are located in Slovakia, one in the Male Karpaty Mountains, the other in the Mala Fatra Mountains (Figure 1). Both Male Karpaty and Mala Fatra Mountains belong to the Tatricum unit of the West Carpathians, Slovakia (Biely et al., 1996). The Tatricum is exposed in the core mountains and represents the deepest exposed tectonic unit of the Inner Carpathians which is an autochthon relative to all overlying units. It consists of a crystalline core and an indigenous Late Paleozoic and Mesozoic envelope. The Tatric crystalline basement is generally made up of medium- to high-grade metamorphic rocks (mica-schist gneisses, gneisses) and granitoids. Some evidence indicates the presence of higher-pressure relics (Janak, 1992). Weakly

Table 1. Relationship of hydraulic conductivity to scale of measurement in crystalline carbonates

Type of Medium	N[a]	Exponent m		95 % CI[b]	Upper Bound[c]	c[d]	
		average	range			average	range
Heterogeneous Porous Flow Media	6	0.51	0.45-0.55	0.03	> 1 - 100,000	-6.2	-7.8 to -5.1
Double-Porosity Media	9	0.71	0.55-0.83*	0.07	> 500 - > 200,000	-6.3	-8.4 to -4.9
Fracture-Flow Media	6	0.94	0.80-1.13	0.10	> 1 - > 100,000	-7.0	-8.5 to -6.2
Conduit Flow Media	4	0.90	0.67-1.11	0.20	> 1,000 - > 10^9	-7.1	-8.7 to -5.9

[a] number of different geological media evaluated
[b] 95 % confidence interval about the exponent
[c] upper bound of the relationships in m^3 of tested material
[d] medium-characteristic parameter c given as log value (m/s)

* after removal of outlier at a 99 % significance level.

metamorphosed complexes also occur in some mountain ranges, including the Male Karpaty Mountains. The boundaries of the complexes of different metamorphic grade are tectonic and probably mostly Hercynian in age (Fritz et al., 1992, Janak, 1992). In the Male Karpaty Mountains, the Tatricum is deformed into complicated slices, recumbent folds and partial nappes that underwent Alpine metamorphism.

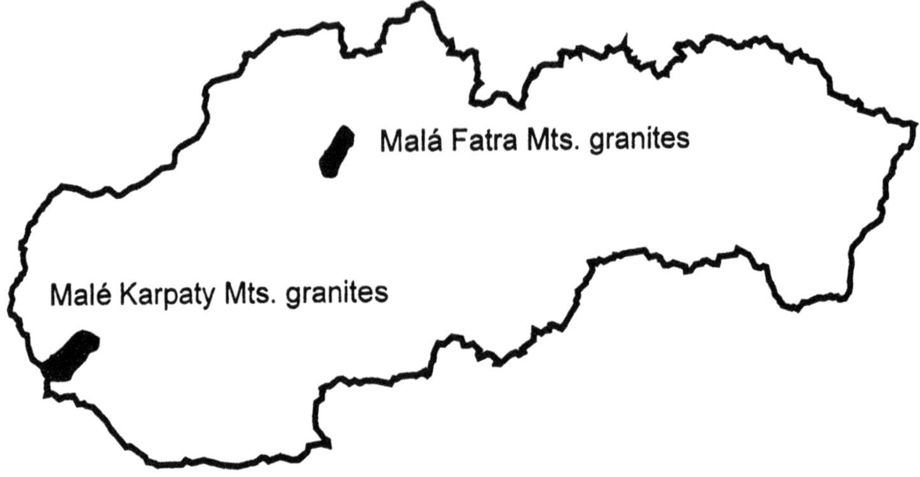

Figure 1. Study sites in Slovakia

Hercynian biotite tonalites to granodiorites are present in both the Male Karpaty and the Mala Fatra Mountains. In the case of the Mala Fatra Mountains they form the whole granitic body, while in the Male Karpaty Mountains they are only found in the northern part of the crystalline core. The biotite tonalites to granodiorites are mostly medium-grained and locally porphyritic (plagioclase phenocrysts). Major minerals are plagioclase (An20-35), quartz, biotite, ± K-feldspar ± muscovite. Amphibole is rarely present. The accessory assemblage is typically dominated by titanite, allanite and magnetite. Mafic microdiorite enclaves are common. The southern part of the crystalline core of the Male Karpaty Mountains ranges in composition from Hercynian biotite-muscovite and biotite granites to granodiorites (Figure 2; type Bratislava; Cambel-Vilinovic, 1987). They are medium- to coarse-grained and locally porphyritic. The basic mineral assemblage includes quartz, plagioclase (oligoclase), K-feldspar (orthoclase and microcline), biotite, and muscovite. The biotite/muscovite ratio is variable. Accessory minerals are typically monazite and ilmenite (Broska and Gregor, 1992). Several radiometric measurements indicate that these granitoids are 360-340 m.y. old (Cambel-Vilinovic, 1987).

The third study site is located in the Mirror Lake area, which is near the lower end of the Hubbard Brook valley in the southern portions of the White Mountains of New Hampshire, USA. In previous studies the granitic rocks of the Mirror Lake area were extensively tested to analyze the hydraulic properties of the granite (Hsieh and Shapiro, 1996). Bedrock underlying the Mirror Lake area is composed of schist, granite, pegmatite and lamprophyre (Tiedeman et al., 1998). The schist is part of the Rangeley formation of early Silurian age (Lyons et al., 1986) and is extensively intruded by the Concord granite of late Devonian age. Both the schist and the granite are intruded by pegmatite dikes, which are possibly a residual differentiate of the Concord granite (Tiedeman et al, 1998). Occasionally, a fine-grained volcanic dike rock (lampophyre) cuts through all three of these rocks. Because the spatial distribution of rock types is very complex, the entire bedrock is referred to as Mirror Lake Granite.

Figure 2. Geology of the Male Karpaty Mountains and location of wells used for the analysis. Well locations are shown as open circle.

Methods

A total of 40 pumping tests was conducted in the granitic rocks of the Male Karpaty Mountains (Figure 2), and 16 pumping tests were conducted in the granitic rocks of the Mala Fatra Mountains (Figure 3). Transmissivity values were computed from the specific yield of the wells using the method described by Jetel (1985, 1995). The method takes into account basic hydraulic resistance of the borehole in steady or unsteady flow conditions, hydraulic resistance of the borehole filter and the partially penetrating well, and also includes an estimate of hydraulic resistance caused by turbulent flow. This method was used for both data sets.

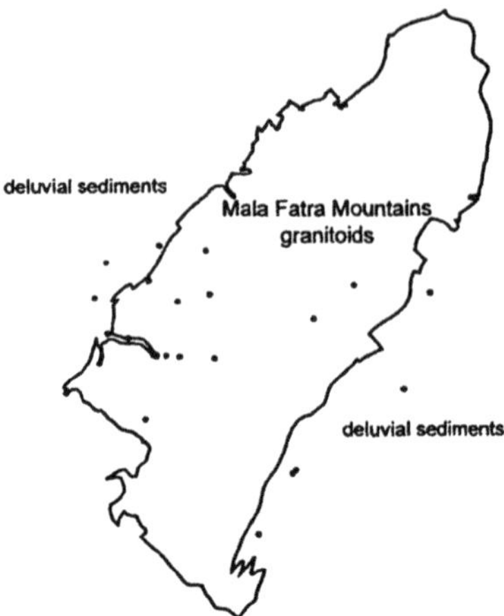

Figure 3. Location of wells in the Mala Fatra Mountains used for the analysis. Wells outside of the granite complex are cased through the cover of the deluvial sediments and screened in the granite.

Hydraulic conductivities were determined by dividing the transmissivity values by the length of the screened section of rock (well screen). Volume of tested rock was used as scale of measurement, because it provides a more accurate and consistent measure than flow distance or radius of influence does (Bradbury and Muldoon, 1988; Schulze-Makuch, 1996; Schulze-Makuch and Cherkauer, 1997). For example, consider comparing piezometer tests (slug- or baildown tests) with permeameter tests. The flow distance in a piezometer test (the dimension parallel to flow) is generally only several cm into the geologic medium, but the procedure is testing several m of material transverse to flow (along the screen). In contrast, the flow distance through a core plug in a permeameter is also several cm, but the transverse radius is only in the cm range. Using a distance parallel to flow approach would essentially project both tests on the same scale even though they are not. With a volume of tested material as scale measure, these two types of tests are separated by orders of magnitude. The volume of tested rock can generally be calculated by dividing the volume of water used in the test by the unit's porosity. In an idealistic, porous homogeneous medium the total porosity of the geologic unit should be used to determine the volume of tested rock. However, because most of the response to stress induced on a granitic rock is transmitted by the fracture network, the use of the fracture porosity to determine the volume of tested rock is more appropriate. Of course, this approach does not account for fractures not connected to the fracture network, but the induced error is minor. For numerical studies the volume of tested rock can not be determined by dividing the volume of water used in the test by the fracture porosity. In that case the volume of tested rock was defined by the product of the average flow distance, a representative transverse distance, and the thickness of the unit.

Before hydraulic parameters can be related to scale of measurement, it must be tested whether there is a correlation between hydraulic conductivity/transmissivity to overburden or internal fluid flow pressure. However, these types of measurements are difficult to obtain and were not available for the study areas. Thus, hydraulic conductivity and transmissivity were related to the depth of the midpoint of the well screen assuming that overburden/internal fluid flow pressures are hydrostatic. A distinct decrease of hydraulic conductivity with increasing depth was found by Stober (1996) for gneissic rocks. However, the same study did not indicate a variation of conductivity with depth for granitic rocks. In Stober's (1996) study, the results of approximately 400 pumping tests from the crystalline rocks of the Black Forest and neighboring regions were analyzed.

Results

The relationship of hydraulic conductivity to subsurface depth for the granitic rocks of the Male Karpaty Mountains is shown in Figure 4. The regression indicates a strong negative correlation with depth. However, if transmissivity is related to depth, no significant correlation is present (Figure 5). Results were similar for the granitic rocks of the Mala Fatra Mountains. Hydraulic conductivity had a strong negative correlation with depth ($K \sim (d)^{-0.95}$, $r = -0.37$), transmissivity did not ($T \sim (d)^{-0.15}$, $r = -0.06$). Thus, a valid relationship of K to scale can only be established for these types of rocks if a correction factor is applied to modify the K values for decreases with depth. The other option is to directly relate transmissivity to scale of measurement (the slight negative correlation of T to depth is not statistically significant). We believe that the latter choice is more appropriate because transmissivity is conceptually the better choice of a hydraulic parameter for granitic rocks. In fractured granitic rocks water flows into the borehole from the whole fracture network during hydraulic testing and is not limited to the section of the rock in which the screen is installed (Figure 6). This concept is valid as long as the well taps into the major fracture network and the fractures of the granitic rock are reasonably well interconnected. However, even though T is believed to be the better conceptual choice, the scaling exponents for K were also calculated for comparison purposes.

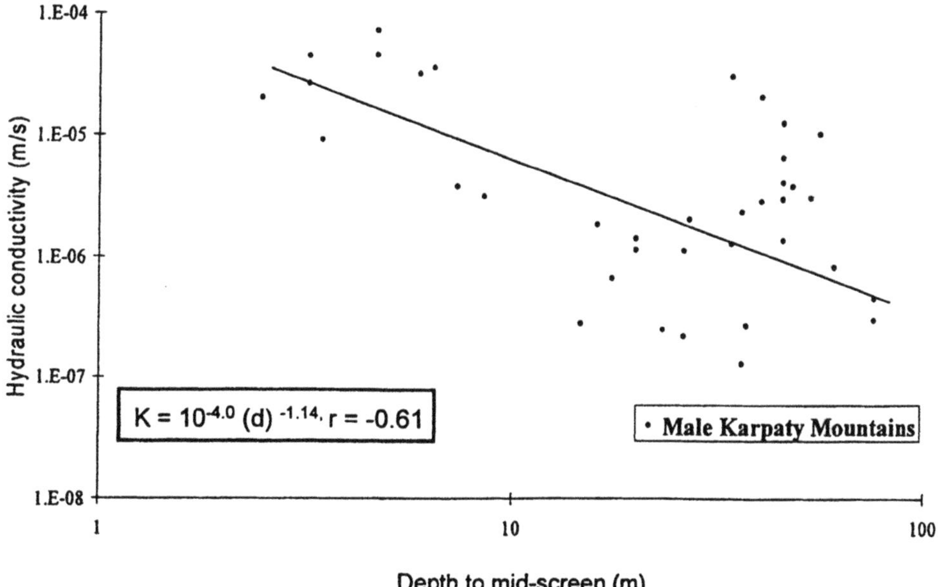

Figure 4. Relationship of hydraulic conductivity to depth for the granitic rocks of the Male Karpaty Mountains based on 40 conducted pumping tests.

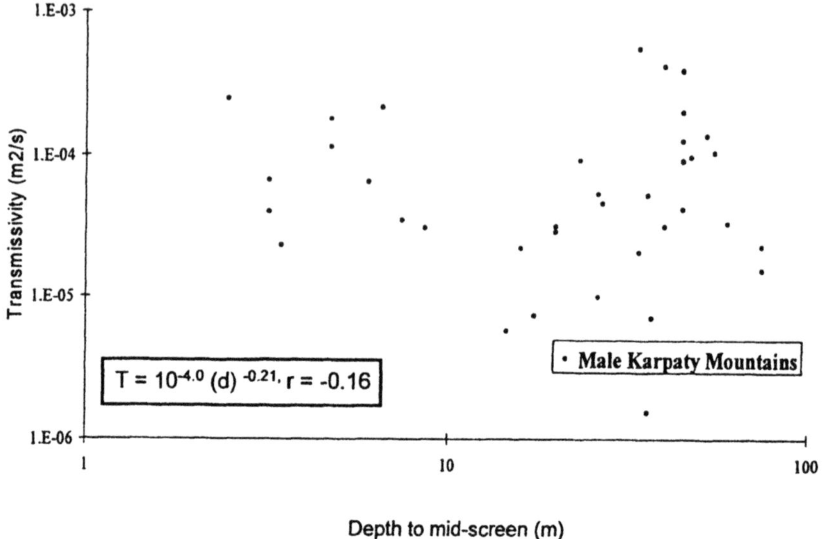

Figure 5. Relationship of transmissivity to depth for the granitic rocks of the Male Karpaty Mountains based on 40 conducted pumping tests.

The relationship of transmissivity to scale of measurement for the granitic rocks of the Male Karpaty Mountains is shown in Figure 7. Based on the graph, two interpretations are possible. Either one may fit a regression line to all data points or one recognizes that there appears to be a break in the scaling relationship. Beyond a certain volume of rock tested transmissivity does not appear to increase with scale of measurement. Based on a greater regression coefficient for interpretation 2 we decided for the latter interpretation and determined the upper bound of the relationship to be at a rock volume of about 250 m^3 . Beyond the upper bound T does not increase with scale of measurement anymore but remains constant. Below the upper bound of the relationship, a regression analysis produces the equation

$$T = 10^{-6.6} (V)^{0.81} \tag{2}$$

where T is transmissivity (m^2/s) and V is the volume of rock tested (m^3).

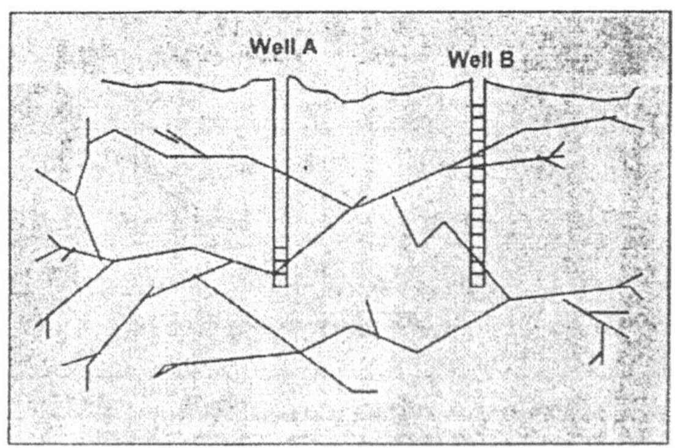

Figure 6. Illustration of a fracture network in a hypothetical granitic rock. Comparable values of transmissivity may be measured in both wells because of the presence of a well- connected fracture network even though the screened sections differ greatly.

Transmissivity values vary spatially over several orders of magnitude even if measurements are taken on a comparable scale. The spread in T values results in the relation's relatively poor correlation coefficient of 0.40. If hydraulic conductivity (corrected for depth) is related to scale of measurement, the scaling exponent becomes 1.0. This exponent is in exact agreement with the exponent proposed for fractured media by Schulze-Makuch et al. (in review) for the fractured crystalline carbonates displayed in Table 1. However, the correlation coefficient of the relationship is lower than that of the scale behavior of transmissivity. Data are insufficient to define the scale relationships with a high certainty at the present time.

The relationship of transmissivity to scale of measurements for the granitic rocks of the Mala Fatra Mountains is shown in Figure 8. A regression line produces the equation

$$T = 10^{-4.8} (V)^{0.22}, \tag{3}$$

where T and V are as above.

No upper bound on the relationship is apparent. The data have a lesser degree of spread around the regression and thus a somewhat larger correlation coefficient (r = 0.58). If hydraulic conductivity (corrected for depth) is related to scale of measurement, no significant relationship of K to scale appears (m = 0.04, r = 0.07). One may speculate that the granitic rocks at the Mala Fatra Mountains are more homogeneous and that they have encountered the upper bound at a scale on which the aquifer tests were conducted. However, to solve this issue more aquifer tests would have to be performed.

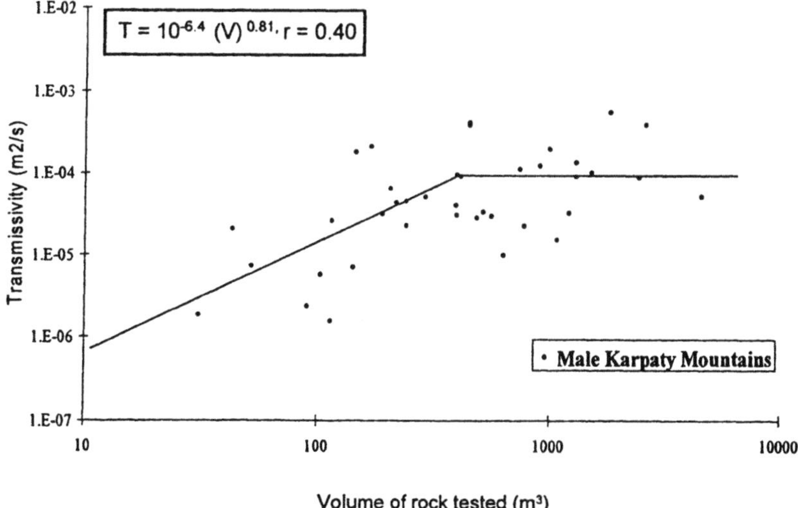

Figure 7. Relationship of transmissivity to scale of measurement for the granitic rocks of the Male Karpaty Mountains based on 40 pumping tests. An upper bound appears to be located at a rock volume of about 250 m^3. If hydraulic conductivity (corrected for depth) is plotted versus scale of measurement, the scaling exponent becomes 1.0.

The results of the third test site are based on a scale analysis of data that were obtained by other researchers (Hsieh, 1996; Hsieh and Shapiro, 1996; Tiedeman et al., 1998). The smallest scale tests available at the test site were packer tests of which a total of 387 were conducted. Packer tests were conducted to a depth of about 100 m in roughly 5 m intervals (Tiedeman et al., 1998). The lower limit of K measurements for the packer test equipment was estimated to be 10^{-10} m/s (Hsieh, 1996). 88 of the 387 tests were no-flow tests and were included in the data as a K-value of 5 x 10^{-11} m/s. A geometric mean K-value of 9.1 x 10^{-9} m/s of the whole data set resulted for the packer test scale when including the no-flow tests. Because packer tests were only conducted in sections of wells that were previously identified by borehole televiewer log and downhole video camera images as fractured (Hsieh, 1996), the mean K-value has to be considered an upper limit for the K characteristic of the Mirror Lake Granite. The average rock volume tested during the packer tests was estimated to be approximately 6 m^3.

On the larger field scale, multiple-well aquifer tests were performed in a 120 m by 80 m well field where 13 wells were drilled to investigate the upper 60 m of rock (Hsieh et al., 1994). Aquifer test results showed that the rock underlying the well field contained fracture networks of variable conductivity. Reported K-values ranged between 3 x 10^{-8} and 6 x 10^{-5} m/s (Hsieh et al., 1994). A numerical simulation of the aquifer tests resulted in a K-value of 2 x 10^{-7} m/s for the same test area (120 m x 80 m x 60 m) (Hsieh et al., 1994). In addition, a regional ground-water flow simulation covering a 3 km by 3 km area to a depth of 150 m resulted in a calibrated K-value of 3.2 x 10^{-7} m/s (Hsieh et al., 1994).

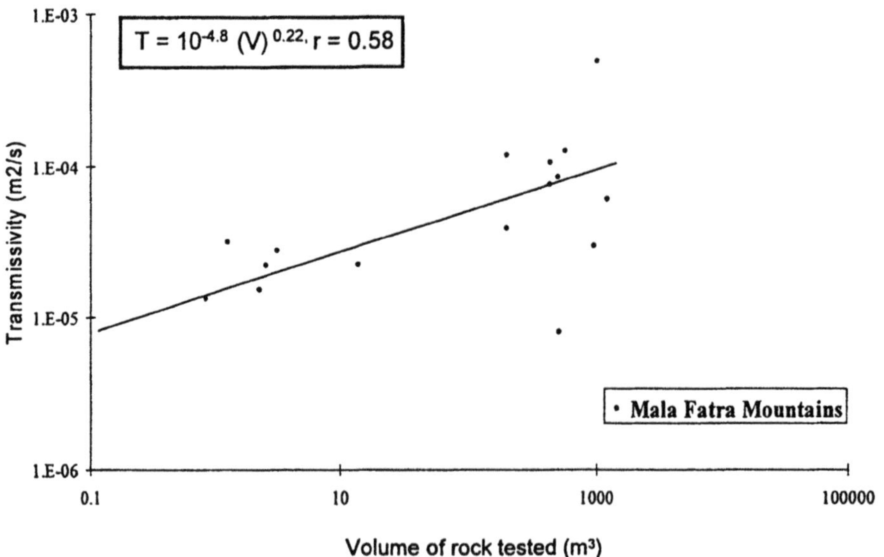

Figure 8. Relationship of transmissivity to scale of measurement for the granitic rocks of the Mala Fatra Mountains based on 16 pumping tests.

Because K was determined at comparable depths on various scales, a possible relationship of K to overburden pressure or depth is not expected to affect results significantly. K appears to remain constant between the large field scale and the regional scale. However, K decreases at a scale below the large field scale with a scaling exponent of at least 0.4. This scaling exponent may be much larger than 0.4 because of (1) the bias in the K-measurements of the packer tests described previously and (2) the location of the upper bound which may be located anywhere between the packer test scale and the aquifer test scale.

Conclusions

Data from granitic rocks of the Male Karpaty Mountains in Slovakia and the Mirror Lake area, USA, indicate that hydraulic properties increase with scale of measurement. The scaling property of the granitic rocks from the Male Karpaty Mountains and from the Mirror Lake area appears to cease at the lower field scale. This may indicate that at these sites no larger fracture networks exist that would increase the transmissivity for a larger portion of rock tested. Data from all sites including the Mala Fatra Mountains in Slovakia indicate large local variations of T and K. 40 pumping tests from the granitic rocks of the Male Karpaty Mountains and 16 pumping tests from the granitic rocks at the Mala Fatra Mountains appear insufficient to adequately define the relationship of transmissivity or hydraulic conductivity to scale of measurement. Thus, at this point it is not clear whether the relationship of K to scale of measurement established for fractured crystalline carbonates is also valid for granitic rocks. Due to the strong correlation of K to depth and the presence of transmissive fractures outside of the well screen that discharge to the borehole, transmissivity appears to be a more suitable parameter than hydraulic conductivity when related to scale (for granitic rocks).

Acknowledgements

We thank Jaromir Svasta , hydrogeologist at the Geological Survey of the Slovak Republic, for his technical support.

References

Biely, A., Bezak, V., Elecko, M., Gross, P., Kaliciak, M., Konecny, V., Lexa, J., Mello, J., Nemcok, J., Potfaj, M., Rakus, M., Vass, D., Vozar. J., and Vozarova, A. 1996. Vysvetlivky ku geologickej mape Slovenska 1:500 000. (Explanations to the geological map of Slovakia in 1:500 000 scale). Dionyz Stur Publishers, Bratislava, 77 p.

Bradbury, K.R. and Muldoon, M.A. 1988. Hydraulic conductivity determinations in unlithified glacial and fluvial sediments: In Groundwater and vadose zone monitoring, ASTM Special Technical Publication, pp. 138-151.

Broska, I. and Gregor, T. 1992. Allanite-magnetic and monazite-ilmenite granitoid series in the Tribec Mts: In Vozar, J. (ed.) Special volume to the problems of the Paleozoic geodynamic domains.,IGCP Project no. 276, Bratislava, Geol. Ústav D. Štúra, pp. 26-36

Cambel-Vilinovic, 1987. Geochemia a petrologia granitoidnych hornin Malych Karpat. Veda, Bratislava, 248 p.

Clauser, C. 1992. Permeability of crystalline rocks, EOS, Transactions American Geophysical Union, vol. 73, no. 21, 233-240.

Fritz, H., Neubauer, F., Janak, M., and Putis, M. 1992. Variscan mid-crustal thrusting in the Carpathians, Part II: Kinematics and fabric evolution of the Western Tatra basement, Terra Abstr., Suppl. 2, vol. 4, no. 24.

Guimera, J., Vives, L. and Carrera, J. 1995. A discussion of scale effects on hydraulic conductivity at a granitic site (El Berrocal, Spain), Geophysical Research Letters, vol. 22, no. 11, 1449-1452.

Hsieh, P. A. 1996. Personal communication, Hydrologist at the U.S. Geological Survey, Menlo Park, California.

Hsieh, P.A., and Shapiro, A.M. 1996. Hydraulic characteristics of fractured bedrock underlying the FSE well field at the Mirror Lake Site, Grafton County, New Hampshire: In Morganwelp, D.W. and Aronson, D.A. (eds.), U.S. Geological Survey Toxic Substances Hydrology Program – Proceedings of the technical meeting, Colorado Springs, September 20-24, 1993, U.S. Geological Survey Water Resources Investigations Report 94-4015, pp. 127-130.

Hsieh, P.A., Shapiro, A.M., Goode, D.J., and Tiedeman, C. 1994. Hydraulic conductivity of fractured crystalline rocks from meter to kilometer scale: Observations from the Mirror Lake Site, New Hampshire, Conference proceedings of abstracts of presentations given at Chapman Conference on "Hydrologic Processes: Building and Testing Atomistic- to Basin-Scale Models", June 6-9, 1994, Lincoln, New Hampshire, p. 21.

Janak, M. 1992. Variscan mid-crustal thrusting in the Carpathians. Part I: Metamorphic conditions and P-T paths of the Tatry Mountains, Terra Abstr., Suppl. 2, vol. 4, no. 35.

Jetel, J. 1985. Metody regionalniho hodnoceni hydraulickych vlastnosti hornin (Methods of the evaluation of the rock regional hydraulical properties), Ustredni ustav geologicky, Praha, 147 p.

Jetel, J. 1995. Utilizing data on specific capacities of wells and water-injection rates in regional assessment of permeability and transmissivity, Slovak Geol. Magazine, 1, 7-18.

Kranz, R.L., Frankel, A.D., Engelder, T. and Scholz, C.H. 1979. The permeability of whole and jointed Barre Granite, The J. Rock Mech. Min. Sci. & Geomech. Abstr., vol. 16, 225-234.

Lyons, J.B., Bothner, W.A., Moench, R.H., and Thompswon, J.B. Jr. 1986. Interim geologic map of New Hampshire, New Hampshire Department of Resources and Economic Development Open-File Report 86-1, 1 sheet, scale 1: 250,000.

Schulze-Makuch, D. 1996. Facies dependent scale behavior of hydraulic conductivity and longitudinal dispersivity in the carbonate aquifer of SE Wisconsin, unpublished Ph.D. dissertation, Univ. of Wisconsin-Milwaukee, Milwaukee, WI, 342 p.

Schulze-Makuch, D., Carlson, D.A., Cherkauer, D.S., and Malik, P. in review. Scale dependency of hydraulic conductivity in heterogeneous media, Ground Water.

Schulze-Makuch, D. and Cherkauer, D.S. 1997. Method developed for extrapolating scale behavior, EOS, Transactions American Geophysical Union, vol. 78, no. 1, p. 3.

Stober, I. 1996. Researchers study conductivity of crystalline rock in proposed radioactive waste site, EOS, Transactions American Geophysical Union, vol. 77, no. 10, 93-94.

Theis, C.V. 1935. The relation between the lowering of the piezometric surface and the rate and duration of discharge of a well using groundwater storage, Transactions American Geophysical Union, vol. 2, 519-524.

Tiedeman, C.R., Goode, D.J., and Hsieh, P.A. 1998. Characterizing a groundwater basin in a New England mountain-and–valley terrain, Ground Water, vol. 36, no. 4, 611-620.

Chapter 3

Hydrochemical properties of water in Crystalline Rocks

THE COMPOSITION OF GROUNDWATER IN THE CONTINENTAL CRYSTALLINE CRUST

Kurt Bucher
Institute of Mineralogy, Petrology and Geochemistry,
University of Freiburg,
Albertstr. 23b, 79104 Freiburg, Germany
bucher@ruf.uni-freiburg.de

Ingrid Stober
Geological Survey of Baden-Württemberg,
Albertstr. 5, 79104 Freiburg, Germany
stober@lgrb.uni-freiburg.de

Abstract

The composition of water stored in the crystalline rocks (basement) of the upper continental crust has in general four components: i) a surface water component derived from rain, snow and other precipitation, ii) a seawater component derived from modern or fossil seawater, iii) an imported component from ongoing magmatic or metamorphic reactions elsewhere in the crust, and iv) a contribution from the reactions between water and the local rock matrix.

Continental crust consists predominantly of granitic and gneissic rocks and water found in the crust reflects the granitic mineralogy of plagioclase, K-feldspar, quartz and mica (± hornblende). Less abundant minerals can be important for deviations from "normal granite water" and also for trace element patterns of groundwater in the basement.

Water is found in cavities, fractures and other water conducting features forming an interconnected pore space that allows for flow and mixing. Flow velocities rapidly decrease with depth and stagnant, density stratified water dominates large areas of the continental basement.

Dissolution rates are very slow for all major rock forming minerals of the granitic basement at the temperatures prevailing in the upper few km of the crust. Water never reaches equilibrium with any of the "granite minerals" with the exception of quartz which rapidly reaches saturation as a result of feldspar and mica dissolution. Dissolution of plagioclase and biotite and precipitation of related

I. Stober and K. Bucher (eds.), Hydrogeology of Crystalline Rocks, 141–175.

secondary minerals as coatings on fracture surfaces control the composition of groundwater in the basement.

Feldspar and mica hydrolysis tends to consume acidity and increases the pH. All basement waters world wide are typically low to moderate pH waters even in brines with very high TDS. This suggests that a mechanism of dessication is primarly responsible for the high concentration of solutes in deep groundwater of the crust.

1. Introduction

The continental crust consists predominantly of granite and gneiss. In a normal and stable tectonic setting the continental crust is about 35 km thick and underlain by subcontinental upper mantle. The crust deforms brittle in the upper and ductile in the lower portion.

Water fills the fracture and cavity related pore space of crystalline rocks. Water filled pore space at great depth has been reported from all deep drill holes in the crust (e.g. Kola, KTB, Urach, Soultz).

The composition of water in the crystalline crust of the continents has received considerable attention during the past decades. Fyfe at al. (1978) in their landmark book "Fluids in the crust" presented a section on chemistry of natural fluids. The only water data available at that time where oil field brines from sedimentary aquifer rocks such as limestones, shales and sandstones. No basement water data were considered in the "Fluids in the Crust" book. The problem was twofold at that time: i) the basement was a priori considered impervious and dry by most scientists, ii) it was also thought that if water is present in fractures it will be weakly mineralized because granite does not dissolve well in water.

Still today, the basement-cover interface is often assumed to be a no-flow boundary in many large scale flow models (e.g. Person and Garven, 1992) allthough typical basement often has a much higher permeablity than many sedimentary units (see Stober and Bucher, this volume).

Over the past two decades much has changed in the perception of water in the basement (Fyfe, 1987). It became increasingly clear that the fractured brittle upper crust is water saturated and behaves like any other aquifer (Gustavson and Krasny, 1993). Except for a zone close to the surface, groundwater in the crystalline basement is typically highly mineralized, a fact that became increasingly evident, first by the most influential book "Saline Water and Gases in Crystalline Rocks" edited by Fritz and Frape (1987). Since then many new analyses of basement water and other related data became available in the literature (e.g. Edmunds et al., 1987; Fritz et al., 1994; Gascoyne and Kamineni, 1993; Stober, 1995; Lodemann et al., 1998; Kamineni, 1987, May et al., 1996). Much data and evidence has been gathered by the nuclear waste disposal programs of various countries, notably Sweden (Stripa, Nordstrom et al., 1985) and Switzerland (NAGRA, Mazurek et al., this volume).

Table 1: composition of average continental crust and of major water controlling minerals

	average crust		andesine		oligoclase		biotite		oligoclase:biotite 5:1		
	mg/kg	wt.% oxide	mg/kg	wt.% oxide	mg/kg	wt.% oxide	mg/kg	wt.% oxide	mg/kg	wt.% oxide	
Na	22182.	2.99	48075.	6.48	73373.	9.89	1483.	0.20	368353.	8.25	Na_2O
K	23743.	2.86	9132.	1.10	415.	0.05	66499.	8.01	68574.	1.37	K_2O
Ca	39380.	5.51	56032.	7.84	23299.	3.26	5288.	0.74	121784.	2.83	CaO
Mg	18397.	3.05	180.	0.03	1508.	0.25	79441.	13.17	86981.	2.40	MgO
Cl	600.	0.06									Cl
HCO_3	16632.	1.20									CO_2
SO_4	1500.	0.05									S
Si	281528.	60.22	271617.	58.10	299667.	64.10	179146.	38.32	1677483.	59.65	SiO_2
Al	80347.	15.18	139946.	26.44	119939.	22.66	80506.	15.21	680203.	21.36	Al_2O_3
Fe	44100.	6.30	280.	0.04	980.	0.14	119000.	17.00	123900.	2.94	FeO
Ti	4088.	0.73	5.		5.		16184.	2.89	16212.	0.48	TiO_2
Mn	1008.	0.14					1584.	0.22	1584.	0.04	MnO
H	1533.	1.37					4520.	4.04	4520.	0.67	H_2O
X_{Na}	0.33		0.46		0.76		0.22		0.75		

Source of data: average crust: Carmichael (1989), Minerals: Deer et al. (1992) andesine plagioclase: DHZ p438a4, oligoclase plagioclase: DHZ p438a3, biotite mica: DHZ p285a5

The contribution of the continental deep drilling programs of Russia (Kola well) and Germany (KTB) to the understanding and composition of crustal fluids has been disappointing. However, the geothermal energy programs and research notably at Soultz-sous-Forêts, Urach, Cornwall and Los Alamos provided a wealth of information and data on water in the wet crystalline crust (ironically enough these projects started out originally as "hot-dry-rock" projects). The importance of crystalline bedrock as an aquifer was also emphazised by the international meetings at Ås (Banks and Banks, eds., 1993), EUG9 (this volume) and EUG10.

In retrospect and now knowing it better, the misconception of a dry crystalline crust and weakly mineralized basement water is surprising in view of the century old tradidion of mineral water production and thermal spas in basement areas (such as the Black Forest and the Rhenish Massif for example). Also the presence of water, often much water, in underground mines, tunnels, galleries and caverns in basement areas is well known in engeneering and mining geology.

In this paper, we present and discuss water data from the Black Forest area and compare them with published water data from central Europe and from the Canadian Shield. The Black Forest represents a surface exposure of the crystalline crust of central Europe that has been affected by the Variscan orogeny during the upper Carboniferous. The basement consists predominantly of granites and gneisses. It was covered by sediments during the Mesozoic and the early Tertiary. However, the cover sediments have been completely removed by erosion in the southern part of the Black Forest during the Tertiary in response to the formation of the Rhine rift system. The southern part of the rift valley has been covered by marine water until the Oligocene (Schreiner, 1991). The Black Forest basement is relatively strongly fractured and shows a distinct fracture porosity. Permeability is high and the typical average K-values of gneiss is $5 \cdot 10^{-7}$ m/s and of granite 10^{-6} m/s (Stober and Bucher, 1999).

2. Composition and mineralogy of basement rocks

Water residing in the fracture porosity of crystalline basement is compositionally strongly related to the rocks and minerals with which it is in contact. Continental crust, in general, consists predominantly of quartz-feldspar rocks, such as granite, granodiorite, tonalite, and their gneissic derivatives including migmatites and meta-sedimentary gneisses. The crystalline rocks are composed of a small number of rock forming minerals and these minerals are of remarkable restricted composition. The inventory of minerals includes quartz (Qtz), K-feldspar (Kfs), plagioclase (Pl), biotite (Bt) and hornblende (Hbl). These five different species of silicate minerals constitute > 90% of continental crust (Carmichael, 1989). The remaining < 10 % of the minerals found in the crust are made up by a very large number of different species. The average rock of the continental crust can be characterized as Bt-Hbl-granodiorite or Bt-Hbl-gneiss including migmatite. The composition of average continental crust is listed in Table 1 in the form of a water analysis and in the conventional oxide component style. The schematic pattern diagram (Fig. 1) of the average crust, plagioclase and a plagioclase + biotite 5:1 mixture has been constructed using chloride as a charge balance anion. Different TDS has been arbitrarily choosen in order to graphically separate the patterns on Fig. 1. The figure shows the water composition pattern that would result from completely dissolving a given amount of crust, plagioclase and Pl+Bt rock.

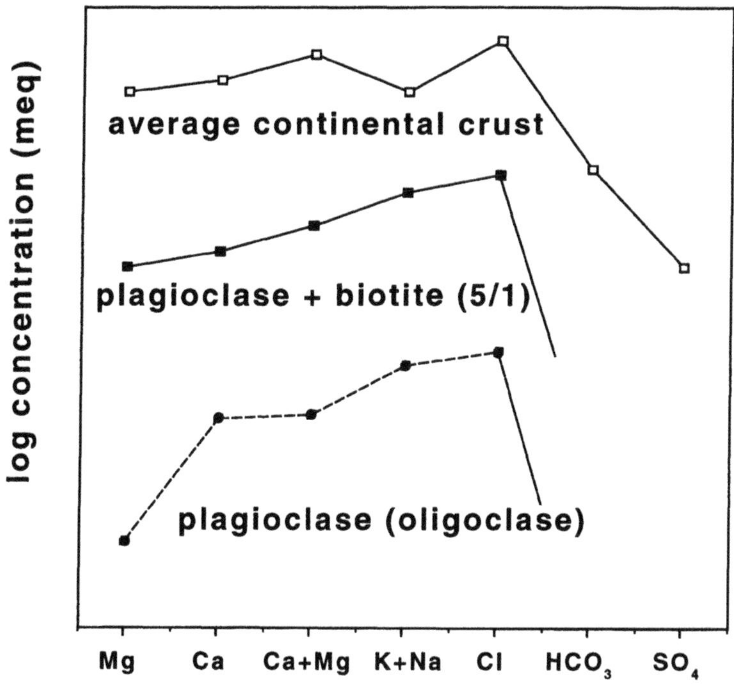

Figure 1: Schematic composition patterns showing the theoretical composition of waters derived from the dissolution of i) average continental crust, ii) a plagioclase (oligoclase) + biotite 5:1 mixture, iii) plagioclase (oligoclase). Cations charge balanced with chloride. Composition of rocks and minerals from Tab. 1.

Dissolution rate constants of the five major silicates in solutions with pH=5 (Lasaga, 1984) show that water composition will largely be controlled by plagioclase and biotite dissolution. The solubility of most of the minor minerals is similar to that of the five principal minerals. This includes muscovite, garnet, chlorite, alumosilicates, epidote and more exotic Al-bearing silicates such as staurolite, cordierite, chloritoid. Some rare minerals of the continental crust are significantly more soluble than the major rock-forming silicates. Of particular interest are various carbonate, sulfate, sulfide, phosphate, halide and fluoride minerals.

Water present in the fracture pores of the crust is predominantly in contact with "Bt-Hbl-granodiorite". This water is never in equilibrium and can never reach equilibrium with the solid phase assemblage of the rock. This is because at low temperature (< 50°C) plagioclase is not stable in the presence of water and biotite is stable in extremely high-pH waters only. As the reaction of water with biotite-hornblende-plagioclase-K-feldspar-quartz rock at low temperature progresses, the water composition changes continuously at the same time as the rock is altered by the process of fluid-rock reaction. The water-rock reaction results in a texturally, chemically and mineralogically visible alteration of the original Bt-Hbl-granodiorite.

Because of the restricted composition of crystalline basement, one would, therefore, expect that the composition of all basement water converges toward a uniform composition with increasing residence time. It will be shown that this is not the case. Reported compositions of deep groundwater in the crystalline basement shows a distinct but restricted variability which suggests that other sources and processes in addition to equilibration with the "crust" are important. Minor and rare minerals can be important for deviations from "granite water" and trace element patterns of groundwater in the basement (Möller, this volume).

3. Composition of water in basement rocks

The composition of water in the fracture pore space of crystalline basement varies widely with respect to total mineralization and relative proportion of the major components (Carlé, 1975; Albu et al., 1997; Michel, 1997). Dominant components of most waters are the cations Ca, Na, Mg and K and the anions HCO_3, SO_4, and Cl. The waters also contain dissolved gasses, predominantly CO_2, but also CH_4 and N_2, in variable amount and composition. The waters do not normally vary much with respect to pH and redox state. CO_2-rich water has its pH controlled by equilibria in the carbonate system and is typically around 5.5 to 6.5. CO_2-poor chloride-rich water has pH of around 7, decreasing with increasing temperature in accord with the temperature dependence of the dissociation constant K_w. Water in the gneiss-granite crust of the continents is remarkably oxidized to very great depth. With very few exceptions, in most waters CO_2 and HCO_3 are the prevailing carbon species, SO_4 is the predominant sulfur species and, consequently, Fe and Mn concentrations are relatively low.

Table 2c: Groundwater in the crystalline basement of the Black Forest (Cl dominant anion, thermal waters). mg/kg

locality	Wildbad 1	Wildbad 2	Lieb.	Berg.	Herr.	Baden-Baden 1	Baden-Baden 2	Säckingen 1	Säckingen 2	Böb.	Schramb.	Ohlsbach
depth m	97	120	77	392	489	154	341	142	488	715	505	59
temp. °C	23.3	38.7	24.2	24.6	26.3	61.2	53.6	30.4	28.3	33.6	27.0	27.3
pH	7.64	7.55	6.90	7.95	7.25	7.06	7.06	6.47	6.6	6.67	7.44	6.50
TDS	474	547	1174	1859	2513	2681	3522	2846	7365	6150	10677	10117
Na	93.	142.	298.8	567.8	640.2	758.1	994.0	844.9	2158.	1919.9	3204.8	2891.3
K	15.7	7.5	16.4	14.8	13.2	52.1	101.2	82.1	149.	92.86	62.95	252.6
Ca	36.5	27.8	48.1	24.0	185.4	110.0	146.0	96.2	328.	176.35	368.34	593.7
Mg	4.03	2.39	10.9	3.89	16.29	3.6	7.29	9.6	27.7	22.48	56.91	34.4
Li			1.6		3.1	10.7	13.46	5.89	26.7			15.4
Sr			1.4		5.3	1.96	1.81	3.4	18.3			20.5
Rb			0.15		0.63	0.77	0.73	0.03				
Fe			0.01	0.68	0.37	0.1	0.13	0.14	1.8	12.44	66.0	1.8
Mn			0.07	0.04	0.33	0.39	0.3	0.03	0.8	0.38	1.07	0.06
Al			0.28		0.29	4.0	5.2	0.7				1.26
Cl	112.7	147.6	344.0	446.7	870.6	1273.7	1775.1	1361.6	3705.	2572.11	3317.6	5357.3
HCO_3	183.0	183.0	347.8	451.5	134.2	159.5	160.7	292.8	519.	838.98	381.25	500.4
SO_4	27.6	34.3	69.1	276.1	609.6	149.2	158.0	99.9	353.	514.01	3206.40	384.6
Br			0.64		1.88	3.7	3.79	3.77	22.5			18.36
F			0.38		1.08	5.6	5.12	4.52	3.3			1.0
SiO_2			18.5	17.6	14.6		105.4	18.46	16.9		8.46	12.5
CO_2			33.		3.3	17.5	18.6	92.4	418.	330.		264.
X_{Na}	0.82	0.90	0.92	0.98	0.86	0.92	0.92	0.94	0.92	0.95	0.94	0.89

Lieb.=Liebenzell, Berg.=Berghaupten, Herr.=Herrenaln, Bad.=Baden-Baden, Böb.=Böblingen, Schramb.=Schramberg

3.1 Black Forest waters

The composition of 28 water samples from the crystalline basement of the Black Forest area shows some characteristic features of groundwater in the continental crust in general (Table 2). The data are from wells in the basement mainly drilled for mineral and thermal water exploitation. The wells range from a few meters to more than 1000 m depth. The depth given on Table 2 is the average depth of all water entering points that contribute to the composition of the sampled water. The reservoir and source area of the sampled thermal water has been shown to be much deeper than sampling depth at many locations. For example, the Ohlsbach water sampled from a 60 m deep well, originates from a reservoir at 3- 4 km depth. Several waters ascend from great depth. The waters follow distinct upwelling paths in response to topographically driven flow (Stober, 1996 and Stober and Bucher, 1999).

Table 2a: Composition of groundwater in the crystalline basement of the Black Forest (HCO_3 dominant anion, mineral waters). mg/kg.

locality	Roth.	Ripp.	Freyersb.	- Griesbach -			
well		1		1	2	3	4
depth m	2	28	88	114	38	103	161
temp. °C	6.5	12.3	11.8	17.3	11.3	16.5	12.6
pH	5.88	6.18	6.10	6.45	6.47	6.46	6.18
TDS	84	1211	1390	1579	1440	1795	3754
Na	4.63	67.	175.7	145.	86.2	202.	245.
K	1.9	9.5	14.0	8.5	14.1	18.1	22.0
Ca	5.8	174.	131.3	240.	238.5	210.	488.
Mg	1.5	32.	25.7	20.3	24.3	27.2	115.
Li				0.4			1.3
Sr	0.03		0.95	2.2			27.2
Fe		13.3	2.94	6.8	1.5	7.1	15.6
Mn		2.3	0.85	1.1	0.96	0.4	0.65
Cl	5.21	19.0	27.2	19.1	26.72	18.4	12.3
HCO_3	20.7	734.8	773.1	722.9	930.3	1010.	2593.
SO_4	2.8	121.7	155.4	368.6	99.58	265.	172.
F	0.28		2.7	2.4		0.8	0.27
SiO_2	24.0	28.39	61.08	28.39	11.54	25.16	45.77
CO_2	18.92	840.4	891.	815.	726.	571.	2485.
X_{Na}	0.58	0.40	0.70	0.51	0.39	0.63	0.47

Roth.=Rothaus, Ripp.=Rippoldsau, Freyersb.=Freyersbach,

Table 2a continued: Composition of groundwater in the crystalline basement of the Black Forest (HCO3 dominant anion, mineral waters). mg/kg.

locality	- Peterstal -				- Teinach -		Ripp.
well	1	2	3	4	1	2	2
depth m	64	70	28	210	752	209	4
temp. °C	15.4	14.5	12.6	17.5	26.3	16.0	9.8
pH	6.05	5.82	5.95	6.24	6.63	6.00	5.70
TDS	2368	2557	2275	3608	1882	3306	3814
Na	236.8	300.	225.3	377.	372.8	436.5	435.
K	19.55	23.6	19.16	28.	10.2	15.9	21.06
Ca	284.57	279.9	280.56	406.	92.2	328.7	492.9
Mg	60.8	61.9	48.64	85.3	21.9	53.5	63.78
Li			0.13		1.7	2.29	1.1
Sr				3.35	1.1	1.61	2.8
Fe	7.76	5.7	0.05	14.0	1.0	12.0	8.5
Mn	0.01	1.3	0.6	0.98	0.12	1.34	2.92
Cl		29.4	31.24	34.1	38.1	54.4	63.4
HCO_3	1382.04	1485.1	1281.	1958.	1009.	2175.	1583.
SO_4	372.91	347.6	319.32	568.	263.4	135.2	1060.
F			0.76	2.06	2.0	0.56	2.1
SiO_2	0.25	10.77	143.33	100.78	52.39	66.93	58.8
CO_2	2552.	3068.	2307.	1610.	262.	2040.	2276.
X_{Na}	0.59	0.65	0.58	0.62	0.88	0.70	0.61

Ripp.=Rippoldsau

The water found in the basement can be grouped into various categories depending on the parameters used to characterize water composition. The 28 analyses readily fall into three groups according to the dominant anion. The composition of all waters is shown on Fig. 2 in meq/kg. In all analyses from Rothaus to Rippoldsau 2, HCO_3 is the abundant anion (Table 2a), whereas SO_4 greatly dominates the water from wells Griesbach 5 and Waldkirch (Table 2b). All waters from Wildbad to Ohlsbach contain predominantly Cl as anion (Table 2c). The total amount of dissolved solids (TDS) varies from 64 (Rothaus) to 10677 mg/kg (Schramberg). For the further discussion of the Black Forest waters, we prefer to collect the HCO_3 and SO_4 waters under a descriptive term "mineral water" (chiefly wells of the mineral water industry) and the chloride-rich waters under the term "thermal water" (mostly wells of the thermal spas).

Table 3: Groundwater in crystalline basement from locations in Switzerland, Germany and France (mg/kg).

locality	Bö NAGRA	Ka NAGRA	Le NAGRA	Sc NAGRA	We NAGRA	Sä 3	Ur 3	Wi KTB-VB	SsF GPK1	Bu B	Sea
depth m	792.4	1271.9	1643.35	1887.85	2218.1	487.5	1774.	4000.	1930.	2535.	
temp. °C	31.3	34.9	34.5	30.7	17.8	28.3	97.5	119.	137.	115.	25.
pH						6.60	6.43	8.28	5.14		8.22
TDS	1324	1439	1052	8554	6484	7365	26775	68260	97913	201348	35159
Na	395.4	385.	295.	2750.	2205.	2158.	9000.	7159.	27900.	63900.	10768.
K	8.3	14.3	8.8	168.0	70.4	149.0	288.1	231.0	3400.0	503.0	399.1
Ca	8.5	30.2	9.4	85.3	151.8	328.0	742.0	15700.0	6930.0	11700.0	412.3
Mg	0.20	2.20	0.10	19.50	0.10	27.70	286.00	2.19	152.00	1900.00	1291.80
Li						26.70		2.41	126.00	41.20	0.18
Sr						18.30		244.00		485.00	8.14
Fe						1.80	66.90	0.28	30.00	36.00	0.0020
Mn						0.80	1.82	0.13			0.0002
Al								0.01			0.0020
Cl	141.8	63.2	125.	3630.	3382.	3705.	11982.	44100.	58500.	120500.	19353.
HCO_3	372.2	366.	273.	866.	76.3	519.	2336.	45.1	255.0		141.7
SO_4	339.0	529.0	263.0	794.0	431.0	353.0	2035.0	307.0	225.0	1525.0	2712.0
Br						22.50		417.00	302.00	726.00	67.30
F						3.30		3.82		31.00	1.39
SiO_2						16.93	23.09	54.0	93.0		4.28
CO_2	4.84	8.00	2.64	290.	18.	418.	5456.				
X_{Na}	0.99	0.96	0.98	0.98	0.96	0.92	0.95	0.44	0.88	0.90	0.98

Locations and source of data: NAGRA drill holes (Swiss Nuklear Waste Disposal Program); Bö:Böttstein, Ka:Kaisten, Le:Leuggern, Sc:Schafisheim, We:Weiach, Sä:Säckingen (Data taken from compilation in Stober, 1995), Ur:Urach (Stober, 1995), Wi:Windischeschenbach, KTB pilot hole 4000m fluid (Lodemann, 1993), SsF:Soultz-sous-Forets (Pauwells et al., 1997), Bu:Buehl (Pauwells et al., 1997), Sea:Seawater (Nordstrom et al., 1979)

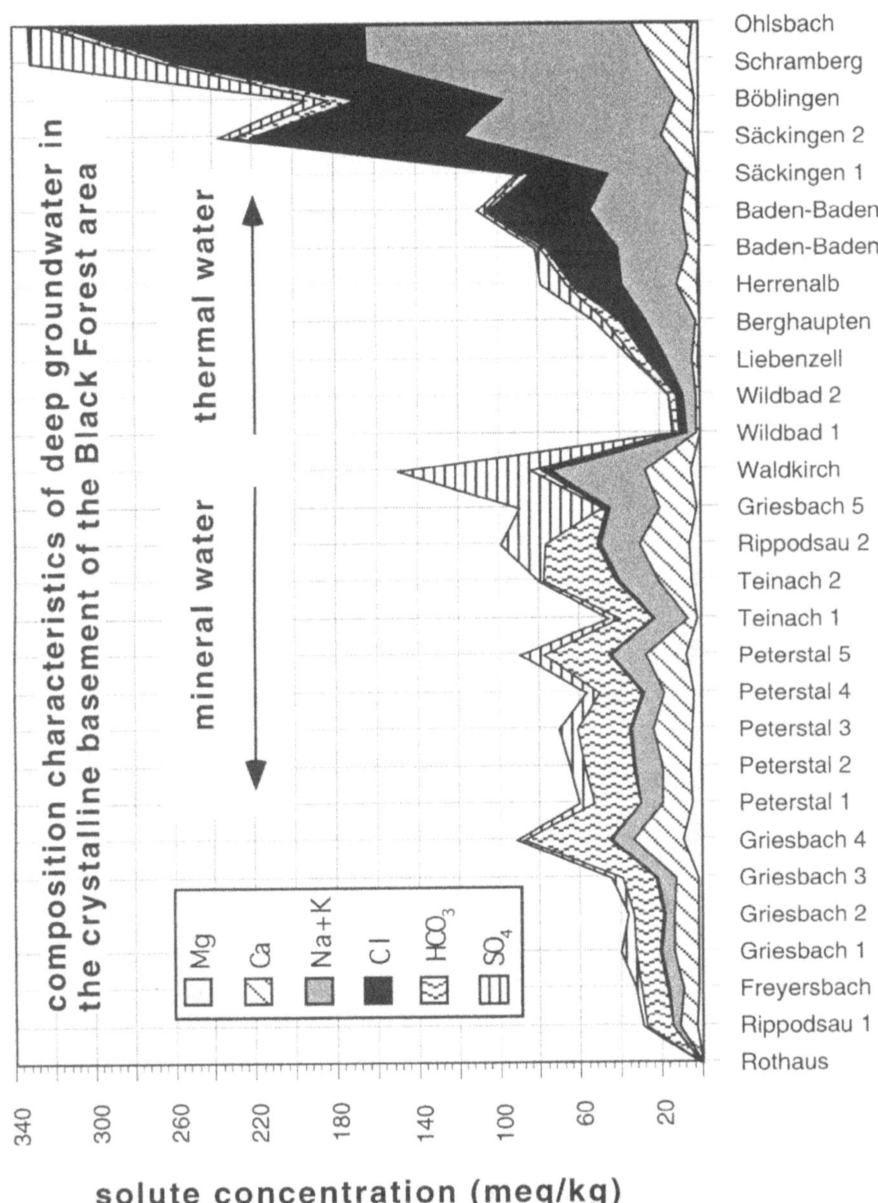

Figure 2: Composition of waters from the Black Forest basement (Table 2).

Table 2b: Composition of groundwater in the crystalline basement of the Black Forest (SO_4 dominant anion). mg/kg.

locality	Griesbach	Waldkirch
well	5	
depth m	412	509
temperature °C	27.9	17.4
pH	7.33	6.61
TDS	2143	5427
Na	566.4	1043.8
K	18.9	39.1
Ca	372.0	458.9
Mg	16.6	59.58
Sr		16.0
Rb		6.51
Fe	1.4	3.8
Mn	0.2	
Al		4.78
Cl	66.0	159.55
HCO_3	112.2	338.55
SO_4	977.6	3134.
Br		0.65
F		2.54
SiO_2	11.15	20.0
CO_2	15.4	44.0
X_{Na}	0.73	0.80

The low-TDS Rothaus water represents perhaps the water type that geologists would normally associate with groundwater from crystalline basement. The weakly mineralized cold Ca-Na-HCO_3 water represents the type water in the granitic crust near the surface. All other water samples have TDS of typically more than several hundred mg/kg. Chemical similarities within each group are significant but distinctive differences among the analyses from the various sites are equally important. Most of the HCO_3 group waters contain Ca as the most abundant cation and the waters are of the Ca-Na-HCO_3 type. The waters of Teinach contain more alkalies than Ca, however, and belong to the Na-Ca-HCO_3 type waters. TDS is weekly correlated with sampling depth at Peterstal and Griesbach but inversely correlated with depth at Teinach. This is strong evidence for a local, shallow level source of solutes. The water flow systems, particularly ascending waters play obviously an important role in the vertical distribution of water compositions in the Black Forest area. The situation is different from the Precambrian shields where the lack of topography and a low stady state heat flow prevents fluid flow and results in a clearer chemical stratification of the water compositions (Gascoyne and Kamineni, 1994).

3.1.1 Mineral water

All HCO_3 -rich mineral waters contain much dissolved CO_2. More than 3000 mg/kg dissolved CO_2 gas were analyzed in the water sample from the well Peterstal 2. There is no clear depth dependence of CO_2 content and depth. It appears, however, that deeper wells, such as the wells Teinach 1 and Peterstal 4 contain less CO_2 than the waters from shallower wells in the same area. It is remarkable that the ascending waters at well Rippoldsau 2 contain more than 2000 mg/kg CO_2 gas just 4 meters below the surface. Because of the high free CO_2 content of these waters, pH is low throughout, often below 6. SiO_2 data are varying widely and no plausible pattern can be seen. Despite a suspicious scatter, Mn and Fe is low throughout and indicates highly oxidizing conditions in all waters. The mole fraction X_{Na} (= Na/(Na+Ca)) in all waters varies between 0.39 and 0.70, the average of 0.56 is clearly much lower than in the other chemical types of water presented below. Ca/Sr ratios of the HCO_3 waters are all high that is 80 - 200, Ca/Sr of all thermal chloride waters is below 60 (with the exception of Säckingen) typically between 20 and 40. The mean Ca/Sr ratio of igneous crustal rocks is about 50.

The two sulfate dominated waters (Table 2b) were sampled at 400 and 500 m depth respectively. Both waters are low in CO_2 gas and consequently pH is higher than in the HCO_3 waters. The TDS of between 3 and 5.5 g/kg and the X_{Na} of about 0.75 is characteristic of basement water that resides at this depth. The waters are essentially mixtures of Ca and Na sulfate waters. The two sulfate waters from the two geographically unrelated locations have a remarkable chemical similarity.

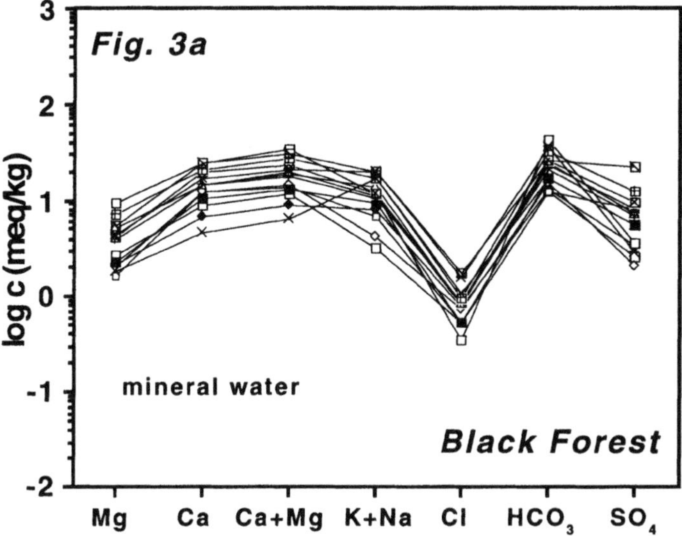

Figure 3a: Composition patterns of mineral waters from the Black Forest..

3.1.2 Thermal water

All Cl dominated waters from the Black Forest area are thermal waters with well head temperatures of up to 60°C. Most wells produce water of around 25 to 30°C. The silica content of the waters is in equilibrium with quartz at temperatures of around 50 - 60°C which suggests that the waters reside typically at 2-3 km depth. Dissolved silica in the chloride waters is much lower than in the HCO_3 waters. The pH in this group is significantly higher than that of the HCO_3 waters. The average of 7.1, however, is low compared with waters that originate predominantly from silicate hydrolysis and water-rock reactions (Bucher and Stober, 1999). The waters are best characterized as neutral sodium chloride solutions. TDS shows a remarkable wide variation from 474 mg/kg to more than 10000 mg/kg. The Schramberg and Ohlsbach water are the only saline waters with a TDS > 10000 mg/kg retrieved from the Black Forest basement. All Cl waters are dominated by NaCl. Waters from Herrenalb and Schramberg are similar and exceptional with significant amounts of SO_4. The sulfate water from Herrenalb contains Ca as major cation, in the Schramberg water sulfate is associated with Na. The waters from Böblingen and Berghaupten are Na-Cl-HCO_3 waters. The typical chloride-rich thermal water is a simple NaCl-rich water, however. Fluorine in Cl water is higher by a factor of two compared with HCO_3 water. Lithium is much higher (2 - 10 times) in Cl water relative to HCO_3 water.

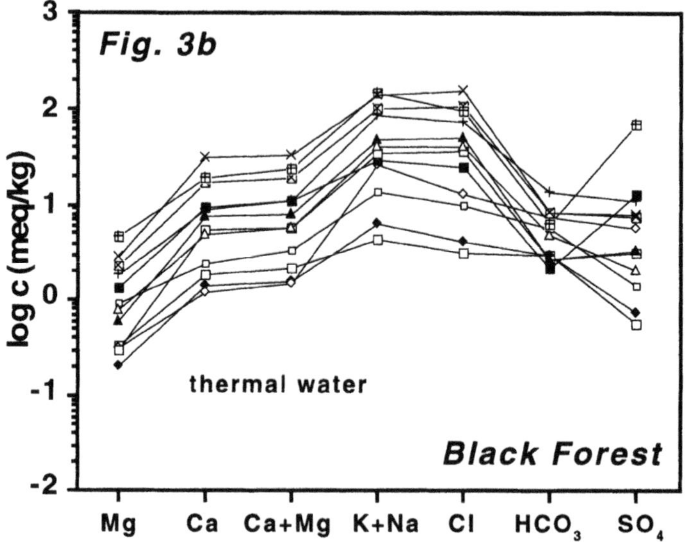

Figure 3b: Composition patterns of thermal waters from the Black Forest basement.

Cl waters contain little dissolved CO_2 gas. The water types in the Black Forest differentiate excellently with respect to the CO_2/TDS ratio, where HCO_3 water has a high CO_2/TDS (> 10 up to 120), whereas in Cl water the CO_2/TDS ratio is low (between 0 and 6). The CO_2/TDS ratio in SO_4 water is similar to that of Cl waters. Calcium is between 4 and 30 times higher than Mg in all water types. Na outweighs K by factors of between 2 and 50, but the Na/K ratios are not correlated with the

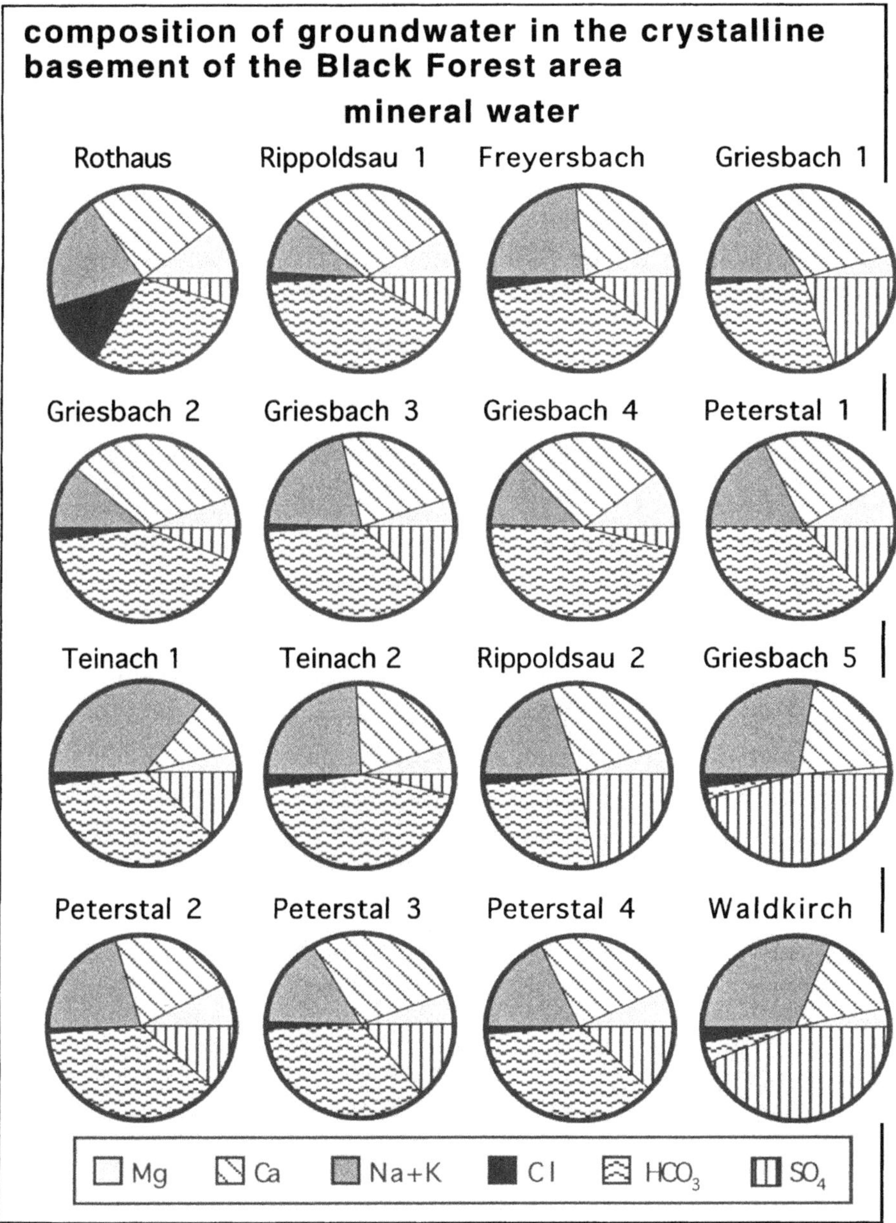

Figure 4a: Pie chart of water composition from mineral water of the Black Forest basement (Table 2).

water type. The Mg/K ratio in contrast is strongly associated with the water type; HCO_3 water has values significantly > 1.5 (2.5 and more), in Cl water Mg/K is normally below 0.3. X_{Na} of the Cl waters varies from 0.82 to 0.98 with an average value of 0.91. It is distinctively higher than in the sulfate waters and dramatically higher than in the HCO_3 waters. Both manganese and iron are slightly higher, on an average, in the HCO_3 waters compared with the Cl waters.

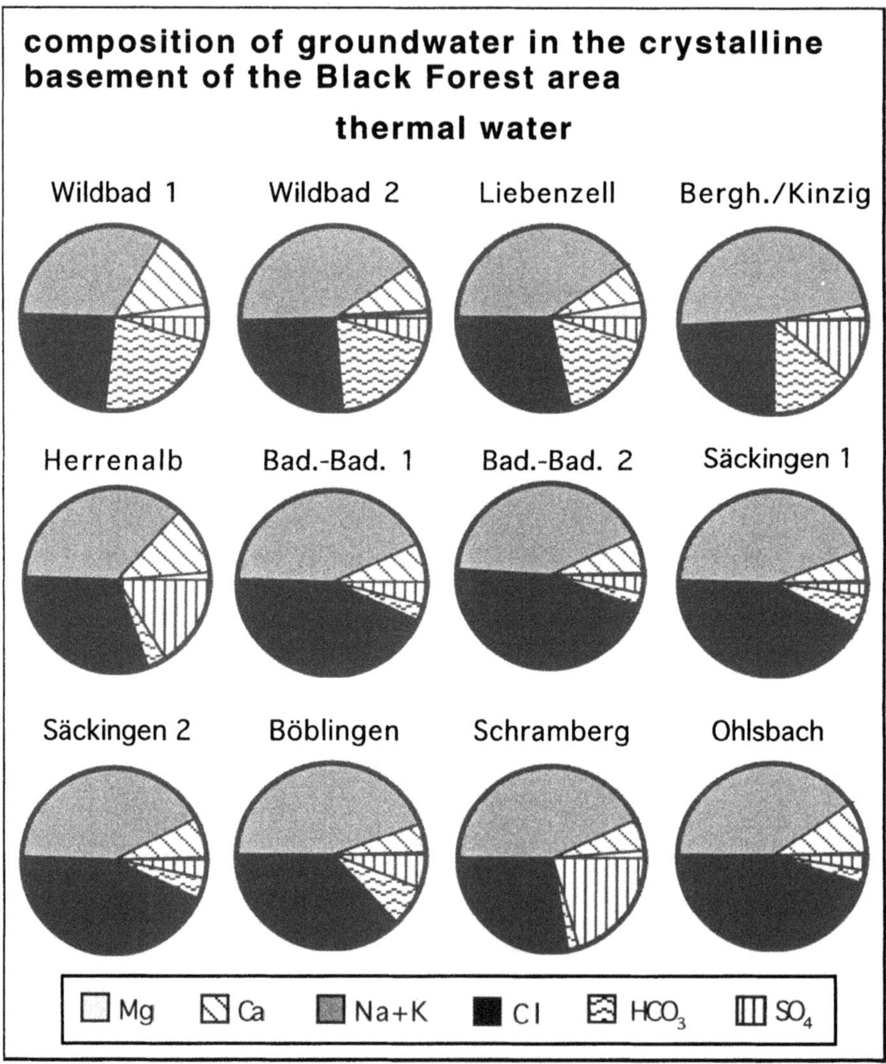

Figure 4b: Pie chart of water composition from thermal water of the Black Forest basement (Table 2).

The general chemical characteristics of groundwater in the Black Forest basement is also clearly depicted on the Schöller Fig. 2 and on the pie chart Fig. 3. The two major water types display very distinct patterns. It can be seen that the TDS variation is smaller for the HCO_3 waters than for the Cl waters. The two sulfate-rich thermal waters from Schramberg and Herrenalb show a distinct deviation from the type pattern of the average Cl-rich deep groundwater. Deviation from the HCO_3 type water pattern is most pronounced for the well Teinach 1 with a high alkali prominence. Also clearly distinct are the relatively SO_4-rich waters Rippoldsau 2 and Griesbach 1.

The compositional characteristics of the Black Forest basement waters as described above are not unique in the world. It appears that all deep groundwater in crystalline basement observes a limited number of characteristic type compositions. This in turn suggests that relatively few typical mechanisms and processes create the water composition in granitic crust worldwide.

3.2. Other Central European basement waters

Additional water compositions from crystalline basement are given on Table 3. The water samples include five analyses from the research wells in northern Switzerland drilled by NAGRA (Swiss Radioactive Waste Disposal Program, see Mazurek et al. this volume), two waters from Hot-dry-Rock (HDR) wells in France (Soultz-sous-Forêts) and Germany (Urach), one water from Bühl (Rhine river valley) and one from Säckingen (southern margin of Black Forest), the fluid analysis from the German continental deep drilling program (Windischeschenbach, KTB-VB) and seawater composition for comparison. The NAGRA wells were drilled into the sediment covered crystalline basement south of the Black Forest area, the HDR well at Soultz-sous-Forêts is located in the Upper Rhine rift valley and reaches well into the crystalline basement, the Bühl well is located at the escarpment of the Rhine rift valley (Münch, 1981) and KTB-VB (continental deep drilling- pilot well) was drilled in SE Germany (Oberpfalz) near the border to Czechia. The compositions on Table 3 are depicted on a Schöller diagram (Fig. 5) which permits an immediate comparison of the most prominent compositional features with the water compositions presented above. The waters can be systematized into four types, represented and defined by distinct Schöller patterns. Fig. 5a shows six waters of the NaCl-type similar to the patterns of the Black Forest thermal waters (Fig. 3b). The waters include two waters from the basement of N-Switzerland beneath the Mesozoic-Tertiary sediment cover, two waters from the margins of the Black Forest in the W and S and one water from the basement beneath the sediment fill of the Rhine rift valley. This later water, the well known Hot-Dry-Rock (HDR) well site Soultz-sous-Forêts, is high in TDS but else the composition is similar to the Black Forest thermal waters in general. Also the research well at Bühl produced an NaCl brine with a very high TDS of more than 200 g/kg.

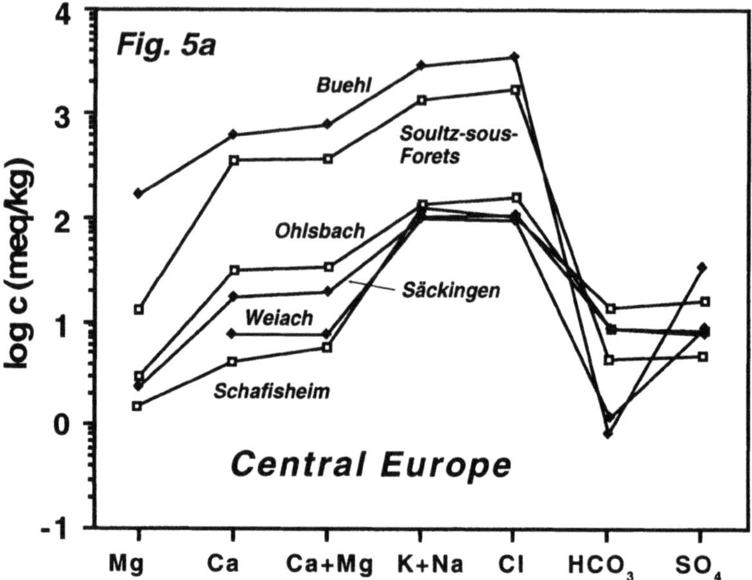

Figure 5a: Composition patterns of high-TDS waters from the Central European basement (Table 3). The patterns resemble the Black Forest thermal water pattern Fig. 3b.

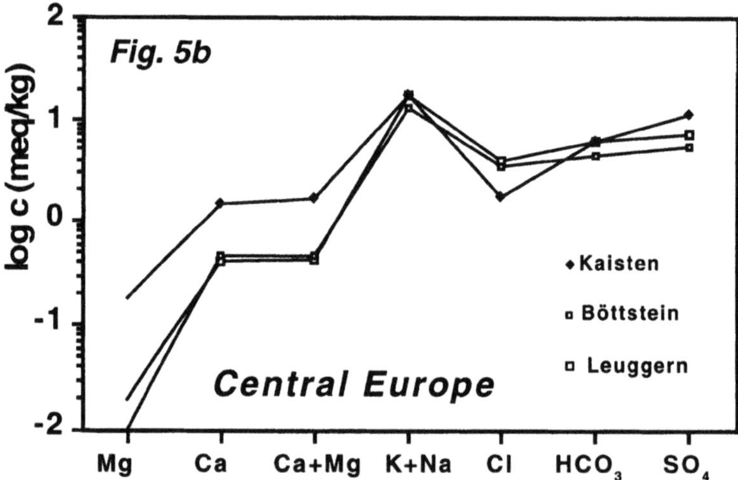

Figure 5b: Composition patterns of high-TDS waters from the Central European basement (Table 3). 3 unique patterns of water from NAGRA boreholes in the basement of northern Switzerland. Compare also Fig. 7d.

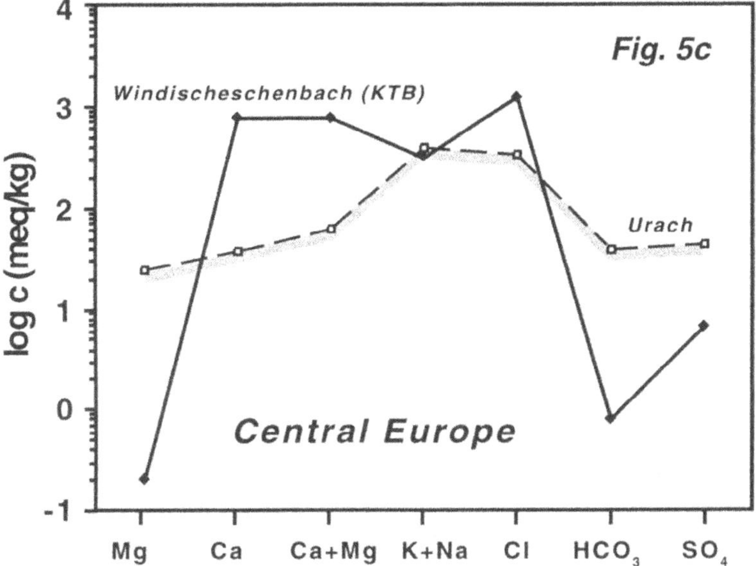

Figure 5c: Composition patterns of water (Table 3) from the German Continental Deep Drilling borehole at Windischeschenbach (KTB) and the Urach 3 HDR research drillhole (Stober and Bucher, this volume).

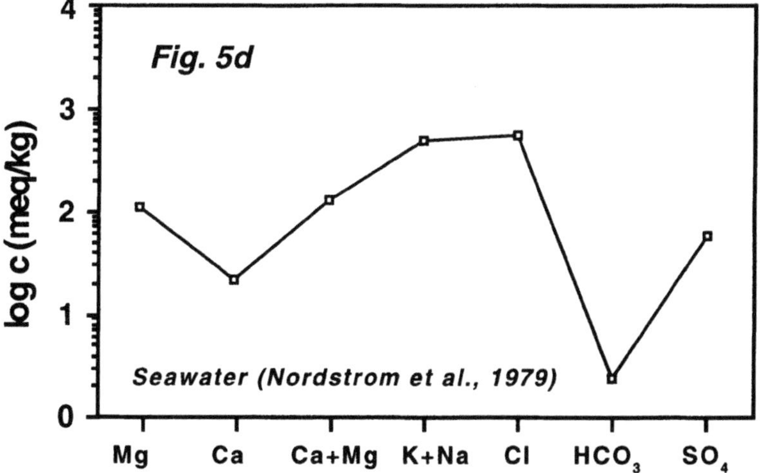

Figure 5d: Composition pattern of standard seawater (Nordstrom et al., 1979; Table 3).

Three other waters from the N-Switzerland basement show distinctly lower TDS and patterns not seen previously in Black Forest waters. The distinct patters (Fig. 5b) suggest a different origin of the solutes and an other chemical evolution of the waters. The cation pattern is identical to the Black Forest thermal waters, the anion pattern differs significantly. Sulfate not Cl is the dominant anion. Also bicarbonate is higher than chloride. All three anions occur at similar concentration levels. Similar water composition patterns were described by Gascoyne et al. (1987) from the Canadian shield at Chalk River. The pattern of the Urach geothermal well (Fig. 5c) is similar to that of the thermal waters of the Black Forest with the most notable exception of an unusually high Mg-content, comparable to that of Ca. In this respect, the Urach pattern is closest to that of seawater (Fig. 5d) which is characterized by much higher Mg than Ca. On the other hand, Urach also contains abundant bicarbonate and CO_2-gas, much more than any other water with a similar TDS.

The Schöller pattern of the water sampled at 4000m in the pilot well (KTB-VB) at Windischeschenbach during the German continental deep drilling program is markedly different from the patterns of all other samples from the Variscian crust in Central Europe (Lodemann et al., 1998). This reported KTB fluid is a Ca-Cl-rich water, whereas all other deep waters of the European crust (thermal waters) are Na-Cl waters. Composition patterns similar to that of KTB were reported from some locations on the Canadian shield (Eye-Dashwa, Lake du Bonnet 2, Gascoyne et al. 1987; the saline waters of the Canadian shield, Frape and Fritz, 1987). This pattern similarity with cold Canadian shield brines is very difficult to explain. The reservoir temperature at 4000 m of the KTB-VB is about 120°C (Lodemann et al., 1998). At this temperature, water in contact with average crustal plagioclase should not be low in X_{Na} (Bucher and Stober, 1999). The problem could be caused by an anknown amount of contamination with bore fluid which has not been marked with a tracer during the experiment (the KTB main well did not yield any water samples with low bore fluid contamination). We remain baffled. The HDR fluids from Soultz and Bühl have a higher TDS than the KTB-VB fluid.

The pie chart diagrams (Fig. 6) of the water compositions of Table 3 clearly show the similarity of the Böttstein, Leugern, and Kaisten waters from the sediment covered basement, all other waters from the Variscian basement of the Black Forest and surrounding areas make up a homogeneous composition group and the Ca-Cl-rich KTB water is a strange water on its own.

With the exception of the KTB water, X_{Na} of all waters is high . All waters from N-Switzerland contain more than 0.96 X_{Na}. Whereas X_{Na} of the KTB water is 0.44, the lowest value of any Cl-water in the central European basement and similar to the average value of 0.47 of the Canadian shield waters (Frape and Fritz, 1987). The Cl/Br ratio of the Ohlsbach water is that of seawater, all other thermal waters is below seawater, KTB has a Cl/Br ratio of 106 which compares to 106 and 95 of average Cl/Br ratios of Canadian shield brines and saline waters respectively (Frape and Fritz, 1987).

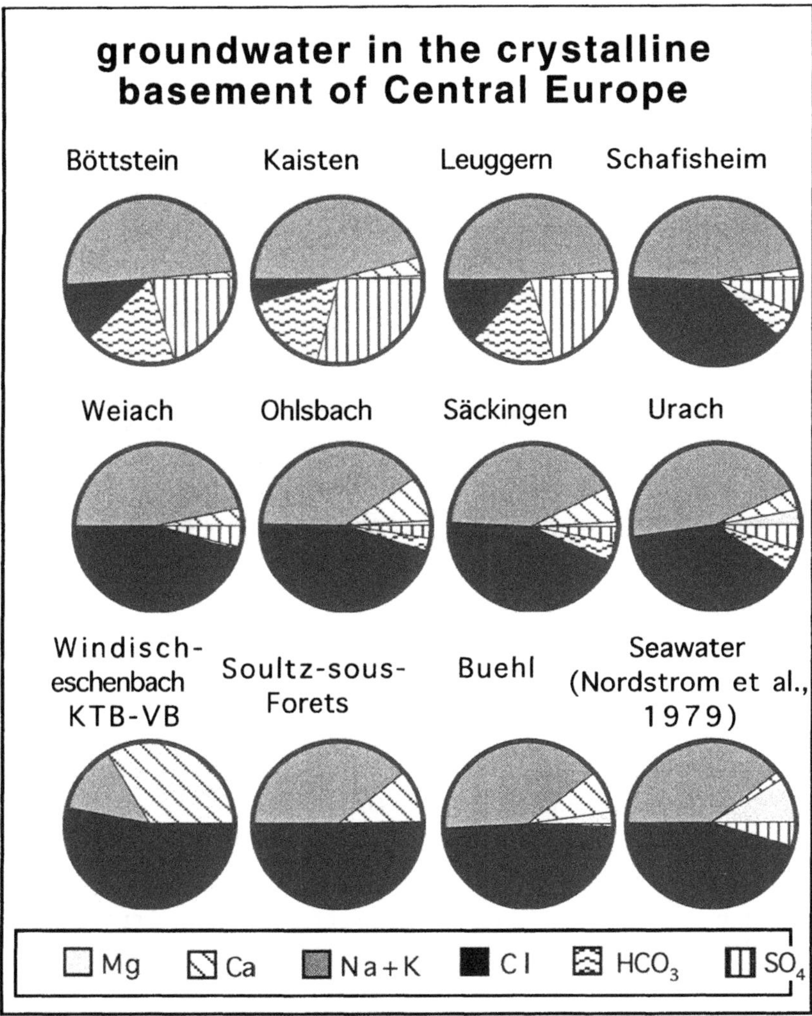

Figure 6: Pie chart of water composition of high-TDS waters from the Central European basement (Table 3).

3.3 Origin of basement water composition

Compositional characteristics of waters from the Canadian shield from Frape and Fritz (1987, their Table 1) and Gascoyne et al. (1987, their Table 3) are shown in the form of Schöller diagrams for comparison with Black Forest and vicinity basement waters (Figs. 7 and 8). It is remarkable that the same type pattern as before can be recognized. This strongly suggests that few principal mechanisms control the composition of water in crystalline basement and that relatively little transition exists between the few (perhaps three) endmember waters. One surprising feature of all waters is that none of them strictly follows a seawater pattern. This indicates that, if seawater was the primary source of water in the fractured basement this water is always chemically modified by reaction with the rock matrix. Another general important feature is that all waters are low in pH. Consequently, the high TDS of all water in the basement has not been acquired by hydrolysis of silicate mineral according to conventional "weathering" reactions of the type:

unstable "primary" silicate mineral + $H^+ \Rightarrow$
residual minerals (kaolinite, quartz) + cations in solution (1)

The production of high-TDS chloride brines is most probably a consequence of desiccation of pore and fracture space by water consuming reactions of the type:

unstable "primary" silicate mineral + H_2O-rich dilute solution (low TDS)
$\Rightarrow$ residual minerals (zeolite, clay, quartz) + H_2O-depleted solution (high TDS) (2)

a specific example:
$Na_4CaAl_6Si_{14}O_{40} + 8\ H_2O \Rightarrow 2\ SiO_2 + CaAl_2Si_4O_{12}{\bullet}4H_2O + 4\ NaAlSi_2O_6{\bullet}H_2O$ (3)
plagioclase + 8 water = 2 quartz + laumontite + 4 analcite

or for the anorthite component alone:
$CaAl_2Si_2O_8 + 4\ H_2O + 2\ SiO_2 \Rightarrow CaAl_2Si_4O_{12}{\bullet}4H_2O$ (4)
anorthite + 4 water + 2 quartz = laumontite

Laumontite stands here as general representative of Ca-zeolite, other zeolites may be formed at low temperature, such as Heulandite and Stilbite. These reactions use up water in the fractures and the residual fluid gets increasingly concentrated in a process that does not affect pH. The solutes initially present in low concentrations from dissolution of soluble minerals such as salts from fluid inclusions, sulfate from sulfide oxidation and small amounts of secondary carbonate increase in concentration as pore water is fixed chemically in hydrous minerals. Reaction 4 is accompanied by a large volume change of the solids (+61.4 cm^3 per 100.79 cm^3 or 1 mole An consumed). This means that the reaction consumes pore space, reduces permeability, clogs fluid pathways and tends to isolate and trap high-TDS water zones.

The effectiveness of the process can be esteemed by carrying through a simple calculation. Consider 1000 cm^3 of a rock with 50% plagioclase of composition An_{20} and 2% water filled fracture pores. 137 cm^3 plagioclase are sufficient to consume (chemically fix) all water in the pore space. 363 cm^3 plagioclase remain untouched,

57 cm^3 zeolite is formed corresponding to about 6% of the rock. Doubling the TDS can be accomplished by producing as little as 25 cm^3 hydrous minerals per 1000 cm^3 rock. The TDS evolution is dramatic at the final stages of the process in a closed system. With increasing TDS the activity of H_2O decreases, which, under certain conditions, may prevent the system from falling completely dry, because low a_{H_2O} water may reach equilibrium with plagioclase. The described desiccation of the pore water leads to the evolution of highly saline brines and to precipitation of solids including zeolite, calcite, gypsum and finally halite. In open systems, low-TDS water is supplied via fractures the water may reach a high mineralization but do not completely desiccate.

The fine details of the water composition are modified by temperature dependent exchange equilibria between water and the solid phase assemblage. An example is the control of the Na-Ca relations in deep groundwater by the presence of plagioclase according to:

$$\text{anorthite (in plagioclase)} + 2\,Na^+ \Leftrightarrow \text{albite (in plagioclase)} + Ca^{2+} \qquad (5)$$

which adjusts X_{Na} in the water in contact with plagioclase of a given anorthite content to low values at low temperature and to higher values at high temperatures. Calculation of the equilibrium conditions of reaction (5) using data and equations of state from Berman (1988) and Helgeson et al. (1981) shows for example that water with a TDS of 1000 mg/kg in equilibrium with An_{20} plagioclase has an X_{Na} of 0.5 at 30°C and an X_{Na} of 0.95 at 120°C which is in perfect agreement with the average X_{Na} of the mineral and thermal waters of the Black Forest area (Stober and Bucher, 1999). The pH-conserving desiccation of crustal aquifers is an universal process leading to highly concentrated brines. It is very similar to evaporation in the surface environment and leads to precipitation of salts and ultimately to a dry crust. Free halite has been reported from grain boundary coatings of gneisses of the Urach well together with zeolites (Althaus et al., 1985), in lower crustal rocks (Markl and Bucher, 1998), amphibolite-facies marbles (Trommsdorff et al., 1985) and from grain boundaries in granites and gneisses of the Black Forest (Liegl et al., 1999).

Plagioclase is not stable in the presence of water at low temperature (< 400°C, Bowers et al., 1984). The process of plagioclase dissolution in the deep crystalline basement aquifer will therefore never stop unless all plagioclase or all water is used up in the reaction. Plagioclase, the most abundant crustal mineral, cannot be used up in this process. The amount of water that is stored in the pore space of the crust or that can be supplied by water reservoirs above or the surface hydrosphere is the limiting quantity that will control the progress of the reaction. If water cannot be supplied to the reacting rocks in the basement at a sufficient rate due to low permeability the reaction may locally use up all water completely.

The consequences of the described hydration reactions include a reduction of the hydraulic gradient with depth that causes water to flow from higher to deeper crustal levels. This supplies fresh low-TDS water to the dessicating deeper parts of the aquifer. A substantial reduction of the hydraulic gradient with depth can be obsereved e.g. in the Urach geothermal well and the KTB wells (Stober and Bucher, this volume).

The continuous consumption of free water in the crust and its chemical fixing into hydrous minerals concentrates solutes in the "residue". In view of this, it is not surprising that chloride is the dominant anion of all high-TDS waters in the basement. Chlorides have the highest solubility of all common salts, carbonate precipitation removes almost all bicarbonate from Ca-rich brines, gypsum precipitation keeps the sulphate concentration low as well compared to Na-Ca-Cl dominated high-TDS waters. Calcite and dolomite are widespread fracture coatings and vein minerals in many basement wells in the Black Forest and e.g. the Canadian shield (Gascoyne et al., 1987). Gypsum has been reported from wells in the granite by Gascoyne et al. (1987). Massive, clear, transparent gypsum (so called Marienglas) has been reported as fracture fillings in gneisses from the Hechtsberg well in the Black Forest (Jenkner et al., 1986). Gypsum of internal origin (as opposed to the dissolution of evaporite in the cover rocks) is evidence of ongoing desiccation in the basement pore space. Abundant laumontite and prehnite were reported from fracture coatings in core samples from the KTB well (Möller et al., 1997). Laumontite, often together with prehnite, has been recognized as the most prominent secondary alteration mineral at all depth levels of the bore hole. The presence of laumontite is independent on lithology, it is abundant in gneissic rocks as well as in amphibolites and other mafic gneisses.

An important modifier of composition patterns is gaseous CO_2. It is locally present in large amounts, for example in the mineral waters of the Black Forest (Tab. 1). Its source and origin is unknown at present although many suggestions were made in the literature (Muffler and White, 1968; Touret, 1986; Behr, 1989; Dai et al., 1996; Kerrick et al., 1994; Griesshaber-Schmal, 1990; see Stober, 1995 for a review). CO_2 of deep seated origin in the basement is an important acid that may be used in reactions of the type (1), the hydrolysis of silicates. It keeps pH low, as observed, and releases solutes to the water. In high TDS brines of the Canadian shield, CO_2 is low because of the saturation conditions with calcite. Much of the CO_2 that may be introduced to $CaCl_2$ rich brines will be removed as calcite. CO_2, therefore may not reach the uppermost crust from deep crustal sources because it would be fixed as calcite in deeper parts of the brittle crust. On the other hand CO_2-streaming of saline brines will result in Na-HCO_3-Cl waters, a water type observed in Chalk river (Fig. 7).

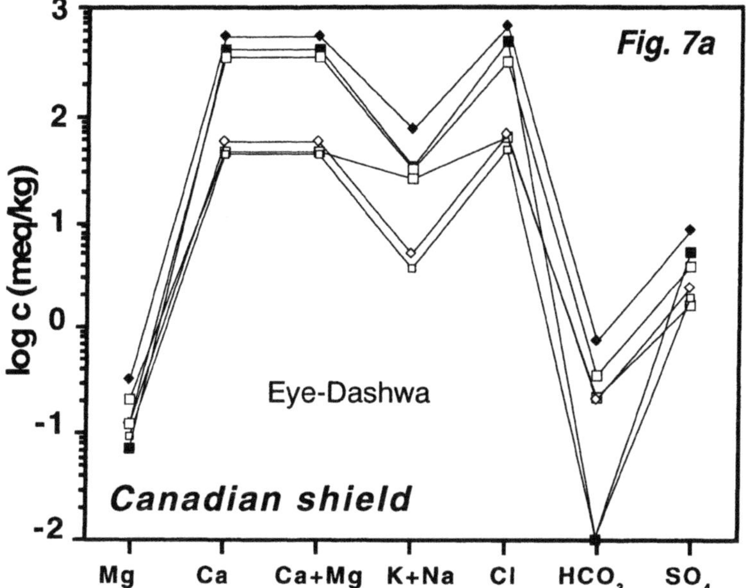

Figure 7a: Composition patterns of waters from the Eye-Dashwa wells on the Canadian shield (Gascoyne et al., 1987).

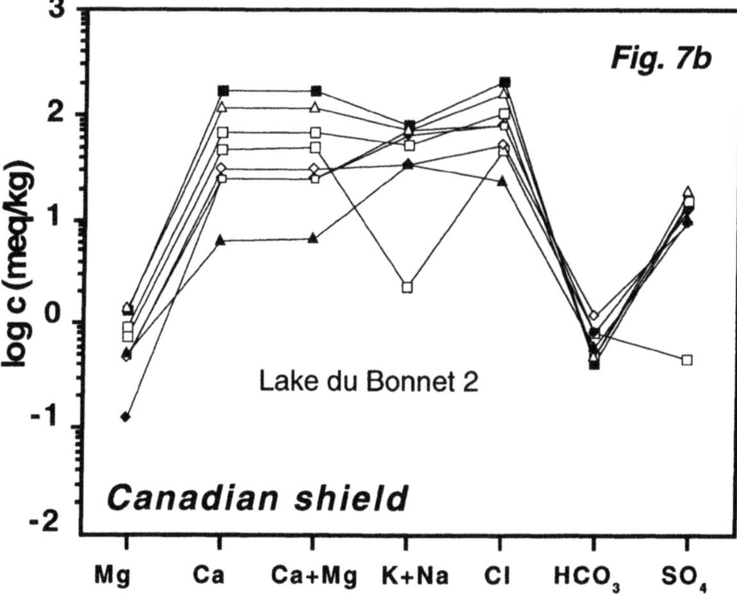

Figure 7b: Composition patterns of waters from wells at Lake du Bonnet on the Canadian shield (Gascoyne et al., 1987)

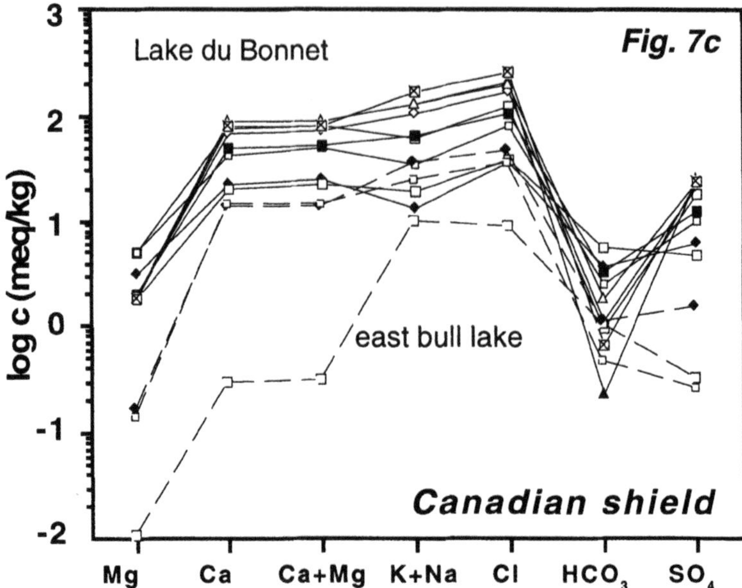

Figure 7c: Composition patterns of waters from wells at Lake du Bonnet and East Bull Lake on the Canadian shield (Gascoyne et al., 1987). The patterns are similar to the petterns on Fig. 3b and Fig. 5a.

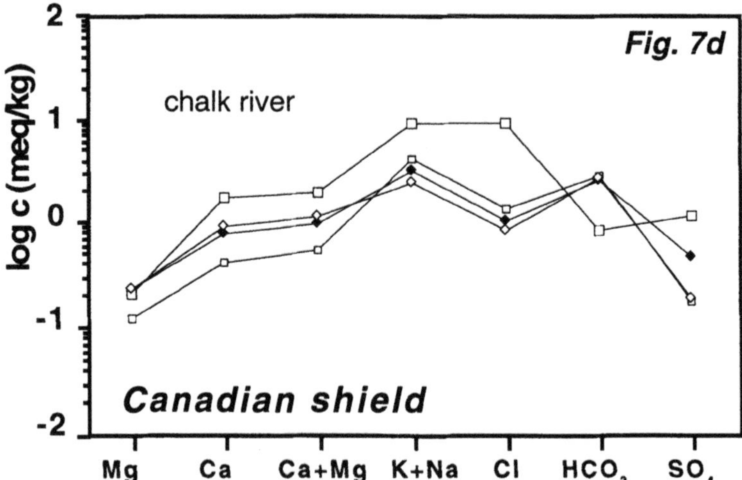

Figure 7d: Composition patterns of waters from wells at Chalk River on the Canadian shield (Gascoyne et al., 1987). Compare Figs. 3b, 5b and 5c.

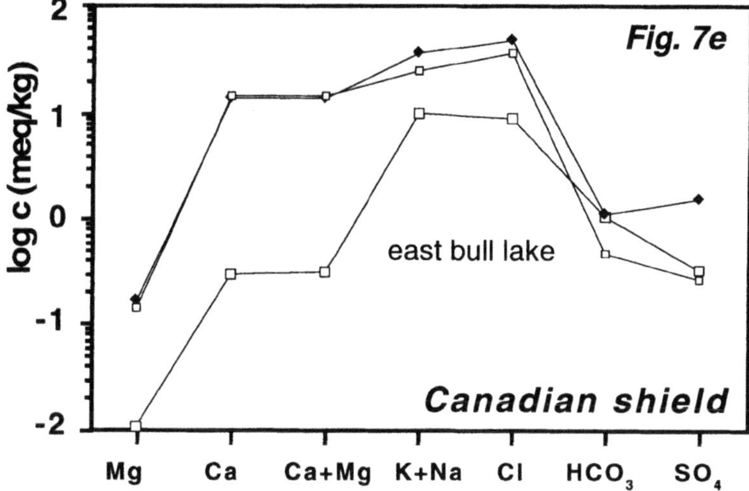

Figure 7e: Composition patterns of waters from wells at East Bull Lake on the Canadian shield (Gascoyne et al., 1987). Compare Fig. 3b. and 5.

4. The water reservoir

The water in the crystalline basement resides in fractures and fissures as well as in open cavities. For a review of water conducing features see Mazurek (this volume). The basement with its water conducting features behaves like any other aquifer (Stober and Bucher, this volume). Its hydraulic properties can be quantified by simple scalar descriptive macroscopic parameters, e.g. a uniform permeability and porosity (Stober, 1986).

4.1 Vertical zonation of water composition

The composition of the water in the crystalline basement shows a striking depth dependence (Gascogne et al., 1987; Gascoyne and Kamineni, 1993; Stober, 1995; Stober and Bucher, 1999) that has been observed worldwide. Near surface meteoric water is weakly mineralized Ca-HCO_3 water that may aquire its composition from dissolution of small amounts of secondary calcite. With increasing depth, TDS increases along with X_{Na} and sulfate. Deep water is chloride water, Ca-rich at low temperature, Na-rich at high temperature. It is expected that the fluid residing near the base of the aquifer (transition zone to the ductile lower part of the crust) is a highly saline brine in equilibrium with calcite, gypsum and various salts.

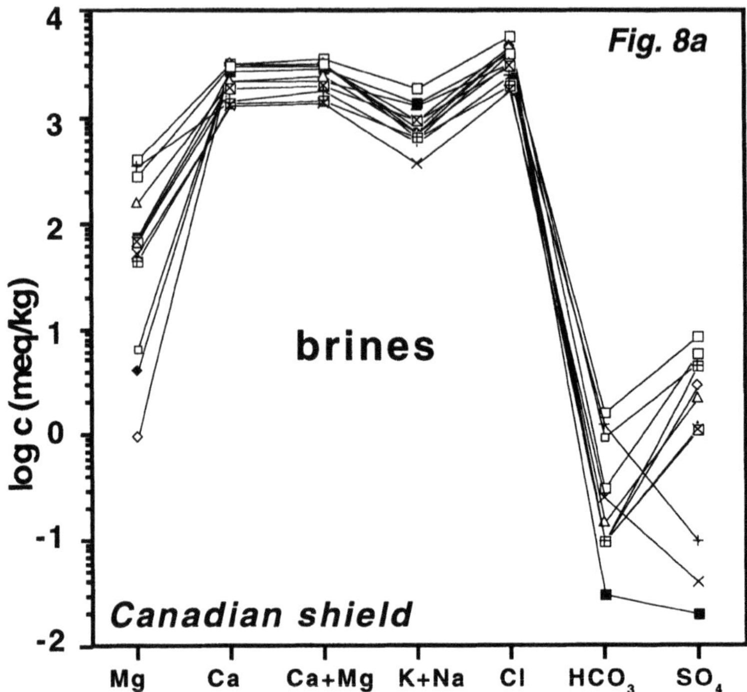

Figure 8a: Composition patterns of brines from mines on the Canadian shield (Frape and Fritz, 1987). TDS > 100 000 mg/kg

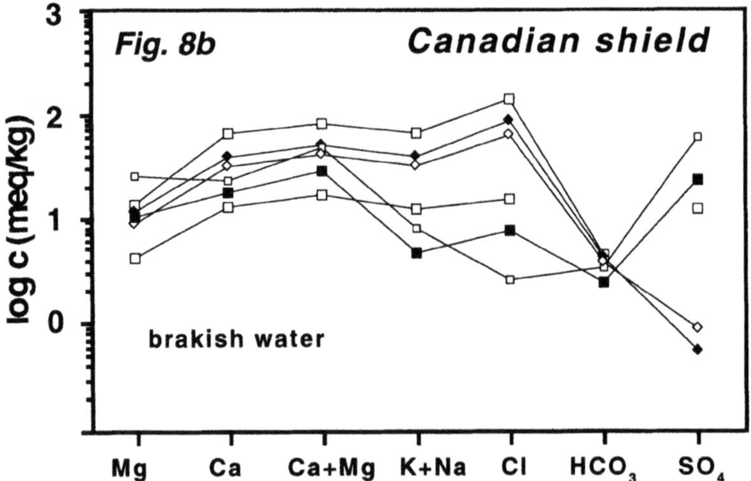

Figure 8b: Composition patterns of water from mines on the Canadian shield (Frape and Fritz, 1987). 10 000 > TDS > 1000 mg/kb

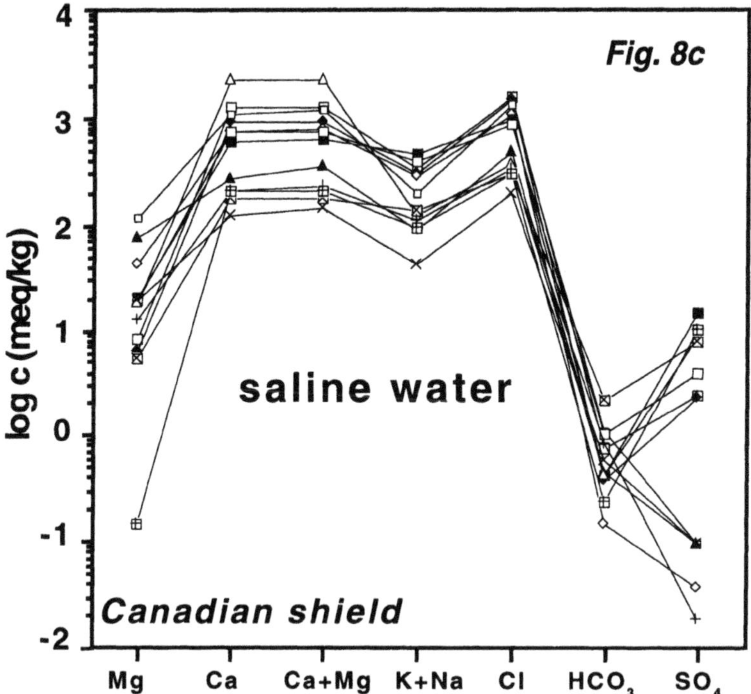

Figure 8c: Composition patterns of saline water from mines on the Canadian shield (Frape and Fritz, 1987). 100 000 > TDS > 10 000 mg/kg

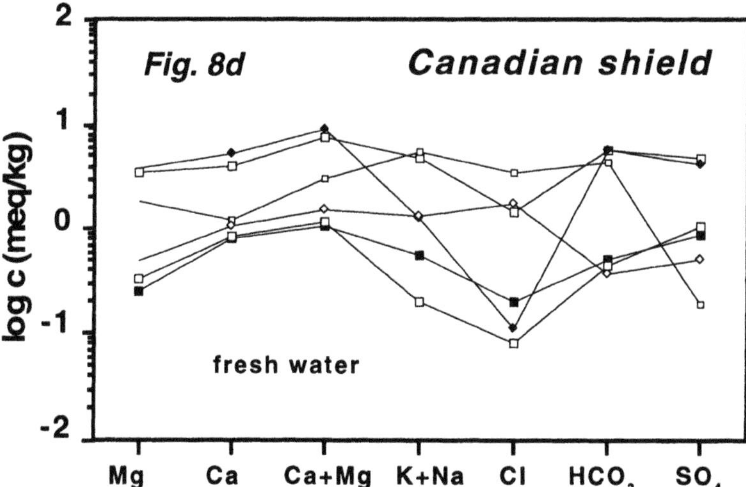

Figure 8d: Composition patterns of waters from mines on the Canadian shield (Frape and Fritz, 1987). TDS < 1000 mg/kg

The average material of the upper continental crust does not change much with depth. The observed depth zonation of groundwater composition in the upper crust (about upper 15 km) is therefore, in general, not related to changes in rock composition. The likely causes of the depth-dependent composition include the progress of the described dessication process and the temperature increase with depth.

The ductile lower crust, which represents the aquiclude in hydrogeological terms, may not contain free water under normal circumstances (Frost and Bucher, 1994). The hydration of minerals that are unstable in the presence of water consumes all water that might be present in isolated pore space of the aquiclude and the process desiccates the lower parts of the crust effectively within geologically short periods of time (Markl and Bucher, 1998).

4.2 Transition to the deeper crust

Conventional groundwater concepts and textbooks appeal to a "base-of-the-aquifer" which corresponds to an impervious no-flow boundary. The continental crust is about 35 km thick and its lower boundary, the MOHO, separates it from the Earth mantle. The nature and properties of the water conducting features in the crustal material changes with the rheology which is controlled predominantly by temperature and deformation rate. At some depth in the crust response to stress on the material changes from brittle to ductile deformation (Etheridge et al., 1983; Stanley et al., 1990). With this fundamental change also the geometry and nature of the water conducting features change from open interconnected fracture systems to isolated pockets and cavities, the pressure on the water changes from hydrostatic to lithostatic. This transition zone is equivalent to the base-of-the-aquifer in a hydrogeological sense. Its depth depends on the temperature and the rate of deformation (Rutter and Brodie, 1992). But in typical continental environments, this transition zone corresponding to the base of the aquifer is at 15 - 20 km depth. It is deeper in cold Precambrian shields where it may be below the MOHO and parts of the upper mantle may be part of the aquifer. In young warm crust the brittle-ductile transition may be not deeper than 5 km (Chester, 1995). Alteration and hydration reactions tend to seal the crustal aquifer which results in an upward migration of the "base-of-the aquifer". However, neotectonics and continuous deformation produces fractures that tend to keep the interconnected flow network open and ties the "base of the aquifer" to the ductile-brittle transition zone.

The significance, nature and composition of aqueous fluids in the deep ductile part of the crust has been described and discussed by Touret (1986), Wickham (1992), Newton (1989), Newton et al. (1998).

Acknowledgments

This research has been supported by the German Science Foundation (DFG, grant: Sto 203/6-1) and by the Rinne Foundation. The financial support is gratefully acknowledged. We also thank Tim Drever and the University of Wyoming for the generous hospitality during our sabbatical visit. The paper benefited from the thoughtful review by V. Dietrich..

References

Albu, M., Banks, D., and Nash, H. (1997) Mineral and thermal groundwater resources. Chapman & Hall, London, 447 p.

Althaus, E., Bauer, F., and Karotke, H. (1985) Bericht des Arbeitskreises Geochemie. In: *3. Zwischenbericht über "Erweiterte Zirkulation wässriger Fluide im Hot Dry Rock-System (Gneisgebirge) der Bohrung Urach 3" (unveröffentlicht).*, 59-61. Stadtwerke Bad Urach: Bad Urach.

Banks, S. and Banks, D., editors. (1993) Hydrogeology of Hard Rocks, *Memoirs of the XXIVth Congress International Association of Hydrogeologists* 28th June - 2nd July 1993, Ås (Oslo), Norway, NGU, Geological Survey of Norway, Trondheim, 1205 p.

Behr, H. (1989) Die geologische Aktivität von Krustenfluiden. *Nds. Akad. Geowiss. Veröfftl.*, 1, 7-43.

Berman, R.G. (1988) Internally-Consistent Thermodynamic Data for Minerals in the System: Na_2O- K_2O- CaO- MgO- FeO- Fe_2O_3- Al_2O_3 - SiO_2- TiO_2- H_2O- CO_2, *Journal of Petrology,* 29, 445-522.

Bowers, T. S., Jackson, K. J., and Helgeson, H. C. (1984) Equilibrium Activity Diagrams for Coexisting Minerals and Aqueous Solutions at Pressures and Temperatures to 5 kb and 600°C. Springer: Berlin. 397 p.

Bucher, K. and Stober, I. (1999) Mass balance versus equilibrium control of groundwater composition in crystalline basement, *J. Conference Abstracts,* 4, 586.

Carlé, W. (1975) Die Mineral- und Thermalwässer von Mitteleuropa. Wissenschaftliche Verlagsgesellschaft, Stuttgart, 419 p.

Carmichael, R.S., (1989) Practical handbook of physical properties of rocks and minerals, CRC Press: Boca Raton, 674 p.

Chester, F.M. (1995) A rheologic model for wet crust applied to strike-slip faults, *Journal of geophysical Research,* 100, 13033-13044.

Dai, J., Song, Y., Dai, Ch. S., and Wang, D.-R. (1996) Geochemistry and accumulation of carbon dioxide gases in China. *American Association of Petroleum Geologists Bulletin,* 80, 1615-1626.

Deer, W.A., Howie, R.A., and Zussmann, J. (1992) An introduction to the rock forming minerals. - 2nd ed. Longman, Harlow, UK. 696 p.

Edmunds, W. M., Kay, R. L. F., Miles, D. L., and Cook, J. M. (1987) The origin of saline groundwaters in the Carnmenellis granites, Cornwall (U.K.): Further evidence from minor and trace elements. In: *Saline Water And Gases In Crystalline Rocks*, 127-143. Fritz, P. and Frape, S. K. (editors), Geological Association of Canada Special Paper, 33. The Runge Press Limited: Ottawa.

Etheridge, M.A., Wall, V.J., and Vernon, R.H. (1983) The role of the fluid phase during regional metamorphism and deformation, *J. Metamorphic Geology,* 1, 205-226.

Frape, S. K., and Fritz, P. (1987) Geochemical trends for groundwaters from the Canadian shield. In: *Saline water and gases in crystalline rocks*, 19-38. Fritz, P. and Frape, S. K. (editors), Geological Association of Canada Special Paper, 33. The Runge Press Limited: Ottawa.

Fritz, P., and Frape, S. K., editors, (1987) Saline Water And Gases In Crystalline Rocks, Geological Association of Canada Special Paper, 33. The Runge Press Limited, Ottawa. 215 p.

Fritz, P., Frape, S. K., Drimmie, J. R., Appleyard, E. C. and Hattori, K. (1994) Sulfate in brines in the crystalline rocks of the Canadian Shield, *Geochimica et Cosmochimica Acta*, 58, 57-65.

Frost, B.R., and Bucher, K. (1994) Is water responsible for geophysical anomalies in the deep continental crust? A petrological perspective, *Tectonophysics,* 231, 293-309.

Fyfe, W. S. (1987) The fluid inventory of the crust and its influence on crustal dynamics. In: *Saline Water And Gases In Crystalline Rocks*, 1-4. Fritz, P. and Frape, S. K. (editors), Geological Association of Canada Special Paper, 33. The Runge Press Limited: Ottawa.

Fyfe, W. S., Price, N. J., and Thompson, A. B. (1978) Fluids in the earth's crust: their significance in metamorphic, tectonic and chemical transport processes. Elsevier: Amsterdam. 383 p.

Gascoyne, M., Davison, C. C., Ross, J. D., and Pearson, R. (1987) Saline groundwaters and brines in plutons in the Canadian Shield. In: *Saline Water And Gases In Crystalline Rocks*, 53-68. Fritz, P. and Frape, S. K. (editors), Geological Association of Canada Special Paper, 33. The Runge Press Limited: Ottawa.

Gascoyne, M., and Kamineni, D. C. (1993) The hydrogeochemistry of fractured plutonic rocks in the canadian shield. In: *Hydrogeology of Hard Rocks*, 440-449. Banks, S. B. and Banks, D. (editors) Geological Survey of Norway: Trondheim.

Gascoyne, M., and Kamineni, D.C. (1994) The hydrogeochemistry of fractured plutonic rocks in the Canadian shield, *Applied Hydrogeology,* 2/94, 43-49.

Griesshaber-Schmal, E., (1990) Helium and Carbon Isotope Systematics in Groundwaters from W.Germany and E.Africa. University of Cambridge, Dissertation (unpublished), 211 p.

Gustavson, G., and Krasny, J. (1993) Crystalline rock aquifers: their occurrence, use and importance. In: *Hydrogeology of Hard Rocks*, 3-20. Banks, S. B. and Banks, D. (editors) Geological Survey of Norway: Trondheim.

Helgeson, H.C., Kirkham, D.H., and Flowers, G.C. (1981) Theoretical prediction of the thermodynamic behaviour of aqueous electrolytes at high pressures and temperatures. IV. Calculation of activity coefficients, osmotic coefficients, and apparent molal and standard and relative partial molal properties to 600°C and 5 kb, *American Journal of Science*, 281, 1249-1516.

Jenkner, B., Jenkner, I., Klein, H., Schädel, K., and Stober, I. (1986) Geothermievorbohrungen im Mittleren Schwarzwald für das Kontinentale Tiefbohrprogramm der Bundesrepublik Deutschland, *Abschlußbericht 1986 des Geologischen Landesamtes Baden-Württemberg vom 15.12.1986 (unpubl.)*, 74 p.

Kamineni, Ch.D. (1987) Halogen-bearing minerals in plutonic rocks: A possible source of chlorine in saline groundwater in the Canadian Shield. In: *Saline Water And Gases In Crystalline Rocks*, 69-79. Fritz, P. and Frape, S. K. (editors), Geological Association of Canada Special Paper, 33. The Runge Press Limited: Ottawa.

Kerrick, D.M., McKibben, M.A., Seward, T.M., and Caldeira, K. (1994) Convective hydrothermal CO_2 emission from high heat flow regions. *Chemical Geology*, 121, 285-293.

Kozlovsky, Ye.A. (1984) The world's deepest well, *Scientific American*, 251, 106-112.

Lasaga, A.C., (1984) Chemical kinetics of water-rock interactions, *Journal of Geophysical Research*, 89, 4009-4025.

Liegl, R., Stober, I., and Bucher, K. (1999) Experimental water-rock reaction of Black Forest gneiss and granite, *Journal of Conference Abstracts*, 4, 590.

Lodemann, M., Fritz, P., Wolf, M., Ivanovich, M., Hansen, B. T., and Nolte, E. (1998) On the origin of saline fluids in the KTB (Continental Deep Project of Germany), *Applied Geochemistry*, 13, 651-672.

Markl, G., and Bucher, K. (1998) Composition of fluids in the lower crust inferred from metamorphic salt in lower crustal rocks, *Nature*, 391, 781-783.

May, F., Hoernes, S., and Neugebauer, H. J. (1996) Genesis and distribution of mineral waters as a consequence of recent lithospheric dynamics: the Rhenish Massif, Central Europe. *Geologische Rundschau*, 85, 782-799.

Mazurek, M. (this volume) Geological and hydraulic properties of water-conducting features in crystalline rocks. In: *Hydrogeology of Crystalline Rocks*. Stober, I. and Bucher, K. (editors) Kluwer: Amsterdam.

Mazurek, M., Gautschi, A., Smith, P. A. and Zuidema, P. (this volume) On the water-conducting features in the Swiss concept for the disposal of high-level radioactive waste. In: Stober, I. and Bucher, K., (editors) *Hydrogeology of Crystalline Rocks*, Kluwer, Amsterdam.

Michel, G. (1997) Mineral- und Thermalwässer - Allgemeine Balneogeologie. In: *Lehrbuch der Hydrogeologie*, 7, Mattheß, G. (editor), Gebr. - Bornträger, Berlin/Stuttgart, 256 p.

Möller et al. (1997) Paleo- and recent fluids in the upper continental crust - Results from the German Continental deep drilling Program (KTB), *Journal of geophysical Research,* 102, 18245-18256.

Möller, P. (this volume) Rare earth elements and yttrium as geochemical indicators of the source of mineral and thermal waters. In: *Hydrogeology of Crystalline Rocks.* Stober, I. and Bucher, K. (editors) Kluwer: Amsterdam.

Münch, H.G. (1981) Zur Geologie des Geothermik-Pilot-Projektes Bühl, *Aufschluss,* 32, 335-344.

Muffler, L.I.P., and White, D.E. (1968) The origin of CO_2 in the Salton Sea geothermal system, south eastern California, U.S.A. *23 Int. Geol. Congr. Prague*, 17, 185-192.

Newton, R.C. (1989) Metamorphic fluids in the deep crust, *Annual Reviews of Earth and Planetary Science,* 17, 385-412.

Newton, R.C., Aranovich, L., Hansen, E. C., and Vandenheuvel, B. A. (1998) Hypersaline fluids in precambrian deep-crustal metamorphism, *Precambrian Research,* 91, 41-63.

Nordstrom, D.K., Andrews, J.N., Carlsson, L., Fontes, J.-C., Fritz, P., Moser H., and Olsson, T. (1985) Hydrogeological and Hydrogeochemical Investigations in Boreholes - Final report of the phase I geochemical investigations of the Stripa groundwaters, *Technical Report STRIPA Project, Stockholm,* 85-06.

Nordstrom, D. K. et al. (1979) A comparison of computerized chemical models for equilibrium calculations in aqueous systems. In: *Chemical modelling in aqueous systems, speciation, sorption, solubility, and kinetics*. S.93, 857-892, Jenne, E. A. (editor) American Chemical Society.

Person, M., and Garven, G. (1992) Hydrologic constraints on petroleum generation within continental rift basins: Theory and application to the Rhine Graben, *American Association of Petroleum Geologists Bulletin,* 76, 468-488.

Rutter, E. H., and Brodie, K. H. (1992) Rheology of the lower crust. In: *Continental Lower Crust*. 201-267. Fountain, D. M., Arculus, R., and Kay, R. W. (editors) Elsevier: Amsterdam.

Schreiner, A. (1991) Geologie und Landschaft. In: *Das Markgräfler Land: Entwicklung und Nutzung einer Landschaft*, 11-24. Hoppe, A. (editor) Naturforschende Gesellschaft Freiburg i.Br., 81.

Stanley, W.D, Mooney, W.D., and Fuis, G.S. (1990) Deep crustal structure of the Cascade Range and surrounding regions from seismic refraction and magnetotelluric data, *Journal of Geophysical Research,* 95, 19419-19438.

Stober, I. (1986) Strömungsverhalten in Festgesteinsaquiferen mit Hilfe von Pump- und Injektionsversuchen. *Geologisches Jahrbuch,* Reihe C, H. 42, 204 p.

Stober, I. (1995) Die Wasserführung des kristallinen Grundgebirges, *Enke-Verlag, Stuttgart,* 191 pp.

Stober, I. (1996) Hydrogeological Investigations in Crystalline Rocks of the Black Forest, Germany, *TERRA Nova,* 8, 255-258.

Stober, I., and Bucher, K. (this volume) Hydraulic Properties of the upper Continental Crust: data from the Urach 3 geothermal well. In: *Hydrogeology of Crystalline Rocks.* Stober, I. and Bucher, K. (editors) Kluwer: Amsterdam.

Stober, I., and Bucher, K. (1999) Deep groundwater in the crystalline basement of the Black Forest region, *Applied Geochemistry,* 14, 237-254.

Touret, J. (1986) Fluid inclusions in rocks from the lower continental crust. In: *The nature of the lower continental crust.* 161-172. Dawson, J. B., Carswell, D. A., Hall, J., and Wedepohl, K. H. (editors) Geological Society of London Special Publication 81. London: United Kingdom.

Trommsdorff, V., Skippen, G., and Ulmer, P. (1985) Halite and sylvite as solid inclusions in high-grade metamorphic rocks, *Contributions to Mineralogy and Petrology,* 24-29.

Wickham, S. M. (1992) Fluids in the deep crust - petrological and isotopic evidence. In: *Continental Lower Crust.* 391-421. Fountain, D. M., Arculus, R., and Kay, R. W. (editors) Elsevier: Amsterdam.

EVOLUTION OF FLUID CIRCULATION IN THE RHINE GRABEN: CONSTRAINTS FROM THE CHEMISTRY OF PRESENT FLUIDS

L. AQUILINA[1], A. GENTER[1], P. ELSASS[2], AND D. PRIBNOW[3]
[1] BRGM, Dir. de la Recherche, 1039 rue de Pinville, 34000 Montpellier
[2] BRGM, SGR Alsace, Lingolsheim, BP 177, 67834 Tanneries Cedex
[3] GGA, Stilleweg 2, 30655 Hannover - Germany

Abstract

A Hot-Dry-Rock geothermal power plant is being built in Soultz-sous-Forêts (Alsace, France), in an area of anomalously high surface heat flow. This anomaly is the result of a re-distribution of heat in the upper crust by convection of fluids. To assess the regional flow field, saline fluids have been collected down to a depth of 3500 m in the Rhine Graben from geothermal boreholes reaching the Triassic Buntsandstein sandstone aquifer and the granitic basement.

The comparison with shallow thermal boreholes located on the western side of the Soultz horst shows mixing effects between meteoric fluids and a saline end-member (80 g/l). This implies (1) that circulation of recent (< 30,000 yr) meteoric fluids occurs at shallow depths (< 1000 m) on the western edge of the graben, and (2) that the Buntsandstein is a brine-bearing reservoir, carrying highly saline fluids (80 to 200 g/l) of the type recovered in the geothermal boreholes. No indication of recent meteoric circulation is observed in the deeper part of the graben, east of Soultz.

The brines originate from the dilution of a sedimentary "primary" end-member by meteoric water, probably during the Eocene-Oligocene period, *i.e.* during the initiation of the graben formation. Subsequently, the fluids have undergone intensive water-rock interactions at high temperature (200-250°C) during burial, except on the eastern edge of the graben where the temperature effect is weaker.

Chemical analysis of the fluid samples indicates that there is no large-scale flow of recent meteoric water from the Black Forest in the east to the Soultz horst through the Buntsandstein aquifer as has been previously suggested by numerical models. The present state of the fluid circulation in the Graben is thought to be divided into three relatively independent circuits. (1) On each border of the graben, meteoric water is rapidly recycled to the surface. (2) Large-scale exchange of saline fluids occurs between the Buntsandstein and the upper part of the granitic section in a succession of convection cells, from the deeper eastern part of the graben up to the Soultz horst. (3) At great depth in the granite (more than 3 to 5 km), there is a very slow but pervasive seepage of fluids from the bottom of the graben up to the Soultz area. This large-scale migration carries heat without transfering large amounts of water, and thus has a limited effect on the composition of the deep saline fluids.

I. Stober and K. Bucher (eds.), Hydrogeology of Crystalline Rocks, 177–203.

1. Introduction

Saline fluids are well known in sedimentary basins where they are related to evaporitic deposits or evaporated seawater relicts, but they have also been discovered at great depth in the basement of Canada (*e.g.* Frape *et al.*, 1984) and Europe (*e.g.* Edmunds *et al.*, 1985; Paces, 1987; Vovk, 1987; Nordstrom *et al.*, 1989; Nurmy *et al.*, 1988; Boulègue *et al.*, 1990; Aquilina *et al.*, 1997a). However, although the understanding of water-rock interaction processes is making progress, the deep saline fluid circulation systems in the continental crust are still poorly known.

Saline fluids have also been collected in the Rhine Graben, within the framework of the Soultz-sous-Forêts European Hot Dry Rock (HDR) geothermal project (Kappelmeyer *et al.*, 1991). The Soultz project is located on a thermal anomaly which was discovered at the beginning of the century (Gérard *et al.*, 1984). The aim of the project is to develop a heat exchanger in the granite basement of the graben, which is overlain by a 1400 m thick sedimentary pile. As part of this project, three boreholes were drilled to depths of 2200, 3600 and 3800 m in the granite (Baria *et al.,* 1995). Fluid circulation was observed in the Buntsandstein aquifer (Lower Triassic sandstone) which directly overlies the granite, but also in the granite itself, down to a depth of 3,500 m (Criaud *in* Kappelmeyer and Gérard, 1989; Vuataz *et al.,* 1990; Aquilina and Brach, 1995). The fluids circulate along major fault zones which are characterized by an increase in permeability.

The fluids which have been collected at the Soultz site are Na-Cl brines with TDS (total dissolved solids) values close to 100 g/l (Pauwels *et al.,* 1991, 1992, 1993). The similarity between the Soultz fluids and the brines of the Buntsandstein aquifer which have been collected in the northern part of the Rhine Graben (Fritz, 1980, 1981; Fritz *et al.*, 1989) indicates a common sedimentary origin (Pauwels *et al.*, 1993). This implies that fluids from the Buntsandstein reservoir penetrate the granitic basement along fault zones.

Several studies of fluid inclusions have been carried out on the cores of one of the geothermal boreholes, both in the granite and in the Buntsandstein (Ledesert, 1993; Dubois *et al.,* 1994, 1996; Ayt Ougougdal *et al.*, 1995; Yardley *et al.*, 1995; Meere *et al.*, 1995). Most of the measurements yield temperatures ranging between 130°C and 180°C, which match the present geothermal gradient (Yardley *et al.,* 1995). However, higher temperatures (180-270°C) have also been measured. Th results also show numerous measurements in good agreement with the salinites of the present fluids, although a wide range of salinities is observed. Independently of the origin of the fluid samples (Buntsandstein or granite), several generations of fluids can be identified: from moderate salinities (10% Eq NaCl or even lower) to higher salinities (20 to 30% Eq NaCl). The general interpretation is that this is the result of mixing processes.

Several numerical models of coupled heat and mass transport (Clauser and Villinger, 1990; Person and Garven, 1992; Flores, 1992; Toth and Otto, 1993 ; Royer *et al.*, 1995; Meere *et al.,* 1995; Royer *et al.*, 1995; Ayt Ougougdal *et al.*, 1995) suggest that a large-scale dilution process should occur in the graben due to mass flow of fluids within the sedimentary aquifers (and the weathered top part of the granite).

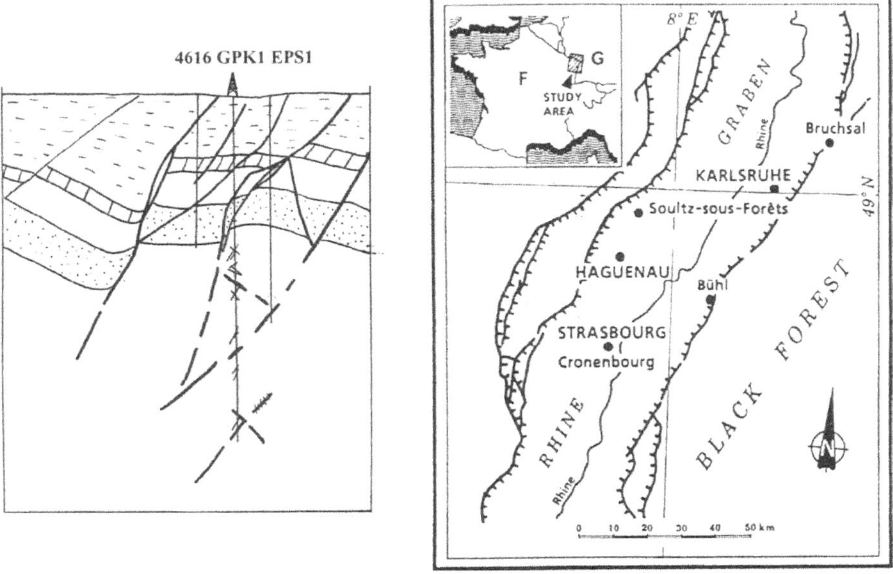

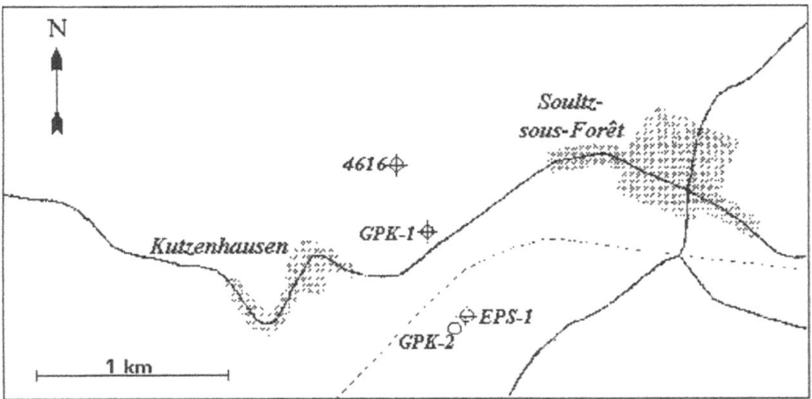

Figure 1 Location map of the Rhine Graben

The present paper discusses the geochemical composition of the fluids collected in the graben in terms of their origin and mixing processes. Other aspects such as water-rock interactions and geothermometry have been studied in other papers (Pauwels *et al.*, 1993; Aquilina *et al.*, 1997b). Our analysis provides several constraints on fluid circulation, which are compared to the existing hydrogeological models and to the results of the fluid inclusion studies. It enables us to propose a new model of the evolution of fluid circulations in the Rhine Graben.

2. Geological setting and fluid-study sites

The Rhine Graben belongs to an extensive rift structure crossing the northwestern European plate (Fig. 1; Villemin and Bergerat, 1987). The Upper Rhine Graben is bounded by N20-N40° faults. It is supposed to present an assymetric structure, the western fault extending with a shallow dip below the graben (Brun *et al.*, 1992). The sediments which filled the graben from Oligocene to present are mainly composed of impervious sediments which seal the Jurassic and Triassic formations. The lower Triassic Buntsandstein sandstone formation (and the middle Triassic Muschelkalk limestone), which unconformably overlies the granite, is a regional aquifer. An east-west section at the Soultz-sous-Forêts level (Fig. 2) shows a horst structure in the vicinity of the Soultz site, along a major west-dipping fault (Kutzenhausen Fault), and an increasing depth of the sedimentary cover towards the east. At the eastern edge of the graben, the Triassic reaches a depth of 5 km. The same pattern appears on a N-S section through the Soultz site. As can be observed on a map of the base of the Triassic formations (Fig. 3), a general trough-like structure extends accross the graben along a NE-SW direction. This trough is induced by a Variscan suture (known as the Lalaye-Lubine suture in the Vosges mountains) which separates two distincts paleozoic domains.

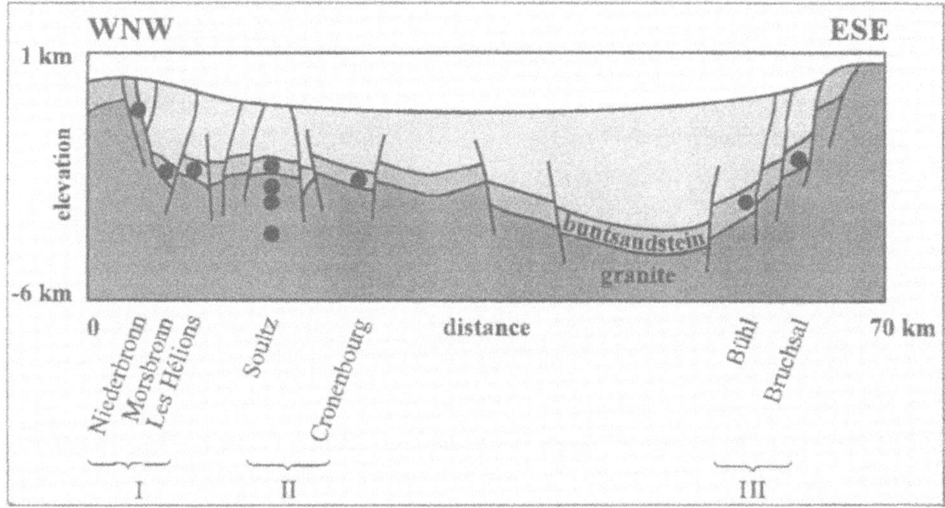

Figure 2 Location of fluid sampling along a simplified cross-section of the Rhine Graben. Sampling positions are projected onto the plane of the cross-section. Bruchsal is about 40 km north of the profile

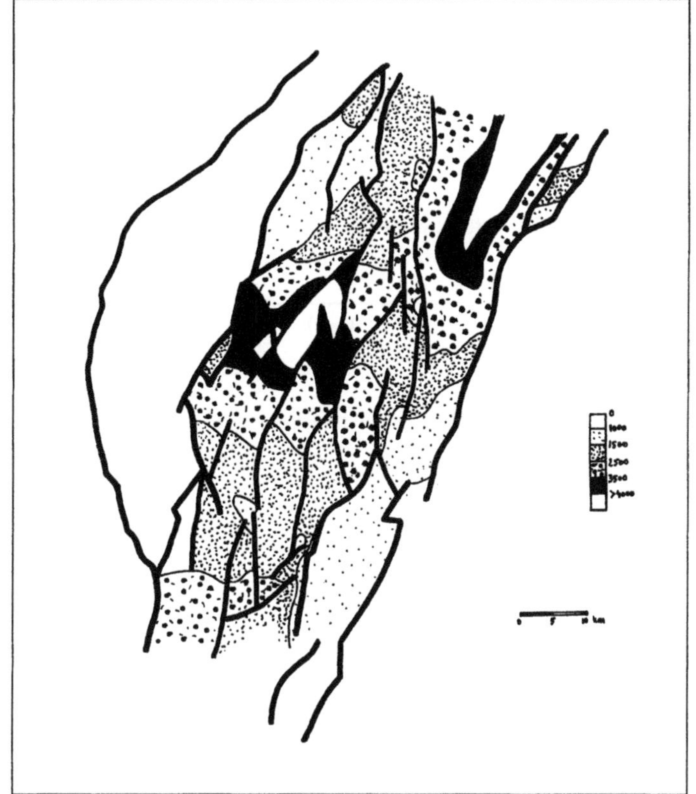

Figure 3 Map of the Triassic-bottom depth (modified from Papillon, 1995)

The granitic section of one of the geothermal boreholes at the Soultz site was entirely cored, which allowed lithology and tectonic features to be extensively investigated (Genter *et al.*, 1991; Traineau *et al.*, 1991; Genter and Traineau, 1992; Ledésert *et al.*, 1993; Ledésert, 1993; Genter *et al.*, 1995). Two major sets of fractures have been recognized: N10° with a westerly dip of 80° and N170° with an easterly dip of 70°. A third E-W set is thought to represents a reactivation of pre-Rhine Graben faulting (Villemin, 1986). The fracture network can induce a porosity of 1 to several percent in the granite. In the granite, fractured zones with extensive alteration haloes are interpreted as the location of major fault structures. Within these zones, a polyphase alteration is observed. It comprises biotite and plagioclase dissolution, and illite, quartz and carbonate precipitation. Within highly altered zones, open fractures partly filled with quartz have been observed. They are the channels of fluid circulation in the granite.

Geothermal fluids of the Rhine Graben have been sampled and analysed by Tardy (1980), Fritz (1980), Fritz (1981); Pauwels *et al.* (1991, 1992, 1993), Royer *et al.*, (1995), and Aquilina *et al.*, (1997b). Pauwels *et al.* (1993) and Aquilina *et al.* (1997b) present complete analytical data sets. The main characteristics of these fluids are summarized in Table 1. Most of the fluid samples originate from the Buntsandstein aquifer at depths ranging from 250 to 2870 m. Only at the Soultz Hot Dry Rock geothermal site have saline fluids been collected in the granitic basement at depths ranging from 1815 to 3500 m.

TABLE 1. Major characteristics of the fluids sampled in the Rhine Graben (data from Pauwels *et al.*, 1993 and Aquilina *et al.*, 1997b)

Sample	Location	Formation	Depth (m)	TDS (g/l)	present T°C	Na/K temp.	SiO_2 temp.
Thermal spas (I)							
Les Hélions	west of Soultz	Bunt.	1100	20.5	72	249	92.5
Morsbronn	west of Soultz	Bunt.	600	6.0	41	250	61
Niederbronn	western edge	Muschelkalk	250	4.7	18	232	51.5
Deep fluids (II-III)							
Cronenbourg (II)	Strasbourg	Bunt.	2870	104	140	262	186.5
4616 (II)	Soultz	Bunt.	1403	103	116	259	175
KS228 (II)	Soultz	granite	1815	99	137	254	162.5
EPS1 (II)	Soultz	granite	2200	101	150	255	150
KF3500 (II)	Soultz	granite	3500	101	165	238	235
Bruhsal (III)	eastern edge	Bunt.	1800	120	114	190	120
Bühl (III)	eastern edge	Bunt.	2655	207	115	71	-

3. Main results of the chemical and isotopic investigations of the fluids

In the following, three classes of fluid samples will be considered (Fig. 2): (I) the fluids from the thermal spas, located on the western border of the graben and characterized by low salinities at shallow depths, (II) the fluids issued from the Soultz site and the Cronenbourg geothermal borehole with intermediate salinities at great depths, (III) the eastern saline fluids (Bühl and Bruchsal), which have the highest salinities at great depth.

The chemical and stable isotopic composition of the fluids is described in Pauwels *et al.* (1993) and Aquilina *et al.* (1997b). However, none of the previous studies has focussed on the origin of the fluids and their possible migration path. The main aspects of these studies, which will help to understand the fluid circulation in the graben, are summarized below.

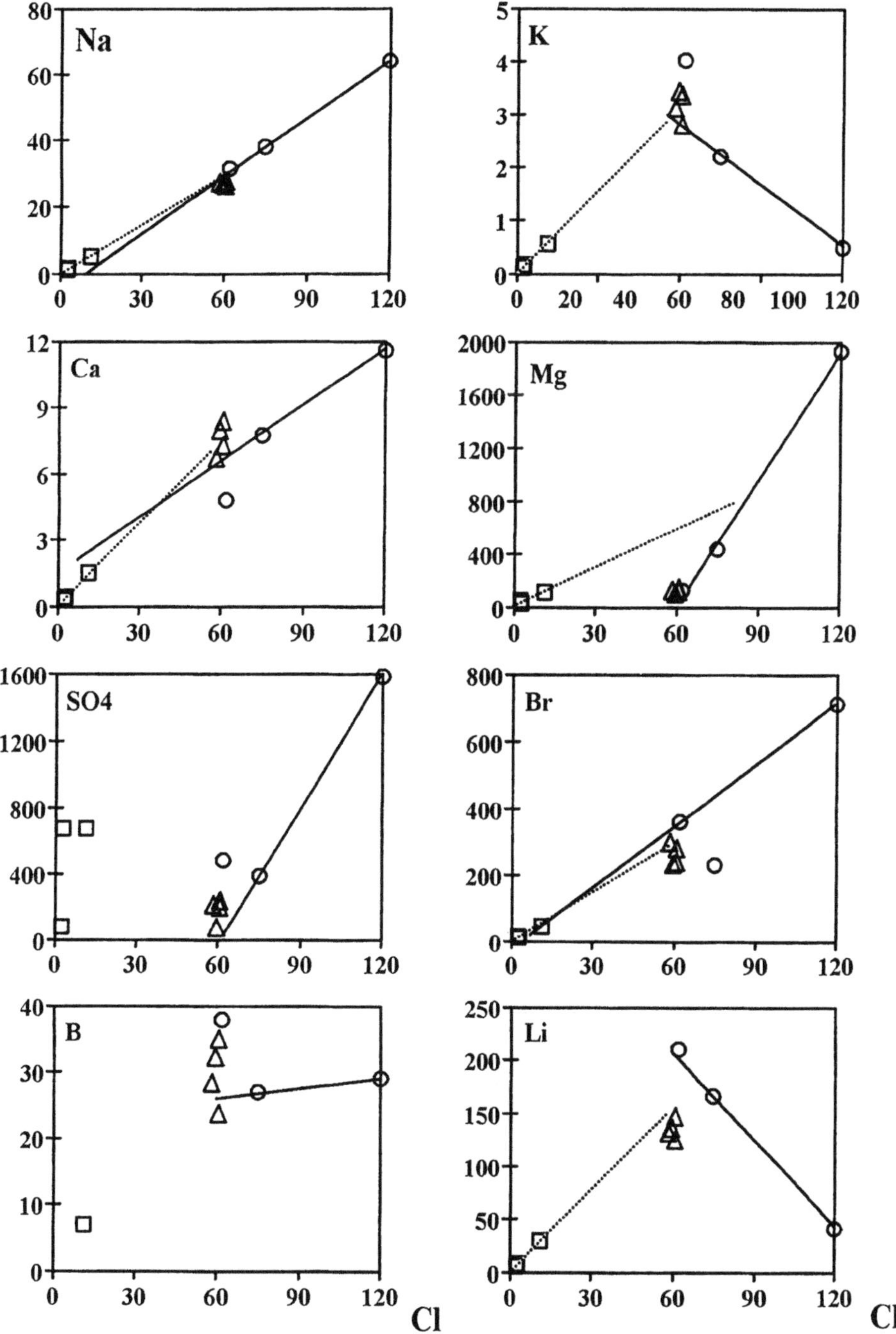

Figure 4 Saline fluids in the Rhine Graben : Na, K, Ca, Mg, SO4, Br, B, Li, Rb, Cs vs Cl circles : Buntsandstein Bühl, Bruchsal, Cronenbourg / triangles : Granite : EPS-1, GPK-1 KS228 (1815m) and KF3500 (3500m) ; Buntsandstein : 4616 plain line joins Bühl and Bruchsal points ; dotted line is a regression for the thermal spa fluids

3.1. CHEMICAL COMPOSITION OF SHALLOW AND DEEP FLUIDS

3.1.1. Thermal shallow spas (I)

The fluid samples from the geothermal spas located to the west of the Soultz site, close to the western border of the graben, have TDS values which are quite different from those of the deep fluids (Tab. 1). TDS increase with increasing distance from the western edge of the graben and increasing depth of the Buntsandstein. Plotting the elements Na, K, Ca, Br, Li versus Cl concentrations (Fig. 4), the datapoints define a straight line which intersects the deep saline data of group II. A similar result is observed for the Sr isotopic ratio, plotted versus Cl concentration. This indicates a mixing between a saline end-member (similar to the Soultz fluids) and a low concentration fluid.

3.1.2. Deep saline fluids (II and III)

Most of the fluid samples originate from the Buntsandstein aquifer. This sandstone formation is similar to water-bearing formations in sedimentary basins which have been investigated for oil resources (*e.g.* Carpenter, 1978; Collins, 1975; Hanor, 1994). Pauwels *et al.* (1993) have suggested that the Br/Cl ratios and stable isotopes of water and dissolved sulfates of the saline fluids from the Buntsandstein aquifer are the result of (1) evaporation of sea-water, (2) mixing with meteoric waters, and (3) dissolution of NaCl. At the Soultz site, sampling of the fluids in the granite allowed a comparison with the fluids from the Buntsandstein (Aquilina *et al.*, 1997b). Although several elements exhibit slight differences, the chemical composition of the fluids is highly similar. This indicates that they have a common origin.

Saline fluids from the same formation as the Triassic of the Rhine Graben are well known in the Paris Basin, on the western side of the Vosges mountains. Geothermal boreholes around Paris have allowed numerous samplings and geochemical investigations (Matray and Fontes, 1990; Fontes and Matray, 1993; Matray *et al.*, 1993, 1994). These studies concluded that the saline fluids from the Paris basin are composed of (1) meteoric water, (2) evaporated seawater, and (3) dissolved halite. The deep geothermal fluids from the Rhine Graben were then compared by Aquilina *et al.* (1997b) to the brines of the Paris Basin and to relicts of evaporated seawater as presented by Fontes and Matray (1993). This comparison shows that the most saline Bühl fluids (group III) have the closest chemical signature to the evaporated seawater. The Bruchsal fluid is intermediate between the Bühl fluid on one hand and the Soultz and Cronenbourg fluids on the other hand (group II). This can be interpreted as an evolution of the most saline Bühl-type fluids: these have the closest marine signature, whereas the Soultz fluids show a strong water-rock imprint at higher temperatures. This view is supported by geothermometry, trace chemistry and the interpretation of strontium isotopic measurements (see next sections).

Although the fluids from the eastern part of the graben (group III) have salinites of 120 and 207 g/l, fluids from both the Buntsandstein and the granite at Soultz show lower salinities of 94-100 g/l. This situation can be explained by two scenarii: (1) dilution of a saline end-member like the Bühl fluid, or (2) different degrees of NaCl dissolution by meteoric waters at different locations in the graben or even a mixing of both solutions.

The relation of several elements versus the chloride content is plotted in Figure 4. For the saline fluids (group II and III), a linear correlation appears for Na and Cl. For Cl and Mg, the lowest Cl points (group II) are only slightly shifted towards higher Mg values. For Li and Cl, the Soultz fluids are shifted from the correlation. The SO_4 content also displays a trend from the most saline to the less saline fluids. The general correlation shown by these elements with Cl is therefore a good argument for producing the Soultz, Cronenbourg and Bruchsal fluids by diluting a brine like the Bühl fluid. In Switzerland, fluids have been sampled in Permian formations and in granite, in a structure similar to the Rhine Graben, known as the Permian trough (Pearson *et al.*, 1991 ; Schmassmann *et al.*, 1992; Michard *et al.*, 1996). These fluids also have salinities in the range of the Bühl fluids, which supports the fact that Bühl-type fluids constitute the "primary" saline solution from which the other fluids originated through dilution.

3.2. OXYGEN AND HYDROGEN ISOTOPIC MEASUREMENTS

3.2.1. Thermal shallow spas (I)

The isotopic composition of the thermal spas shows an evolution from the less saline to the most saline, which defines a mixing line (Fig. 5a). This trend goes from the Global Meteoric Line to the Soultz fluids. It indicates that the shallow thermal fluids result from the mixing of deep brines, of the type found at Soultz, with meteoric water, which is in good agreement with the chemical composition presented in the previous section.

3.2.2. Deep saline fluids (II and III)

No dilution trend in the deep saline fluids can be seen in the oxygen-deuterium plot (Fig. 5b). Pauwels *et al.* (1993) interpreted the $\delta^{18}O$ as the result of a temperature-dependent equilibration of the oxygen isotopes with the host rock. The relation of the $\delta^{18}O$ to the Na/K temperature (Fig. 5c) supports the fact that a thermo-dependant equilibration of the O of the fluid with the minerals of the surrounding rock has occured. On the other hand, the Bühl fluid does not fit the temperature relation. The large negative value of $\delta^{18}O$ obtained for the Bühl fluid although it has the lowest temperature (Fig. 5c) might imply a wide isotopic variation of the fluids. Such a variation, however, seems quite unlikely when compared to the natural variation of the isotopic ratios of the saline fluids from the Paris Basin or the Swiss basement (Schmassmann *et al.*, 1992; Matray *et al.*, 1994).

Another interpretation of the data can be given if one considers the Bühl fluid as the "primary" brine from which the other fluids originated. Indeed, the isotopic position of the Bühl fluid is similar to that observed for many sedimentary brines (Kharaka and Carothers, 1986 ; Knauth and Beeunas, 1986). A dilution process will shift the points from this origin to the left, as shown in Figure 5d, until they reach 40 to 50 % of dilution. Equilibration at high temperatures would then shift the points back to the right (arrow in Fig. 5d), along line *a* in Fig. 5b.

A large difference in deuterium values is also observed for the samples from Soultz and Cronenbourg which have a similar salinity. Although a natural variability cannot be ruled out, it seems that an enrichment in deuterium has occurred. Such an enrichment could have occurred along line *b* in Fig. 5b, with a simultaneous slight decrease in oxygen, due to slow cooling of the fluids from 230°C (cation geotemperature) to 140°C (present temperature). Hydrogen exchange with clays can provide deuterium enrichments at quite low temperatures in a closed system (O'Neil and Kharaka, 1976; Graham, 1981). This has been observed for several fluids encountered in the basements of Canada (Frape *et al.*, 1984), Finland (Nurmi *et al.*, 1988) and France (Aquilina, 1997c), and has also been inferred for saline fluids from the Tanganika Lake basement in Africa (Pflumio *et al.*, 1994). Such fluids are characterized by long-lasting water-rock interaction. This is also the case for the Soultz fluids as indicated by the intense water-rock interaction and the ^{36}Cl data.

3.3. GEOTHERMOMETRY

Geothermometric relationships commonly used in geothermal investigations have been examined in the deep geothermal brines of the Rhine Graben (Pauwels *et al.*, 1993). At Soultz, for the fluids located either in the Buntsandstein or in the upper part of the granite (1815-2200 m depth), the cation geothermometers indicate temperatures above 200°C, while the silica and $\delta^{18}O(SO_4)$ data indicate temperatures only slightly higher than the present temperature (140°C). This has been interpreted as evidence that the fluids had equilibrated in a high temperature reservoir and then migrated towards the Soultz horst. They would have slowly cooled down during the migration, allowing silica to re-equilibrate.

Aquilina *et al.* (1997b) also studied the fluids that were sampled at a depth of 3500 m in the granite. Only slight differences in major elements could be observed between the deepest fluid in the granite (3500 m) and the other fluids at shallower depth in granite and in the Buntsandstein. However, clear differences were observed in trace elements and in the organic content: (1) a higher degree of water-rock interaction for the deepest fluid, supported for example by the Sr isotopic measurements, and (2) the lack of organic matter in the deepest fluid, whereas organic materials are present in the shallower Soultz fluids. Another striking difference is the fact that the cation and the silica geothermometers both give a similar value of about 230°C for the deepest fluid.

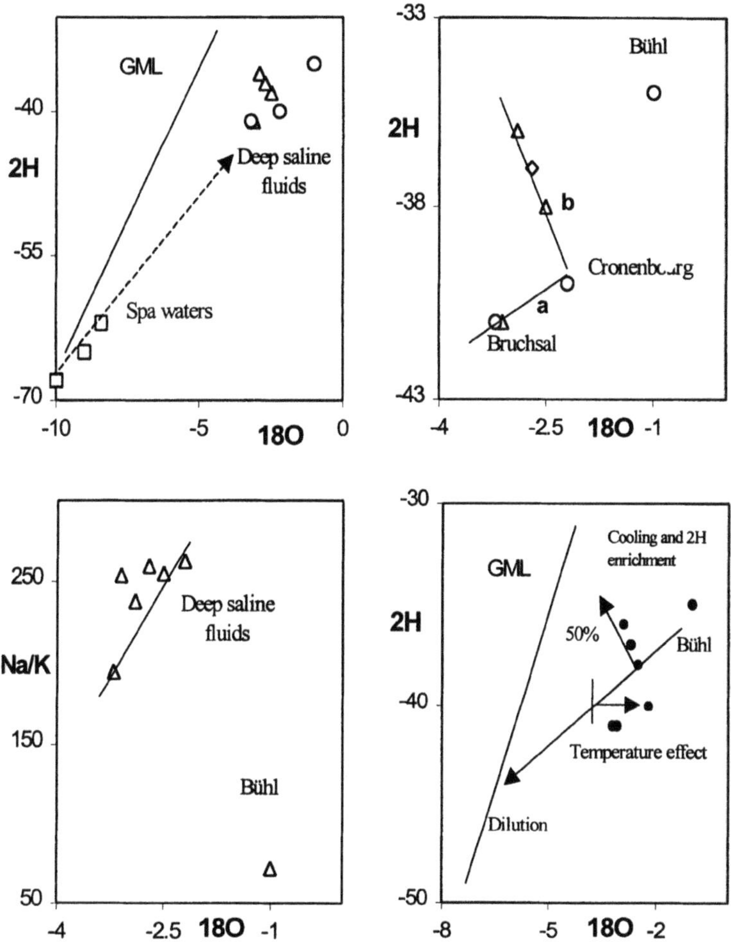

Figure 5 Saline fluids in the Rhine Graben : oxygen and hydrogen isotopic relations
a : δ18O-δD relation for all the fluids ; same symbol as Figure 4
b : δ18O-δD relation for the deep fluids; same symbol as Figure 4 except 4616 : diamond
c : δ18O-Na/K temperature relation ; d : interpretation of the isotopic relations

3.4. STRONTIUM ISOTOPIC RATIO

The Sr isotopic ratio of the Soultz and Bühl fluids is compared to the minerals of the Soultz granite (from EPS1 cores) in a Rb/Sr vs $^{87}Sr/^{86}Sr$ diagram (Fig. 6), revealing two different relationships. One is defined by the Buntsandstein fluids, the other one by the fluids from the granite. Each of them evolves between two end-members, two of them being constituted by minerals from the Soultz granite. The Buntsandstein-fluid correlation leads from the Bühl fluid which has the lowest radiogenic Sr content, *i.e.* the most marine signature, to the composition of biotite. The granite-fluid correlation diverges from this trend along a line correlated to plagioclase.

These trends are interpreted as the dissolution of biotite and plagioclase. This is in good agreement with observations of the trace-element chemistry, especially Rb and Cs which originate from the dissolution of biotite and show a linear increase with the degree of water-rock interaction (Aquilina *et al.*, 1997b). The fact that the plagioclase-Soultz fluids relation intersects the middle of the Bühl-biotite relation is interpreted as a succession in the dissolution processes, biotite preceeding plagioclase. It is thought that this succesion is related to the increase of temperature during burial (Aquilina et al., 1997b).

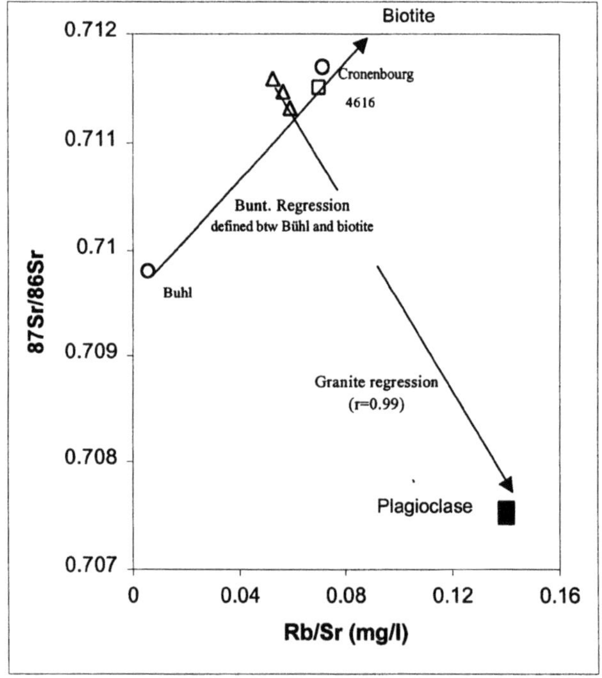

Figure 6 Saline fluids in the graben : $^{87}Sr/^{86}Sr$ vs Rb/Sr
Buntsandstein : circles : Bühl and Cronenbourg, square : 4616
Granite : triangles : EPS-1, GPK-1 KS228 (1815m) and KF3500 (3500m)

3.5. CARBON ISOTOPIC RATIO

The activities of carbon have been measured in the spas of the western part of the graben (Toth and Otto, 1993; Royer *et al.*, 1995). The ages that have been deduced from the ^{14}C concentrations range from several hundred years to 25-30,000 years and increase from west to east. They are correlated to the increase in salinity with depth of the fluids of the Buntsandstein.

3.6. CHLORINE ISOTOPIC RATIO

The chemical and isotopic analyses previously described have been complemented by ^{36}Cl measurements on some of the fluids (Tab. 2). ^{36}Cl concentrations were measured by G. Vourvopoulos (Department of Physics and Astronomy, Western Kentucky University).

^{36}Cl has various origins including cosmic ray production, nuclear tests, subsurface spallation, and deep production induced by neutrons produced by nuclear reactions. In the case of the Soultz saline fluids, the surface contribution can be considered as negligible. The ^{36}Cl evolution law is:

$$RC = R_0C_0\, e^{-\lambda t} + R_{eq}C_0\,(1-e^{-\lambda t}) + R_{eq}(C-C_0) \qquad (1)$$

R : $^{36}Cl/Cl$ ratio of the sample
R_0 : initial $^{36}Cl/Cl$
R_{eq} : equilibrium value (production equilibrated by decay)
C : Cl concentration of the sample
C_0 : initial Cl concentration
λ : radioactive decay constant

If dissolution of Cl occurs in the aquifer, the term $C- C_0$ will become dominant and no age can be estimated from the ^{36}Cl measurement. If the Cl produced in situ is negligible with respect to the initial Cl concentration, due to the high salinity of the fluids, we find:

$$RC = R_0C_0\, e^{-t} + R_{eq}C_0\,(1-e^{-\lambda t}) \qquad (2)$$

If the initial ratio and the equilibrium value can be determined, the residence time can be estimated. Such a method has been applied to saline fluids from the granitic basement of Switzerland (Balderer *et al.*, 1987).

TABLE 2 U, Th content of fresh granite and altered zones, equilibrium values of granite and computation of residence times

Fluid location	Cl (mg/l)	$^{36}Cl/Cl$ 10^{-15}	Petrography	U (ppm)	Th (ppm)	Req 10^{-15}	Residence time (ky)
Bühl	120300	bdl					
GPK1	58500	17	fresh granite	8.3	35.7	82.7	43.5
	id	id	altered zone				
			1815 m	4	20	42.5	96.5
	61000	17	altered zone				
			3500 m	0.7	11	17.5	> 600
EPS1	59820	15	altered zone	5	15	39.7	105
4616	60900	bdl					

bdl : below detection limit ; U, Th conc. from Traineau *et al.*, 1991; Chevremont *et al.*, 1992

The measurements for the Soultz brines yield two different kinds of results. In the Buntsandstein aquifer, the $^{36}Cl/Cl$ ratio is low (below detection limit). In the granite, the three samples (GPK-1 at 1815 and 3500 m depth and EPS-1) give a similar ratio of $17*10^{-15}$. In the Buntsandstein, the very low ratio is probably due to the dissolution of salts in the formation, the ^{36}Cl concentration being extremely low in marine evaporites ($R = 0.02*10^{-15}$). It is thus impossible to estimate residence times.

On the contrary, the fluids which have been sampled in the granite have been kept for a sufficent time without any Cl addition for the ratio to increase. The results indicate that the initial ratio of the fluids when they were introduced in the granite was close to 10^{-15}, as in the Buntsandstein presently. An estimation of the equilibrium value in the Soultz granite is required to compute the residence time in the granite using equation (2).

The equilibrium value can be computed from the neutron production which has to be estimated by the empirical formula of Feige *et al.* (1968, *in* Balderer *et al.*, 1987), since the natural neutron flow is not known :

$$P = p(0.4764\ CU + 1.57\ CU) + 0.7\ CTh \tag{3}$$

P : neutron per kg per year CU, CTh : U and Th conc. (ppm)

In the Soultz granite, the U and Th concentrations which were determined from the physical logs and from chemical analyses are presented in Table 2 (Traineau *et al.*, 1991; Chevremont *et al.*, 1992). The U concentration is 8.3 ppm, the Th concentration is 36 ppm. With such concentrations, the equilibrium of fluid-rock value is $68*10^{-15}$. However, the neutron production which may influence the fluids circulating in the fractured zones is restricted to the few centimeters around the fluid pathways. The fluids circulate in the central part of the altered zones, characterized by quartz veins with open channels. In the three fracture zones where fluids have been collected, U and Th concentrations are much lower than in fresh granite (Tab. 2). With these concentrations, the equilibrium values are also lower (Tab. 2).

The residence times which are computed from the different equilibrium values are also presented in Table 2. When the U and Th concentrations of fresh granite are used in equation (3), the residence time is 43,500 years. Concentrations of the upper part of the granite provide similar residence times of 96,500 and 105,500 years. For the lower part of the granite, the measured $^{36}Cl/Cl$ ratio is in equilibrium with the host rock of the fracture at 3500 m. In that case, the fluid is likely to have been present in the granite for at least one million years.

4. Discussion

The thermal anomaly of the Rhine Graben has been investigated by several authors using simulations by 2D numerical models of coupled heat- and fluid-flow (Clauser and Villinger, 1990; Person and Garven, 1992; Flores, 1992). In these models, the system is recharged in the Black Forest, along the fractures which limit the border of the graben (Fig. 7). Meteoric water infiltrates down to the bottom part of the graben where the sedimentary cover reaches a thickness of 5 km. The fluid flows from the deeper eastern part of the graben towards the Soultz horst where the basement is closest to the surface, mainly along the Triassic and Jurassic geological formations which have a higher permeability than the overlying Tertiary formations or the underlying granite basement. The western border faults of the graben or the faults limiting the Soultz horst are supposed to allow the outflow of the system. The velocities implied by these models are in the order of centimeters per year to meters per year.

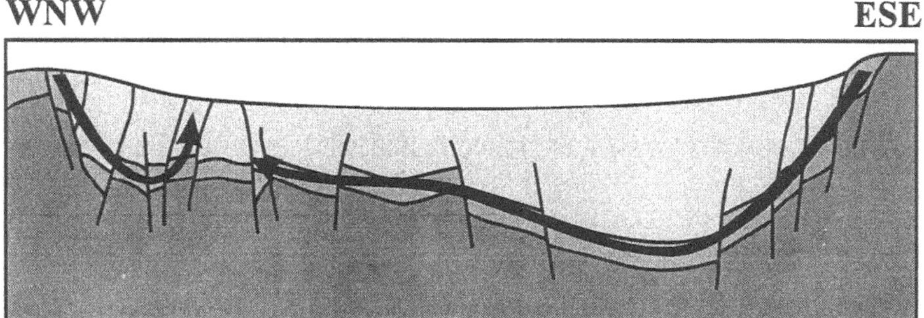

Figure 7 Flow scheme according to previous numerical models of coupled heat- and fluid-flow (Clauser and Villinger, 1990; Person and Garven, 1992; Flores, 1992)

Several aspects of the present-fluid chemical analyses yield hydrogeological constraints which can be compared to the hydrological models. These aspects are presented in the next sections and summarized in Table 3.

TABLE 3 Major chemical characteristics of the fluids sampled in the Rhine Graben

	western spas (I)	**Soultz and Cronenbourg (II)**		**Bruchsal, Bühl (III)**
TDS	low salinities (4 - 20 g/l)	middle range (94 - 106 g/l)		high salinities (120 - 200 g/l)
residence time	^{14}C residence time < 20 ky, increase from W to E	^{36}Cl residence time > 100 ky		
Dilution effect	recent dilution of a saline (80 g/l) end-member with meteoric waters	dilution of a saline end-member like Bühl with palaeometeoric waters		No dilution for Bühl fluids which result from palaeodissolution of NaCl and residual evaporated sea-water
		1815-2200 m	3500 m	
WRI processes	saline end-member exhibits the same WRI signature as the Soultz fluids	increase of Rb, Cs and ^{87}Sr through biot. and plagio. dissolution, organic matter	increase of Rb, Cs and ^{87}Sr through biot. and plagio. dissolution, no organic matter	low Rb, Cs and ^{87}Sr no biot. and plagio. dissolution, sedimentary signature
		1815-2200 m	3500 m	
WRI temperature	50 -250	150 - 260	230	70 - 120

WRI : Water-Rock-Interaction

4.1. CHEMICAL AND STABLE ISOTOPIC COMPOSITION OF THE FLUIDS

4.1.1. Shallow thermal spas (I)

The stable isotopic signature of the water sampled on the western border of the graben defines a mixing line which goes from the Global Meteoric Line towards an end-member corresponding to the deep saline fluids involved in the mixing process. The chemical composition of this saline end-member can be estimated by correction for dilution, using the isotopic composition of the water to estimate the degree of dilution. The calculation indicates that this end-member has a salinity (about 80 g/l) and a chemical composition in the range of the Soultz fluids. It also has a high Sr isotopic ratio, and Rb and Cs contents indicating that it is highly similar to the Soultz brines.

4.1.2. Deep saline brines (II and III)

The composition of the deep geothermal brines indicates that a large-scale dilution of a Bühl-type brine with palaeometeoric waters is more likely than a process of spatially heterogeneous dissolution of NaCl in the Buntsandstein formation by palaeometeoric waters. Independently of the processes which both imply meteoric fluid circulation, the chemical composition of all the samples appears rather homogeneous and only slightly modified by water-rock interactions. Although they have different salinities, they have rather similar element-to-Cl ratios. This is true for the eastern samples, the Soultz fluids and the brines contained in the fluids from the spas from the western border of the graben (after correction for dilution). At the scale of the graben, only one chemical family can be identified for these brines, particularly when one considers the very contrasting compositions encountered in other sedimentary basins. As stated in the previous section, these brines are present in the whole Buntsandstein formation, which indicates that the palaeometeoric fluid circulation was a large-scale process.

It can be concluded from the chemical and stable isotopic analyses that (1) deep saline fluids of the same family are present in the Buntsandstein sandstone in the whole northern Rhine Graben. They have salinities ranging from 80 to 207 g/l and most of them show high water-rock interaction signatures. (2) These deep fluids result from the paleo-dilution of a "primary" sedimentary brine (Bühl-type) at the scale of the graben. (3) In the shallow western part of the graben, these saline fluids have been recently diluted by meteoric waters.

4.2. CARBON ISOTOPIC COMPOSITION

Since ^{14}C datations are available for the thermal spas on the western side of the Kutzenhausen fault, the hydrogeological regime there can be clearly defined. The distribution of the residence times indicates that recharge takes place in the outcrops of the Buntsandstein sandstones. Meteoric water flows through this formation towards the east at least until the Kutzenhausen fault is reached. Part of the water returns to the surface, probably along faults, as several hot springs seep in this area, or by slow drainage through the Oligocene sedimentary filling of the graben. This dilution process is recent with respect to the paleo-dilution process that affected the whole Buntsandstein. The fluid velocity which can be defined from the ^{14}C data is approximately 1 meter per year. This value is of the same order of magnitude as the velocities obtained by numerical modelling. Chemical data from the eastern boundary of the graben suggest that a similar process is highly likely (Kanglin *et al.*, 1995).

4.3. TRACE ELEMENTS AND STRONTIUM ISOTOPIC COMPOSITION

Trace-element and strontium isotope analyses indicate that water-rock interaction has taken place between the granite and the saline end-member. The geochemical signatures correspond to interaction with biotite and plagioclase minerals at temperatures above 150°C (Aquilina *et al.*, 1997b). This strongly suggests that the large-scale paleo-dilution process had occurred before hot water-rock interaction. Thus, the most likely hypothesis of evolution from a geological point of view is that an invasion of meteoric waters, at the scale of the graben, occurred during the Cretaceous-Eocene continental period. After this time, the Oligocene extension faulting led to rapid subsidence and burial under 1500 m of sediments. The tectonic structuration isolated several blocks and allowed an independant evolution without any connection to the surface, the Tertiary deposits sealing the Triassic-granite system. In the central part of the graben and due to deep burial, the temperature increase created the conditions for the equilibration of the waters with the host rock. On the borders of the graben, on the contrary, the fluids have remained relatively unchanged.

4.4. GEOTHERMOMETRY

Analysis of trace elements shows a clear difference between the deeper fluids encountered at 3500 m at Soultz and the upper fluids collected from the top part of the granite basement. The deeper fluid shows (1) a higher degree of water-rock interaction (for example from the Sr isotopic measurements) and (2) the lack of organic matter. Organic materials originate from the sedimentary cover and are a signature of the sedimentary origin of the upper fluids. These observations indicate that the upper and lower part of the granite basement investigated at Soultz do not belong to the same circulation system.

The geothermometers computed for the fluids also indicate a strong difference between the fluids at 1815-2200 m and those at 3500 m. For the deeper fluids at 3500 m, the geothermometers are in better agreement and yield values around 230°C. This can be taken as evidence that a hot geothermal reservoir is present at depth and that the fracture zone allowing fluid flow at 3500 m is directly connected to this reservoir. The independence of the upper and lower zones can be explained by the existence of a zone without fractures of large aperture, between 2800 and 3200 m, which would separate two independent fracture systems in the granite (Genter *et al.*, 1995).

Both the geothermometers and the trace element contents indicate that the upper and the deeper fluids do not belong to the same circulation pathway. The 1815-2200 m fracture zones (in wells GPK-1 and EPS-1) are linked to the Buntsandstein aquifer as demonstrated by hydraulic tests (Baria *et al.*, 1995), which is in good agreement with their common geothermometer pattern. We have yet to identify a link to a high temperature reservoir for the 3500 m fracture zone. Our hypothesis is that such a link can occur in a convection cell within the granite of the Soultz horst. This could allow a short distance from the hot deeper part of the granite to the location of sampling at 3500 m. The deeper part of the granite, due to circulation in the fracture zones could act as a hot reservoir for the fluids. A geothermal gradient of 6°C/km (as determined in the bottom part of well GPK-1) implies a depth of 5 km for a temperature of 250°C. Thus, the bottom of the well (at 3.9 km) is not very far from this hypothetic reservoir and drilling to a depth of 5 km as planned for the period 1998-1999 could give interesting results.

4.5. COMPARISON OF THE FLUID INCLUSION DATA WITH THE HYDROGEOLOGICAL MODELS

The analyses of fluid inclusions and present day fluids both indicate that a large-scale dilution process has occurred during the early stage of the Rhine Graben formation. Although the main trend observed in the fluid inclusions is temperatures and salinities in good agreement with the present fluids, inclusions with a wide range of salinities and temperature have also been observed. The present geothermal fluids, on the contrary, have rather homogeneous compositions and their salinities are not lower than 80 g/l (taking out the recent dilution phenomena on the western border of the graben). The discrepancy between the two sets can easily be explained if we assume that some of the fluid inclusions record the begining of the fluid circulation. The intrusion of meteoric waters created local inhomogeneities which have had time to disappear before injection of the fluids in the granite and interaction with the host rocks.

According to the previous discussion, dilution might be the mechanism which allowed the fluids to flow within the graben, but then this large-scale fluid circulation stopped and the fluids slowly interacted with their host rock to produce the different signatures observed. This means that fluid circulation in the Rhine Graben cannot be explained by a large-scale fluid transfer from the bottom of the graben to the Soultz horst with recharge in the Black-Forest, as suggested by most numerical models (Clauser and Villinger, 1990; Person and Garven, 1992; Flores, 1992). In these models, reproducing the observed geothermal gradient and temperature field requires flow velocities on the order of 0.1 to 1 meter per year. The models thus imply that the regional circulation loop, from the Black-Forest to the Soultz site or to the western border of the graben, should have a duration of less than 50 thousand years. These ages are not in agreement with the ^{36}Cl ages and the fast flow rate does not agree with the observed high water-rock interaction degree. Comparison with the fluid evolution in the Paris Basin supports this interpretation. A dilution process occurrs in the Triassic rocks since Eocene period and the effect is clearly observed in the evolution of the salinity from the outcrops close to the Vosges to the center of the basin (Matray and Chery, 1997). Nothing comparable can be observed in the Rhine Graben.

However, the large thermal gradient in the graben and particularly at the Soultz site is due to fluid circulation, as demonstrated by the correlation between gradient increase and fracture zones in the granite (Aquilina et al., 1997b). Temperature measurements from oil wells reaching the top of the Muschelkalk (just above the Buntsandstein aquifer) in the Soultz area have shown that local high temperature anomalies are aligned along the major faults, which is interpreted as an evidence of channeling of fluids in the faults and of the existence of thermal convection cells (Benderitter et al., 1995 ; Benderitter and Elsass, 1995). A succession of this type of convection cells in fault zones linking the granite basement to the Buntsandstein aquifer could allow heat transfer from the deeper part of the graben to the Soultz horst structure, without fresh water input from the Black Forest. The Buntsandstein aquifer would thus mainly act as a captive aquifer.

Although this process allows transfer of some heat to the Soultz horst, preliminary modelling (Pribnow and Clauser, 1998) indicates that it does not seem to be a sufficient driving force to reproduce the temperature field. A pervasive flow of fluid at a depth of more than 3 to 5 km in the granite could carry huge amounts of heat without necessitating large amounts of fluids. This fluid flow would be limited upwards by the convection level integrating the Buntsandstein and the upper part of the granite. The fluids would be channeled towards the Soultz horst where they would slowly seep through the Oligocene cover. Such a process would allow heat transfer without a large chemical impact on the saline fluids of the Buntsandstein aquifer.

4.6. COMPARISON WITH THE PETROGRAPHIC DATA

Several petrographic studies have been carried out on the Soultz granite (Genter, 1989; Traineau *et al.*, 1991; Ledésert, 1993). Four phases of hydrothermal alteration have been distinguished: (1) dissolution of biotite and plagioclase and precipitation of illite, (2) formation of tosudite (interstratified illite/smectite), associated with organic matter; (3) precipitation of illite and carbonates; and (4) precipitation of quartz in the inner part of the altered zone. Most of the reactions occur without transport but rather as isochemical precipitation (Giggenbach, 1984). The first phase is clearly the main phase of alteration of the granite. The relatively simple relations which can be observed in Figure 6 between the minerals and the fluids are related to this general alteration of biotite and plagioclase. The precipitation of the other minerals did not disturb the relations induced by the dissolution, as Sr isotopes do not fractionate in such processes.

These petrographic alteration phases can only be assigned a relative chronology. However, it is clear that the high water-rock interaction degree of the fluids is also related to the major alteration process of the fractured zones in granite. Such a process is likely to progress during a large part of the graben evolution, and has probably started in Oligocene times with the development of the extension-faulting. The petrographic observations are thus in good agreement with our hypothesis which relates the creation of the deep brines to a large-scale invasion of meteoric water during the Cretaceous-Eocene period, and their water-rock interaction signature to the rapid burial of the Buntsandstein during the Oligocene.

Figure 8 Interpretation of the fluid circulation in the Rhine Graben

Time	Geology and Hydrology	Structural evolution	Petrography	Fluids	Fluid inclusion
Lower Cretaceous	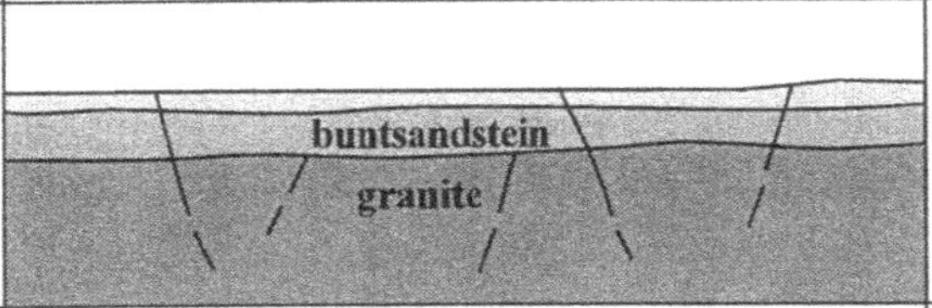 buntsandstein granite	Shallow submarine conditions	precipitation of illite in Buntsandstein	Residual Bühl-like sedimentary brines	?
Cretaceous-Eocene	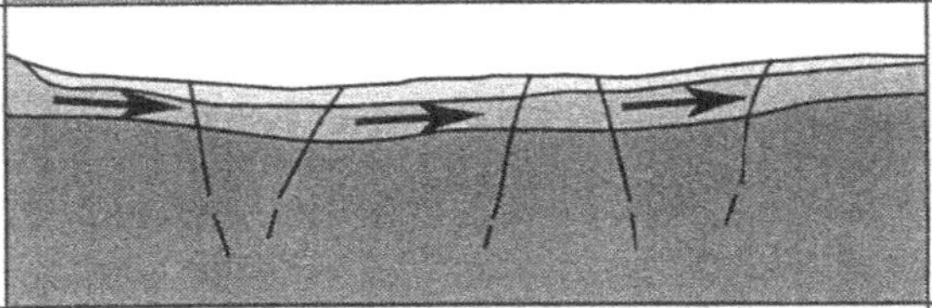	Emertion, rift shoulders uplift (Pyrenean compression)	?	Invasion by meteoric fluids from the west to the east	High salinity, low T ?
Oligocene	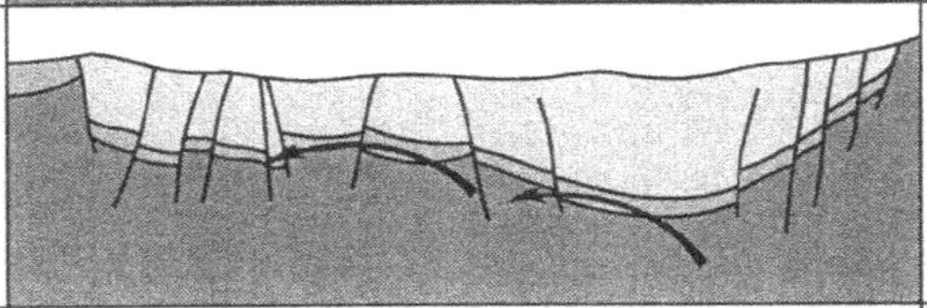	Fracturation and subsidence of the rift (alpine extention)	dissolution of quartz, biotite and plagioclase precipitation of carbonate and tossudite (organic matter)	Mixing processes circulation in the granite, water-rock interaction	large range of salinity >200°C
present	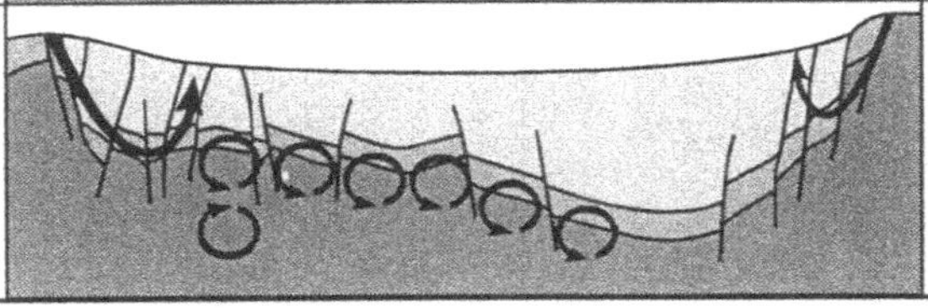	Subsidence and burial	quartz precipitation	Short circulation loops of meteoric fluids at the edges of the graben, deep circulation of saline fluids	homogeneous salinities 130 - 180°C

4.7. A MODEL OF FLUID CIRCULATION

The chemical and isotopic analyses of geothermal fluids collected in the Upper Rhine Graben from deep wells and surface springs, compared to the petrographic and fluid inclusion data, allow us to reconstruct the following phases of evolution of the fluids which are summarized in Figure 8 :
(1) Invasion of meteoric waters into the Buntsandstein aquifer during the Cretaceous-Eocene continental period; mixing of fresh water with the existing sedimentary brines.
(2) Faulting of the graben at the beginning of the Oligocene; intrusion of fluids into the faults of the granite basement.
(3) Rapid burial of the Triassic formations and sealing by the mostly impervious Oligocene deposits. Water-rock interaction in the sedimentary aquifer and granite: dissolution of biotite.
(4) Continuation of burial and increase of temperature in the Triassic formations. Water-rock interaction: dissolution of biotite and plagioclase.

In our model, present-day fluid-circulation in the graben is dominated by three different systems (Fig. 8):
(1) On the western border of the graben, there is mixing of the evolved saline fluids with surface waters infiltrated in the Vosges. The mixing process is clearly identifiable on the western faulted border and fluid velocities are on the order of 1 meter per year. No direct evidence of mixing has been found on the eastern border, close to the Black-Forest, however several springs with low salinities (Baden-Baden) are well known and indicate that short and shallow (less than 2 - 3 km) circulation circuits are also present there.
(2) Large-scale exchange of saline fluids occurs between the Buntsandstein and the upper part of the granitic section in a succession of convection cells. Fluids circulate in the fractured zones which affect both sedimentary cover and granite and link the upper part of the granite to the Buntsandstein reservoir. This circulation results in the observed temperature field at Soultz-sous-Forêts. The differences between the fluids at 1815-2200 m and the fluids at 3500 m (in particular the geothermometry which indicates the proximity of a hot reservoir) suggest that these convection cells might be relatively complex with part of them located entirely within the granite between (at least) 5 to 3 km depth.
(3) A pervasive transfer of fluids through the granitic medium at depths of more than 3 to 5 km is also suggested. Permeabilities of the granite (in the order of 10-16 m/s) only allow slow seepage of these fluids but a large transfer of heat. This transfer could explain the temperature field. The limited inflow of meteoric water on the Black-Forest border of the graben would not noticeably modify the composition of the saline fluids in the Buntsandstein aquifer which would remain a relatively closed system, as indicated by the present-day fluids and fluid inclusions. Further modelling will test the evolution of such a process (Pribnow and Clauser, 1998).

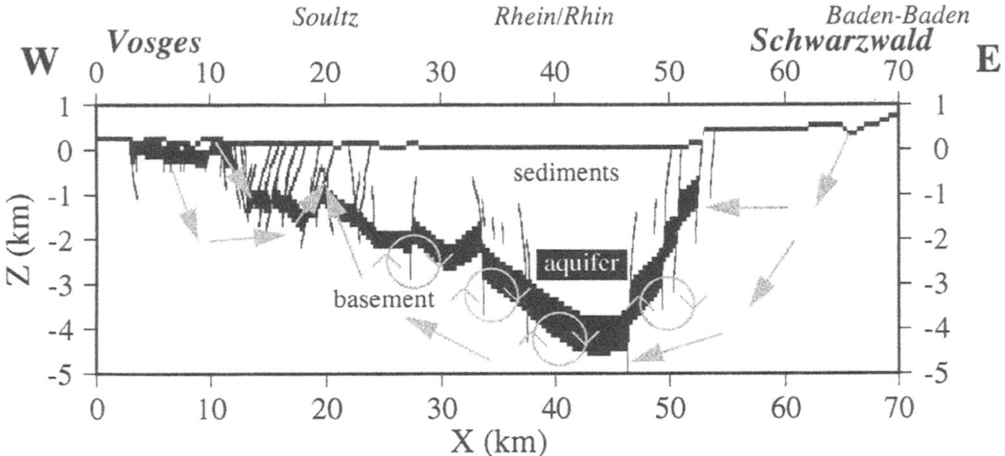

Figure 9 Flow scheme according to recent numerical models of coupled heat- and fluid-flow

5. Conclusion

The first phase of the European Hot Dry Rock project has led to intensive research that has improved the understanding of natural fluid circulations in the Rhine Graben, especially at the Soultz site. The chemical characteristics of the geothermal fluids which have been sampled in the Rhine Graben suggest that (1) these fluids have evolved over a very long geological period, probably since the burial of the graben, and (2) that the reservoir formations have remained mostly preserved from surface water intrusion. The history of the fluids is therefore best explained by a dilution process by meteoric waters which took place under continental conditions at the beginning of the evolution of the graben. After the burial in Oligocene times under a thick pile of impervious sediments, no further meteoric water intrusion is likely to have occurred. Such a process is at the scale of the geological history of the graben and is a good example of what Bethke and Marshak (1990) have described as the "plate tectonics of groundwater".

Present-day infiltration of meteoric water and mixing with brines is observed only on the western edge of the graben, with apparent fluid velocities on the order of 1 meter per year. The fluids located east of the Kutzenhausen fault, either in the Buntsandstein or in the granite, show evidence of intense water-rock interaction (biotite and plagioclase dissolution, temperature equilibration) and ^{36}Cl enrichment. Both these processes indicate time scales of millions of years. It therefore appears that the Buntsandstein sandstone reservoir behaves as a captive aquifer where the fluids remain trapped. Meteoric recharge is mostly recycled to the surface through short circulation loops on the borders of the graben and no evidence of large-scale fluid transfer as suggested by the hydrogeological models of Person and Garven (1992) or Clauser and Villinger (1990) is observed.

Preliminary results of on-going modelling reproduce the flow scheme suggested by the chemical data by allowing flow in the granitic basement (Fig. 9). Convection cells in the deeper part of the graben can develop (i.e. the critical Raleigh number is exceeded) because the viscosity of the fluid decreases with increasing temperature. These convection cells act as a resistance to meteoric water that has penetrated deeper into the granitic basement trying to reach the Buntsandstein aquifer. These fluids are forced to slowly flow below the deepest sediment cover and seep westwards up to the Soultz horst, carrying absorbed heat from the deeper to the shallower parts. The flow field shown in Figure 9 is not only in agreement with the chemical data but also results in a temperature field close to the observed one. Details on these new modelling results will be published soon in a separate paper.

The chemical analysis of the present-day fluids cannot explain the whole complex evolution of the system. It does provide some constraints on fluid evolution in relation with the history of the graben and the hydrothermal alteration of the granite. The model we suggest is a framework which has to be discussed from a hydrogeological point of view and tested by modelling. On-going research on the Soultz site will provide more data and help us to refine our views on fluid circulation in the Rhine Graben.

Acknowledgements

We are thankful to the Socomine team: R. Baria, J. Baumgartner, A. Gérard, F. Kieffer and C. Kléber for site facilities. Juliane Herrmann helped with some figures. This research is part of the European HDR Project funded by BMBF, ADEME, BRGM, and supported by CEC DGXII. Tony Hoch and Ingrid Stober are thanked for reviews and editing.

6. REFERENCES

Aquilina, L., Sureau, J.F., Steinberg, M. and the GPF team (1997a) Comparison of surface-, aquifer-, and pore waters from a Mesozoic sedimentary basin and its underlying Paleozoic basement, southeastern France : chemical evolution of waters with diagenesis and relationship between aquifers, *Chemical Geology* **138**, 185-209.

Aquilina, L., Pauwels, H. and Fouillac, C. (1997b) Water-rock interaction processes in the Triassic sandstone and the granitic basement of the Rhine Graben: geochemical investigation of a geothermal reservoir, *Geochim. et Cosmochim. Acta* **61**, 4281-4295.

Aquilina, L. (1997c) Les circulations de fluides actuelles dans la croûte continentale supérieure (0.5-15 km). *BRGM Report* R39497, 66 p.

Aquilina, L. and Brach, M. (1995) Characterization of Soultz hydrochemical system: WELCOM (Well Chemical On-line Monitoring) applied to deepening of GPK-1 borehole, *Geotherm. Sci. and Tech.* **4**, 239-251.

Ayt Ougougdal, M., Cathelineau, M., Pironon, J., Boiron, M.C., Banks, D. and Yardley, B. (1995) Diagenetic salt-rich and organic-rich fluid migration in the Rhine Graben Triassic sandstones (Soultz deep drilling), European Union of Geosciences, April 9-13 Strasbourg, *Terra Abstracts* **7**, 198.

Balderer, W., Fontes, J.C., Michelot, J.L. and Elmore, D. (1987) Isotopic investigations of the water-rock system in the deep crystalline rock of northern Switzerland, *in:* Saline Waters and gases in crystalline rocks (P. Fritz and S.K. Frape Eds.) *Geol. Assoc. Canada* 33, 175-195.

Baria, R., Garnish, J., Baumgartner, J., Gérard, A., Jung, R. (1995) Recent developments in the European HDR research programme at Soultz-sous-Forêts, *World Geothermal Congress*, May 18-31, Florence, Italy.

Brun, J.P., Gutsher, M.A. and Dekorp-Ecors teams (1992) Deep crustal stucture of the Rhine Graben from Dekorp-Ecors seismic reflection data: a summary, *Tectonophysics* **208**, 139-147.

Benderitter, Y., Tabbagh, A., Elsass, P. (1995) Calcul de l'effet thermique d'une remontée hydrothermale dans un socle fracturé. Application à l'anomalie géothermique de Soultz-sous-Forêts, *Bull. Soc. Géol. France*, **166**, 1, 37-48.

Benderitter, Y. and Elsass, P. (1995) Stuctural control of deep fluid circulation at the Soultz HDR site, France: a review, *Geotherm. Sci. and Tech.* **4**, 227-237.

Bethke, C.M. and Marshak, S. (1990) Brine migration across North America – The plate tectonics of groundwater, *An. Rev. of Earth Plan. Sci.* **18**, 287-316.

Boulègue, J., Benedetti, M., Gauthier, B. and Bosch, B. (1990) Les fluides dans le socle du sondage GPF Sancerre-Couy, *Bull. Soc. Geol. France* **5**, 789-796.

Carpenter, A.B. (1978) Origin and chemical evolution of brines in sedimentary basins, *Oklahoma Geological Survey circular* **79**, 60-77.

Chevremont, Ph., Thieblemont, D., Laforet, C., Genter, A. and Traineau, H. (1992) Etude pétrologique du massif granitique recoupé par le forage EPS1 (Soultz-sous-Forêts), *BRGM Report* RCS 92 T15 SGN/IRG.

Clauser, C. and Villinger, H. (1990) Analysis of conductive and convective heat transfer in a sedimentary basin, demonstration for the Rheingraben, *Geophys. J. Int.* **100**, 393-414.

Collins, A.G. (1975) Geochemistry of Oilfield brines, Elsevier Amsterdam, 496 p.

Dubois, M., Royer, J.J., Zimmermann, J.L. and Cheilletz, A. (1994) Paléothermicité et évolution de la composition des fluides hydrothermaux au cours du temps dans le granite de Soultz : étude des inclusions fluides (Graben du Rhin, Alsace), *Réunion des Sciences de la Terre*, Nancy, France.

Dubois M., Ayt Ougoudal M., Meere P., Royer J.J., Boiron M.C. and Cathelinau M. (1996) Temperature of paleo- to modern self-sealing within a continental rift basin : the fluid inclusion data (Soultz-sous-Forêts, Rhine Graben, France). *Eur. J. Mineral.* **8**, 1065-1080.

Edmunds, W.M., Kay, R.L. and McCartney, R.A. (1985) Origin of saline groundwaters in the Carnmenellis granite (Cornwall, England): natural processes and reaction during hot dry rock reservoir circulation, *Chemical Geology* **49**, 287-301.

Flores, E.L. (1992) Transferts de chaleur et de masse en milieu sédimentaire et fracturé. Modélisation numérique de la convection naturelle autour du site de Soultz (Graben du Rhin), Ph D Thesis Univ. Nancy, 230 p.

Fontes, J.C. and Matray, J.M. (1993) Geochemistry and origin of formation brines from the Paris Basin, France 1. Brines associated with Triassic salts, *Chemical Geology* **109**, 149-175.

Frape, S.N., Fritz, P. and McNutt, R.H. (1984) Water-rock interaction and chemistry of groundwaters from the Canadian shield, *Geochim. et Cosmochim Acta* **48**, 1617-1627.

Fritz, B. (1981) Etude thermodynamique et modélisation des réactions hydrothermales et diagénétiques, *Sci. Géol. Mém.* 65.

Fritz, B. (1980) Analyse des eaux du forage géothermique de Cronenbourg, *Rapport au projet de géothermie Cronenbourg.*

Fritz, J., Eberwein, P., Hackl, S., Hornberger, R.and Schaumburg, D. (1989) Geothermal project Bruchsal Phases 5 and 6, circulation and reinjection tests - Final report, Contract EEC n° GE 265/85-DE.

Genter, A. (1989) Géothermie Roches Chaudes Sèches, le granite de Soultz-sous-Forêts (Bas Rhin, France), Ph D Thesis, Univ. Orléans, 201 p.

Genter, A., Martin, P. and Montaggioni, P. (1991) Application of FMS and BHTV tools for evaluation of natural fractures in the Soultz geothermal borehole GPK1, *Geotherm. Sci. and Tech.* **3**, 189-214

Genter, A. and Traineau, H. (1992) Borehole EPS-1, Alsace, France: preliminary geological results from granite core analysis for Hot Dry Rock research, *Scientific Drilling.* **3**, 205-214.

Genter, A., Traineau, H., Dezayes, C., Elsass, P., Ledésert, B., Meunier, A. and Villemin, T. (1995) Fracture analysis and reservoir characterization of the granitic basement in the HDR Soultz project (France), *Geotherm. Sci. and Tech.* **4**, 189-214.

Gérard, A., Menjoz, A., Schwoerer, P. (1984) L'anomalie thermique de Soultz-sous-Forêts, *Géotherm. Actualités* **3**, 35-42.

Giggenbach, W.F. (1984) Mass transfer in hydrothermal alteration systems - A conceptual approach, *Geochim. Cosmochim. Acta* **48**, 2693-2711.

Graham, C.M. (1981) Experimental hydrogen isotope studies, III. Diffusion of hydrogen in hydrous minerals and stable isotope exchange in metamorphic rocks, *Contrib. Mineral. Petrol.* **76**, 216-228.

Hanor, J.S. (1994) Physical and chemical controls on the composition of waters in sedimentary basins, *Mar. and Petrol. Geol.* **11**, 31-46.

Kanglin, H.E., Stober, I. and Bucher, K. (1995) Hydrochemical characterization of thermal waters from the upper Rhine rift valley, European Union of Geosciences, April 9-13 Strasbourg, *Terra Abstracts* **7**, 204.

Kappelmeyer, O. and Gérard, A. (1989) The European project at Soultz-sous-Forêts. Proc. IV Int. Seminar on the results of EC geothermal energy research and demonstration, April 27-30, Florence, Kluwer Academic Publishers, pp. 283-334.

Kappelmeyer, O., Gérard, A., Schloemer, W., Ferrandes, R., Rummel, F. and Benderitter, Y. (1991) European HDR project at Soultz-sous-Forêts : General presentation, *Geotherm. Sci. and Tech.* **2**, 263-289.

Kharaka, Y.K. and Carothers, W.W. (1986) Oxygen and Hydrogen isotope geochemistry of deep basin brines, *in:* Handbook of Environmental Isotope Geochemistry (P. Fritz and J.C. Fontes Eds.) **2**, 305-360.

Knauth, L.P. and Beeunas, M.A. (1986) Isotope geochemistry of fluid inclusions in Permian halite with implications for the isotopic history of ocean water and the origin of saline formation waters, *Geochim. et Cosmochim. Acta* **50**, 419-433.

Ledésert, B. (1993) Fracturation et paléocirculations hydrothermales. Application au granite de Soultz-sous-Forêts, Ph D Thesis, Univ. Poitiers.

Ledésert, B., Dubois, J., Genter, A., and Meunier, A. (1993) Fractal analysis of fracture applied to Soultz-sous-Forêts Hot Dry Rock geothermal program, *J. Volcanol. Geotherm. Res.* **57**, 1-17.

Matray, J.M. and Fontes, J.C. (1990) Origin of the oil-field brines in the Paris basin, *Geology* **18**, 501-504.

Matray, J.M., Coleman, M.L. and Eggenkamp, H.G.M. (1993) Origin of the Keuper formation waters of the Paris Basin, *Applied Geochem.* **3**, 119-127.

Matray, J.M., Lambert, M. and Fontes, J.C. (1994) Stable isotope conservation and origin of saline waters from the Middle Jurassic aquifer of the Paris basin, *Applied Geochem.* **9**, 297-309.

Matray, J.M. and Chery L. (1997) Origine et âge des eaux profondes du bassin de Paris par l'utilisation des traceurs chimiques et isotopiques, *in* Hydrology and isotope geochemistry, Proc. Int. Symp. In memory of J.C. Fontes. ORSTOM Editions, Paris ; p. 117-136.

Meere, P.A., Cathelineau, M., Dubois, M. Ayt Ougougdal, M. and Royer, J.J. (1995) Are quartz veins forming under Strasbourg today ? A fluid inclusion study, European Union of Geosciences, April 9-13, Strasbourg, *Terra Abstracts* **7**, 185.

Michard, G., Pearson, J.R. and Gautschi, A (1996) Chemical evolution of waters during long term interaction with granitic rocks in northern Switzeland, *Applied Geochem.* **11**, 757-774.

Nordstrom, K.D., Ball, J.W., Donahoe, R.J. and Whittemore, D. (1989) Groundwater chemistry and water-rock interactions at Stripa, *Geochim. et Cosmochim. Acta* **53**, 1727-1740.

Nurmi, P., Kukkonen, I. and Lahermo, P. (1988) Geochemistry and origin of saline groundwaters in the Fennoscandian Shield, *Applied Geochem.* **3**, 185-203.

O'Neil, J.R. and Kharaka, Y.K. (1976) Hydrogen and oxygen isotope exchange reactions between clay minerals and water, *Geochim. et Cosmochim. Acta* **40**, pp. 241-246.

Paces, T. (1987) Hydrochemical evolution of saline waters from crystalline rocks of the Bohemian massif, *in:* Saline Waters and gases in crystalline rocks (P. Fritz and S.K. Frape Eds.) *Geol. Assoc. Canada* **33**, 145-156.

Pauwels, H., Criaud, A., Vuataz, F.D., Brach, M., Fouillac, C. (1991) Uses of chemical tracers in HDR reservoir studies. Example of Soultz-sous-Forêts (Alsace, France), *Geotherm. Sci. and Tech.* **3**, 83-103.

Pauwels, H., Fouillac, C., Criaud, A. (1992) Water-rock interactions during experiments within geothermal Hot Dry Rock borehole GPK-1, Soultz-sous-Forêts, Alsace, France, *Applied Geochem.* **7**, 243-255.

Pauwels, H., Fouillac, C., Fouillac, A.M. (1993) Chemistry and isotopes of deep geothermal saline fluids in the Upper Rhine Graben: origin of compounds and water-rock interactions, *Geochim. et Cosmochim. Acta.* **57**, 2737-2749.

Papillon, E. (1995) Traitements et interprétations des cartes d'anomalies magnétiques et gravimétriques du Fossé rhénan supérieur, Diplôme d'ingénieur de l'Ecole de Physique du Globe, Strasbourg, 95 p.

Pearson, F.J., Balderer, W., Loosli, H., Lehmann, B., Matter, A., Peters, T., Schmassmann, H. and Gautschi, A. (1991) Appplied isotope hydrogeology, A case study in northern Switzerland, *Studies in Environmental Sci.* **43**, Elsevier.

Person, M. and Garven, G. (1992) Hydrologic constraints on petroleum generation within continental rift basins: theory and applicaion to the Rhine Graben, *Am. Assoc. Petrol. Geol. Bull.* **76-4**, 468-488.

Pflumio, C., Boulègue, J. and Tiercelin, J.J. (1994) Hydrothermal activity in the Northern Tanganyka rift, east Africa, *Chemical Geology* **116**, 85-109.

Pribnow, D. and Clauser, C. (1998) Heat- and fluid-flow in the Rhine Graben : regional and local models for a hot-dry-rock system, *Proc. 4th Hot Dry Rock Forum*, Strasbourg, Sept. 28-30, 1998.

Royer, J.J., Le Carlier de Veslud, C. and Gérard, B. (1995) Convective heat and mass transfer around the geothermal site at Soultz-sous-Forêts (Rhine Graben, France), European Union of Geosciences, April 9-13, Strasbourg, *Terra Abstracts* **7**, 194.

Schmassmann, H., Kullin, M. and Schneemann, K. (1992) Hydrochemische Synthese Nordschweiz: Buntsandstein-, Perm- und Kristallin-aquifere, NAGRA Tech. Rep. 91-30.

Tardy, Y. (1980) Rapports sur les géothermomètres chimiques en terrains granitiques et sédimentaires (Plombières et Alsace). Action indirecte dans le domaine de l'énergie, Contract EEC n° 629-78-67-EGF.

Toth, J. and Otto, C. (1993) Hydrogeology and oil deposits at Pechelbronn-Soultz-Upper Rhine Graben, *Acta Geol. Hungarica* **36-4**, 375-393.

Traineau, H., Genter A., Cautru, J.P., Fabriol, H., Chevremont, P. (1991) Petrography of the granite massif from drill cutting analyses and well log interpretation in the geothermal HDR borehole GPK-1 (Soultz, Alsace, France), *Geotherm. Sci. and Tech.* **3**, 1-29.

Villemin, T. (1986) Tectonique en extension, fracturation et subsidence: le Fossé rhénan et le bassin Sarre-Nahe, Ph D thesis, Univ. P. & M. Curie.

Villemin, T. and Bergerat, F. (1987) L'évolution du Fossé rhénan au cours du Cénozoique : un bilan de la déformation et des effets thermiques de l'extension, *Bull. Soc. géol. France* **8**, t III-2, 245-255.

Vovk, I.F. (1987) Radiolytic salt enrichment and brines in the crystalline basement of the east European platform, *in:* Saline Waters and gases in crystalline rocks (P. Fritz and S.K. Frape Eds.) *Geol. Assoc. Canada* **33**, 197-210.

Vuataz, F.D., Brach, M., Criaud, A. and Fouillac, C. (1990) Geochemical monitoring of drilling fluids. A powerful tool to forecast and detect formation waters, *Soc. of Petrol. Eng., Form. Eval.* June, 177-184.

Yardley, B.W., Cathelineau, M., Boiron, M.C., Dubessy, J., Ayt Ougoudal, M., Pironon, J., Landais, P., Noronha, F. , Guedes, A., Doria, A., Vindel, E. and Lopez, J.A. (1994, 1995) Fluid behavior in the upper crystalline crust : a multidisciplinary approach, Report to the ECC, Project N° JOU-CT 93-0318.

Occurrence and origin of Cl-rich amphibole and biotite in the Earth's crust - implications for fluid composition and evolution

Kåre Kullerud
Department of Geology, University of Tromsø, N-9037 Tromsø, Norway

Key words: Chlorine, amphibole, biotite, fluid evolution, fluid immiscibility, fluid filtration

Abstract: Analyses of Cl-bearing amphiboles and biotites from more than 20 occurrences around the world have been reviewed. The Cl-content of amphibole ranges up to about 6 wt%, while the most Cl-enriched biotite contains about 7 wt% Cl. For the individual occurrences of amphibole and biotite, systematic compositional variations, correlated to the Cl-contents of the minerals can be observed. It is argued that these variations were governed by variations in the fluid activity ratio a_{Cl^-}/a_{OH^-} during mineral growth. The most Cl-enriched amphiboles and biotites formed in equilibrium with highly saline solutions.

Several mechanisms are possible for the formation of saline brines in the Earth's crust. High-Cl fluids may originate during dissolution of Cl-rich minerals (e.g. amphibole and biotite) in low-Cl crustal fluids. In many cases, however, highly saline fluids evolve from low-saline fluids (e.g. marine waters) during preferential extraction of volatile components (e.g. H_2O). Several mechanisms may be responsible for a fractination between Cl and H_2O during fluid-rock interactions. Preferential incorporation of OH^- relative to Cl^- in amphibole and biotite during hydration of anhydrous rocks is probably one of the most important mechanisms for the formation of saline brines in the crust. Other mechanisms involve fluid immiscibility and fluid filtration.

I. Stober and K. Bucher (eds.), Hydrogeology of Crystalline Rocks, 205–225.

1. INTRODUCTION

The composition of a fluid in the Earth's crust is subjected to continual changes in response to reactions between fluid and rock along the transport path of the fluid. The Cl-content of the fluid can be altered during dissolution and precipitation of Cl-bearing minerals. Chlorine may be an important component of hydrous silicates, where it substitutes for the hydroxyl group. In particular amphibole and biotite, which are common rock-forming minerals, are important crustal reservoirs of Cl. Thus, much of the Cl which occur in groundwater may have been liberated to the fluid phase during alteration of Cl-bearing amphibole and biotite.

It is, however, important to realise that the Cl-content of a fluid also may change in response to the exchange of other volatile components between the fluid phase and the rock, e.g. by addition or extraction of H_2O or CO_2. Saline solutions may be generated during preferential extraction of H_2O from Cl-bearing hydrous fluids. During the formation of hydrous silicates, OH^- is strongly partitioned into the solid phases relative to Cl^-, resulting in an increase in the Cl-content of the remaining fluid. Fyfe (1987) suggested that large amounts of highly saline solutions are generated in the oceanic lithosphere by this mechanism when the peridotitic component of the lithosphere is serpentinised during reactions with marine water. During hydration of the continental crust and the basaltic component of the oceanic lithosphere, amphibole and biotite are among the most important minerals that form. Although these minerals may incorporate Cl during formation, the element is normally preferred by the fluid phase. Thus, hydration reactions in the crust are probably of major importance for the formation high-Cl geological fluids. Other mechanisms that may be responsible for the formation of Cl-rich fluids in the crust include fluid immiscibility and fluid filtration processes.

In this paper the occurrence, the compositional variations and the mechanisms of formation of Cl-bearing amphibole and biotite will be addressed. Further, it will be focused on the different mechanisms that may be responsible for the evolution of highly saline brines from relatively low-Cl hydrous fluids.

2. COMMON OCCURRENCES OF CL-BEARING AMPHIBOLE AND BIOTITE

Cl-rich amphibole and biotite are known from various geological environments and rock types around the world (Table 1). Clearly, several sources of Cl are possible. Cl-rich minerals have frequently been reported from sub-seafloor metamorphosed oceanic crust (e.g. Jacobson 1975, Ito & Anderson 1983, Vanko 1986). In this setting, seawater is the most probable source of Cl. The presence of Cl-rich minerals in association with skarns (e.g. Dick & Robinson 1979), may suggest a magmatic origin of Cl-bearing fluids. Formation of Cl-bearing silicates from magmatic volatiles have been reported from many localities, e.g. the Skaergaard intrusion (Sonnenthal 1992) and from the Stillwater and Bushveld Complexes (Boudreau et al. 1986). In some cases, the Cl incorporated in silicates was not externally derived, but it was an original constituent of the rock. Mora & Valley (1989) suggested that Cl incorporated in silicates in metasedimentary rocks, was locally derived from halides present in the original sediment.

For many occurrences of Cl-bearing minerals the source of Cl, whether it is derived from the crust or the mantle, is unclear. Recent studies on the isotope systematics of Cl (e.g. Magenheim et al. 1995, Boudreau et al. 1997, Markl et al. 1997) shows that this method has large potential for future studies on the distribution and evolution of Cl in the Earth's crust.

Table 1. Occurrences of amphiboles (Am) and biotites (Bt) used in Figs. 1, 2 and 3

	Locality	Setting/rock type	Am	Bt	
1	Transcaucasia, Russia	Skarn	x		Krutov (1936)
2	Transcaucasia, Russia	Skarn	x		Jacobson (1975)
3	St. Pauls Rocks, equatorial Atlantic	Ultrabasic intrusion	x		Jacobson (1975)
4	Santa Rita Stock, New Mexico, USA	Porphyry copper deposit		x	Jacobs, 1976
5	Southern Yukon	Sphalerite skarn	x		Dick & Robinson (1979)
6	Rajastan, NW India	Calcareous meta-sediment	x	x	Sharma (1981)
7	Visakpatnam, India	Hyperstene-garnet granulite	x	x	Kamineni et al (1982)
8	Pyrénées, France	Intermediate charnockite	x		Vielzeuf (1982)
9	Mid-Cayman Rise, Caribbean Ocean	Sub-seafloor hydrothermally altered gabbro	x		Ito & Anderson (1983)
10	Mathematician Ridge, East Pacific Ocean	Oceanic metabasic rocks	x		Vanko (1986)
11	Eastern Ghats, India	Charnockites	x		Rao & Rao (1987)
12	West Ongul Island, East Antarctica	Amphibolite and carbonate-pegmatite	x		Suwa et al. (1987)

	Locality	Setting/rock type	Am	Bt	
13	Sesia-Lanzo, Italy	Marbles	x		Castelli (1988)
14	Idaho, USA	Carbonate-bearing granofels		x	Mora & Valley (1989)
15	Adirondack Mountains, USA	Marcy anorthosite massif	x		Morrison (1991)
16	Sterling Hill, New Jersey, USA	Skarn in association with Zn-ore		x	Tracy (1991)
17	Salton Sea, California, USA	Metabasic rocks	x		Enami et al (1992)
18	Bergslagen, Sweden	Meta-exhalites	x	x	Oen & Lustenhouwer (1992)
19	Skaergaard intrusion, Greenland	Anorthosites and pegmatites	x	x	Sonnenthal (1992)
20	Quinling, China	Pb-Zn deposit, volcanoclastic rocks	x	x	Jiang et al. (1994, 1996)
21	Lofoten, Norway	Ductile shear zones in gabbro	x	x	Kullerud (1995, 1996)
22	Black Rock Forest, New York, USA	Amphibolite to granulite facies gneisses	x	x	Léger et al. (1996)
23	Ramnes Cauldron, Norway	Porphyritic alkali granitic rocks	x		Sato et al. (1997)

3. COMPOSITIONAL VARIATIONS OF CL-BIOTITE

A selection of analyses of Cl-bearing biotite from the literature is plotted in Fig. 1. All biotite analyses were recalculated on the basis of a total cation charge of 22 [i.e. $\Sigma O=10$, $\Sigma\,(OH,Cl,F) = 2$], except from the data of Léger et al. (1996) which were redrawn from their figure 3. Table 2 gives a selection of biotite analyses from the literature.

The majority of the biotites shown in Fig. 1 have less than 0.3 Cl per formula unit (pfu) (≈ 2 wt% Cl). In a few samples, however, considerably higher Cl-contents are reported (Tracy 1991, Oen & Lustenhouwer 1992, Léger et al. 1996). The biotites of Jacobs (1976) show a negative correlation between Cl and Mg and a positive correlation between Cl and Fe. Biotites from the other localities show similar correlations, however, the slopes of the individual compositional trends vary. It should also be noted that although several of the trends in the Mg-Cl and Fe-Cl diagrams are sub-parallel, they are vertically displaced relative to each other. Similarly, a positive correlation can be observed between Cl and Al^{IV} for several occurrences (Fig. 1c), but the individual variation trends are displaced relatively to each other. For most of the reported Cl-rich biotites the F-content is low. When F is abundant, negative correlations between Cl and F are generally observed (not shown). For Ba-bearing biotites (e.g. biotites from Bergslagen, Lofoten, Sterling Hill and Quinling, see Table 1), Ba generally correlates positively with Cl (not shown).

Biotite from several localities show extreme compositional variations on thin section scale. The biotite analyses reported from Sterling Hills, showing Cl-content in the range 0.89 wt% - 7.15 wt%, were all obtained from one thin-section (Tracy 1991). Léger et al. (1996) describe biotites from a single specimen ranging in compositions from 1.02 wt% Cl to 4.64 wt% Cl, while the biotite analyses from Lofoten (Kullerud 1995) with Cl-content in the range 0.1 wt% - 1.5 wt% were sampled from two specimens.

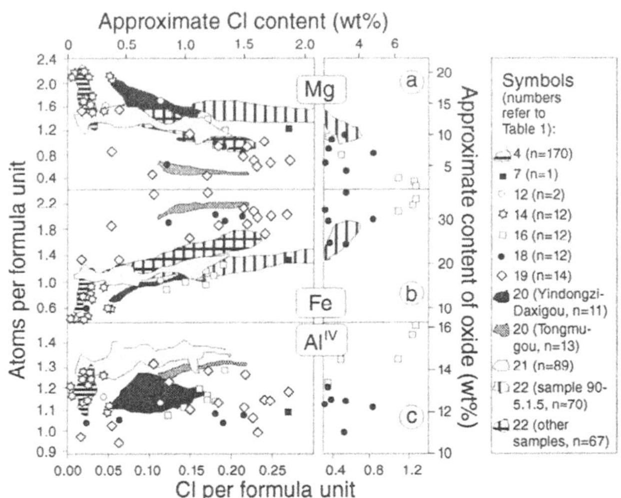

Figure 1. Compositional variations of Cl-bearing biotite from a selection of occurrences described in the literature

4. COMPOSITIONAL VARIATIONS OF CL-AMPHIBOLE

A selection of analyses of Cl-bearing amphibole (Cl > 0.2 wt%) from the literature is plotted in Figs. 2 and 3. Structural formulas of amphibole were recalculated on the basis of total cations - (Ca + Na + K) = 13. Table 2 gives a selection of amphibole analyses from the literature.

Table 2. Selected analyses and calculated structural formulas of biotite and amphibole

Reference (see Table 1)	21	22	19	18		2	13	22	21
	Bt	Bt	Bt	Bt		Am	Am	Am	Am
SiO_2	34.40	35.71	35.58	32.90	SiO_2	33.90	36.77	35.65	32.53
TiO_2	1.74	0.55	3.38	0.37	TiO_2	1.01	0.25	0.45	0.88
Al_2O_3	19.36	11.32	12.65	13.30	Al_2O_3	11.57	12.28	13.02	18.93
FeO	16.44	28.49	30.13	27.70	FeO	31.91	27.68	27.33	22.46
MnO	0.00	0.10	0.09	0.29	MnO		0.19	0.21	0.17
MgO	11.89	7.95	5.87	6.52	MgO	1.81	2.20	3.73	3.53
CaO	0.06	0.03	0.00		CaO	11.16	10.67	10.89	11.09
BaO	3.85		0.24	2.84					
Na_2O	0.09	0.07	0.06		Na_2O	0.74	0.96	1.31	1.82
K_2O	7.42	9.58	9.01	8.34	K_2O	3.34	3.59	2.74	2.73
F	0.00	0.48	0.14		F		0.00	0.14	
Cl	1.18	4.64	2.02	5.50	Cl	4.95	4.09	5.00	3.78
Structural formulas									
Si	2.637	2.926	2.825	2.796	Si	5.673	6.127	5.798	5.308
Al^{IV}	1.363	1.074	1.175	1.204	Ti	0.127	0.031	0.055	0.108
Ti	0.100	0.034	0.202	0.024	Al	2.281	2.411	2.495	3.638
Al^{VI}	0.385	0.019	0.009	0.128	Fe	4.466	3.857	3.718	3.064
Fe	1.054	1.953	2.001	1.969	Mn		0.027	0.029	0.023
Mn	0.000	0.007	0.006	0.021	Mg	0.452	0.547	0.905	0.859
Mg	1.359	0.971	0.695	0.826	Ca	2.001	1.905	1.898	1.939
Ca	0.005	0.003	0.000		Na	0.240	0.310	0.413	0.575
Ba	0.116		0.007	0.095	K	0.713	0.763	0.569	0.568
Na	0.014	0.011	0.009		F		0.000	0.072	
K	0.726	1.002	0.913	0.904	Cl	1.404	1.155	1.379	1.045
F	0.000	0.124	0.035						
Cl	0.153	0.645	0.272	0.792					

The Cl-content of amphibole ranges almost up to 2 atoms pfu. (Fig. 3), i.e. the OH-sites in the most Cl-rich amphiboles, are almost totally occupied by Cl. Amphibole from Dashkesan, Transcaucasia (Krutov 1936) contain above 2 Cl pfu (Cl = 7.24 wt%). However, more recent analyses of amphibole from Dashkesan (Jacobson 1975), indicate that the Cl-content was over-estimated in Krutov's analysis.

The Cl-bearing amphiboles show considerable A-site occupancies. Amphiboles from Lofoten (Kullerud 1996) show a clear positive correlation between A-site occupancy and Cl-content (Fig. 3b). Amphiboles from other localities show similar compositional trends, however, the slopes of the individual trends may vary, and they may be vertically displaced relatively to each other. Further, a general negative correlation between A-site occupancy and Si-content can be observed (Fig. 2a), indicating that the incorporation of Cl in amphibole is coupled with edenite substitution $[Si^{IV} + []^{A} = Al^{IV} + (K,Na)^{A}$, where [] denotes vacancy]. The negative correlation between Mg/(Mg+Fe) and Cl (Fig. 3a), suggests that $FeMg_{-1}$-exchange also is coupled with $ClOH_{-1}$-exchange.

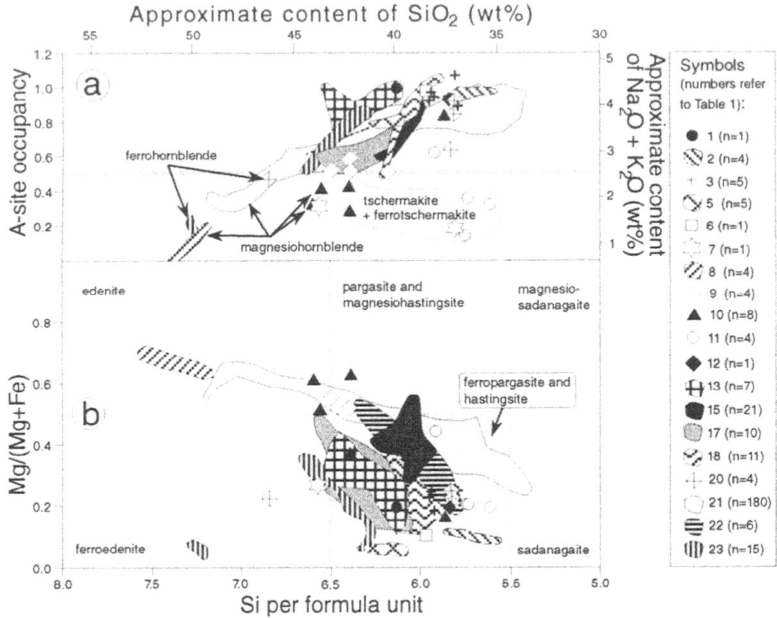

Figure 2. Classification of amphibole analyses selected from the literature. Mineral names are from Leake et al. (1997). See text for discussion.

In the calcic amphibole classification diagram of Leake et al. (1997), the majority of the amphibole data plot in the field of hastingsite/ferropargasite (Fig. 2b). Note that the classification diagram is valid only for amphiboles with A-site occupancy > 0.5 (see Fig. 2a). Names of amphiboles with A-site occupancy < 0.5 are indicated in Fig. 2a. Amphiboles from Lofoten (Kullerud 1996) show a continuos compositional trend from Cl-rich sadanagaite, through the field of hastingsite/ferropargasite, to Cl-poor magnesiohornblende. Kullerud (1996) showed that the compositional variations of the Lofoten amphiboles could be described by the complexely coupled exchange vector $(Si_{-1.6}Al_{1.6})^{IV}(Fe_{1.3}Mg_{-2.1})^{VI}K_{0.4}Cl_1OH_{-1}$. This exchange vector is clearly not universally valid. From the data presented in Figs. 2 and 3, however, there seem to be a general rule that $ClOH_{-1}$-exchange in amphibole is coupled with $FeMg_{-1}$-exchange and an increase in the edenite-component.

Similarly to biotite, amphibole may show extreme compositional variations on thin section scale. Léger et al. (1996) described amphiboles from two samples ranging in compositions from 0.35 Cl pfu to 0.85 Cl pfu and 0.9 Cl pfu to 1.4 Cl pfu, respectively, while Kullerud (1996) reported amphiboles with Cl-contents in the range 0.05 Cl pfu to 1.10 Cl pfu from two samples.

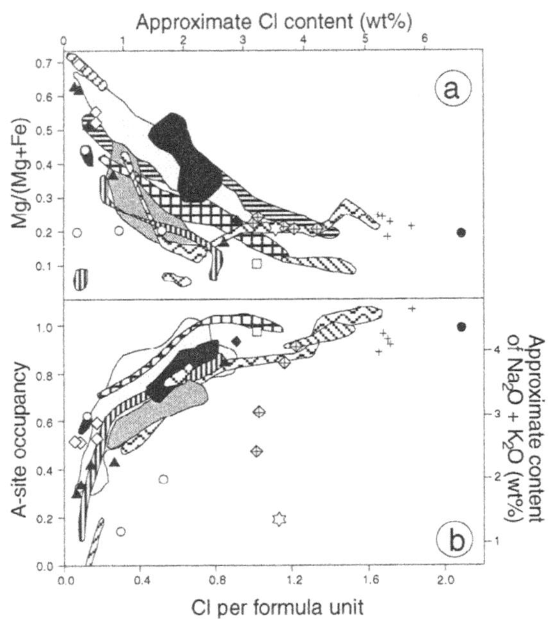

Figure 3. Compositional variations of amphibole selected from the literature. Symbols as in Fig. 2

5. MECHANISMS OF CL-INCORPORATION IN AMPHIBOLE AND BIOTITE

Observations from Fig. 1 and 3 suggest that for individual occurrences of Cl-bearing amphibole and biotite, the Cl-contents of the minerals generally correlate positively with the Fe-contents, and negatively with the Mg-contents. In addition, several other components of amphibole and biotite (e.g. A-site occupancy in amphibole, Al^{IV} in biotite, see also Kullerud 1995 and 1996) are correlated to the Cl-contents of the minerals.

Positive correlation between Fe and Cl and negative correlation between Mg and Cl in amphibole and biotite are in accordance with the Mg-Cl (or Fe-F) avoidance rule (e.g. Ramberg 1952). Ramberg (1952) explained Fe-F avoidance in silicates as an effect of the strong bonding forces between Mg and F and corresponding weak bonding forces between Fe and F. By analogy, Fe-Cl bonds are strongly preferred over Mg-Cl bonds in silicates (e.g. Munoz & Swenson 1981, Munoz 1984). Rosenberg & Foit (1977) gave an alternative explanation of Fe-F avoidance by crystal field theory.

Munoz & Swenson (1981) argued that the incorporation of Cl in biotite is principally controlled by the Mg/Fe-ratio of the mineral, i.e. the incorporation of Cl is controlled by internal crystal chemical constraints, related to differences in the strength of Fe-Cl bonds and Mg-Cl bonds. On the basis of experimental data and the data on the natural biotites of Jacobs (1976, see Fig. 1), Munoz & Swenson (1981) expressed the F- and Cl-contents in biotite as functions of temperature, X_{Mg} and fluid composition. Implicit in their model is the assumption that the biotites of the phyllic + potassic alteration zone of Jacobs (1976), all equilibrated with the same hydrothermal fluid (constant a_{Cl^-}/a_{OH^-}) and were formed at the same temperature.

Volfinger et al. (1985) related the frequently observed positive correlations between Fe and Cl in amphibole and biotite to structural differences between the Fe- and Mg-endmembers of the minerals. They pointed out that substitution of Fe for Mg in amphibole results in a slight rotation of the tetrahedra, and an increase in the sizes of the OH- and A-sites. This allows replacement of OH by the relatively larger Cl-anion, and incorporation of K on the A-site of amphibole.

One problem with the model of Volfinger et al. (1985) is that although the substitution of Fe for Mg does enlarge the octahedral strip of amphibole, the size of the OH-site is not enlarged enough to incorporate the large Cl-anion (Oberti et al. 1993). On the basis of structural refinement studies of Cl-amphibole, Makino et al. (1993) and Oberti et al. (1993) showed that the incorporation of Cl on the OH-site, induces a deformation of the crystal lattice and an increase in cell volume. Oberti et al. (1993) suggested local ordering of Fe and Cl, dictated by the preference of Cl for Cl-Fe rather than Cl-Mg bonds. Further, Oberti et al. (1993) suggested that the incorporation of Cl in amphibole promote incorporation of K on the A-site and substitution of

Al for Si on tetrahedral site. This is in good agreement with the general compositional variations of Cl-bearing amphibole (Figs. 2 and 3).

The implications of the models of Cl-incorporation in amphibole proposed by Makino et al. (1993) and Oberti et al. (1993) are completely different from the implications of Munoz & Swenson (1981) and Volfinger (1985). Following Munoz & Swenson (1981) and Volfinger (1985), the principal control on the variations of Cl in amphibole and biotite is the Mg/Fe-ratios of the minerals. A consequence of the models of Makino et al. (1993) and Oberti et al. (1993), on the other hand, is that the variations in Cl-content of amphibole and biotite are principally controlled by the a_{Cl^-}/a_{OH^-}-ratio of the equilibrium fluid, while the variations in the Mg/Fe-ratios of the minerals are, to a large extent, internally dictated by crystal chemical constraints.

The assumption that the biotites of the phyllic + potassic alteration zone of Jacobs (1976), all equilibrated with the same hydrothermal fluid (constant a_{Cl^-}/a_{OH^-}) and were formed at the same temperature, is fundamental for the model of Munoz & Swenson (1981). According to Jacobs & Parry (1979) the majority of the biotites represents primary igneous biotite and secondary biotite that formed during the potassic alteration (at T=500-600°C, Jacobs & Parry 1979). Munoz & Swenson (1981) suggest that the biotites subsequently reequilibrated with a fluid phase of constant composition during the phyllic alteration (at T=300-400°C, Jacobs & Parry 1979), and that the present Cl-content was dictated by the Mg/Fe-ratios of the minerals. With reference to the strong positive correlation between $\log(X_F/X_{Cl})$ and X_{Mg} of biotite, their main argument was formulated: "The fact that the Santa Rita biotites produce a good linear trend argues persuasively that they all equilibrated at the same temperature and fluid composition." They argue further that: "Biotite assemblages within the same deposit which equilibrated either with markedly different hydrothermal fluids or at markedly different temperatures could not produce a consistent pattern when their halogen contents are plotted against X_{Mg}." The arguments of Munoz & Swenson (1981) have been adopted by several other authors, e.g. Zhu & Sverjensky (1992) state that: "This is a reasonable postulate since the systematic halogen contents for these biotites are difficult to interpret otherwise." An alternative explanation for the formation of the biotites was given by Jacobs & Parry (1979) who suggested that the variations in Cl-contents of biotite were related to variations in the pH and composition of the equilibrium fluid.

The implications of correlations between chemical components of a mineral (e.g. X_{Fe} and X_{Cl}) was addressed by Kullerud (1995). He pointed out that high correlation between two chemical components implies that the variations of the two components are related to the same geochemical process. It is, however, not possible from the correlation alone to decide whether the concentration of one component is dependent on the concentration of an other (e.g. X_{Cl} is dependent on X_{Fe}, or X_{Fe} is dependent on X_{Cl}), or whether the concentrations of the components are dependent on an external factor. The composition of a mineral (e.g. its Fe- and Cl-contents) is, in general, dependent on a complex relationship between pressure, temperature, chemical potentials and internal constraints. Negative correlation between Fe and Mg in a mineral, for example, may be related to variations in any of those variables. Similarly, the compositional variations of the biotites described in Jacobs (1976), in particular the systematic variations of Fe, Mg, Cl and F, may be related to gradual variations in P, T and fluid composition during biotite growth.

Several observations from occurrences of Cl-bearing minerals suggest that the variations in Cl-content of the minerals are controlled by variations in the fluid composition (i.e. the activity ratio a_{Cl^-}/a_{OH^-}), rather than gradients in the activities of Fe and Mg. Vanko (1986) reports replacement of Cl-poor actinolite (Cl < 0.05 wt%) by Cl-rich hastingsite (Cl up to 4.0 wt%) during greenschist facies metamorphism. For these amphiboles, Vanko (1986) argues that the variations in Cl-content were dictated by variations in the Cl activity of the coexisting fluid phase, and not by X_{Mg} of amphibole or temperature. Further, the formation of chemically zoned Cl-bearing amphibole (e.g. Kullerud 1996, Sato et al. 1997) cannot have been formed in equilibrium with a fluid of constant activity ratio a_{Cl^-}/a_{OH^-} (see also discussion in Kullerud 1996). Thus, both structural refinement studies (see above) and natural mineral data suggest that variations in the Cl-contents of amphibole and biotite depend on the variations in the activity ratio a_{Cl^-}/a_{OH^-} of the equilibrium fluid.

On basis of the discussion above it is reasonable to assume that the systematic compositional variations of amphibole and biotite observed at several localities (Figs. 1 and 3), are related to gradual variations in the composition of the equilibrium fluid during growth. It should be noted that gradual changes in temperature will result in similar compositional variations, since both the pH and the speciation of the Cl-fluid are strongly dependent on the temperature. For several of the occurrences of Cl-bearing amphibole and biotite, however, the presence of the Cl-bearing fluid was very transient (e.g. Kullerud 1996), and the Cl-bearing fluid was expelled long before the temperature changed significantly. The differences in the slopes of the individual compositional trends in Figs. 1 and 3, and their vertical displacement relative to each other, probably reflects significant differences in P or T conditions during the formation of the Cl-minerals for the various occurrences. However, the metamorphic conditions during the formation of the Cl-minerals shown in Figs. 1 and 3, have not been sufficiently constrained to confirm this.

The extreme compositional variations in Cl-contents of amphibole and biotite on thin-section scale that occasionally are observed (see above), indicates that extreme gradients in fluid composition may occur locally. Whether these gradients are functions of distance or time will be discussed in the next section.

6. FORMATION OF SALINE FLUIDS DURING FLUID-ROCK INTERACTION

Exchange of Cl^- for OH^- between amphibole or biotite and a coexisting fluid is commonly expressed as:

$$AB(Cl) + OH^-_{fluid} = AB(OH) + Cl^-_{fluid} \qquad (Eq.\ 1)$$

where AB represents amphibole or biotite. The equilibrium constant of the exchange reaction can be expressed as:

$$\ln K = \ln(a_{AB(OH)}/a_{AB(Cl)}) + \ln a_{Cl-}^{fluid} - \ln a_{OH-}^{fluid} \qquad (Eq.\ 2)$$

Since $\ln a_{OH-} = \ln K_w + pH$, where K_w is the dissociation constant of water, Eq. 2 can be reformulated:

$$\ln K + \ln K_W - \ln(a_{AB(OH)}/a_{AB(Cl)}) = \ln a_{Cl-}^{fluid} - pH \qquad (Eq.\ 3)$$

This equation relates the Cl-content of amphibole or biotite to the Cl activity and the pH of the equilibrium fluid. In order to calculate the Cl vs. pH function quantitatively at the pressure and temperature of interest, the values of ΔG of the Cl-endmembers of amphibole and biotite must be known. Until such data are sufficiently constrained, however, only qualitative considerations can be performed.

Cl is a very common constituent in fluids throughout the crust. The rare occurrence of Cl-rich silicates, however, suggests that such minerals form in equilibrium with very saline solutions only. For comparison, Vanko (1986) suggested that “typical” oceanic amphibole, i.e. amphibole in equilibrium with seawater of normal chlorinity, contains a few tenths of a weight percent Cl. In the Salton Sea geothermal system amphibole with up to 2.7 wt% Cl formed in equilibrium with highly saline geothermal fluids, with 15-20 wt% total dissolved Cl (Enami et al. 1992). Further quantification of fluid salinity on basis of the Cl-contents of minerals is difficult, due to the lack of experimental data. Munoz & Swenson (1981) carried out quantitative experiments on the partitioning of Cl between biotite and fluid, however, they succeeded in only one reversal bracket. Further, with reference to the discussion above on the generation of the biotites of Jacobs (1976), it is clear that the equations of Munoz & Swenson (1981), and also the thermodynamic calculations of Zhu & Sverjensky (1992), which were based on the same data, are associated with large uncertainties.

Anyhow, the distribution of Cl-rich amphibole and biotite in nature suggests that Cl-bearing fluids, which are quite common, occasionally evolve to highly saline brines. Several potential mechanisms of Cl enrichment of fluids have been proposed in the literature:

1) Fluid immiscibility - the Cl-bearing fluid separates into two phases, one Cl-rich fluid and one Cl-poor (e.g. Bowers & Helgeson 1983).

2) Fluid filtration - H_2O is preferentially escaping from the Cl-bearing fluid through natural filter membranes (e.g. Hanshaw & Coplen 1973).

3) Preferential extraction of H_2O from the fluid during hydration reactions (e.g. Trommsdorff et al. 1985, Fyfe (1987) Kullerud 1995, 1996).

4) Dissolution of Cl-rich minerals (e.g. halite).

5) Mixing - the Cl-concentration of the original fluid may increase during mixing with saline brines.

The latter of the mechanisms above cannot explain the formation of the most saline brines. Mechanism 4 can explain gradual increase in the Cl-contents of the fluid as long as a solid source of Cl is present (e.g. halite), however, in originally Cl-free rocks this mechanism does not work. Mechanisms 1, 2 and 3, however, can explain the evolution of Cl-bearing fluids to a highly saline solution, even when the fluids are transported through Cl-free rocks. These mechanisms will be discussed further below. It should be emphasised that the following discussion considers mass-balance between fluid and rock, only. Despite of its limitations, this approach may be of value for a qualitative understanding of fluid evolution.

6.1 Extraction of H_2O from the fluid phase

Fluid transport in the crust occur over a range of scales and velocities by various mechanisms. Diffusion along crystal defects and grain boundaries are very slow processes, while fracture zones and shear zones may provide rapid transport of large volumes of fluids. The large extent of non-equilibrium textures that commonly are observed in rocks which have been infiltrated by externally derived fluids, may suggest that equilibrium generally is not attained between fluid and rock, before the fluid escapes or is expended through fluid-consuming reactions (e.g. hydration reactions). Most likely, reactions occur continuously between fluid and rock along the transport path of the fluid, resulting in continual changes in fluid composition. As pointed out above, Cl is strongly partitioned into the fluid phase relative to H_2O during fluid-rock interaction. A consequence of this is that H_2O preferentially will be extracted from a Cl-bearing hydrous fluid during hydration reactions, while the Cl-content continuously increases along the transport route (Fig. 4).

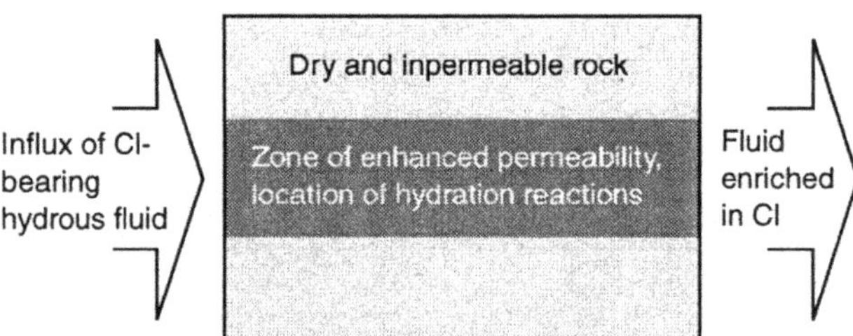

Figure 4. Evolution of saline solutions from less saline solutions. H_2O is preferentially extracted from the fluid through hydration reactions along the fluid transport path

Kullerud (1995, 1996) suggested that the large gradients in the compositions of amphibole and biotite from Lofoten (see above) were related to the differential partitioning of Cl and H_2O between fluid and minerals (Fig. 5, see also Markl & Bucher 1998, Markl et al. in press). He suggested that after an initial flushing of the rock by a Cl-bearing fluid, the fluid-rock system closed. The free fluid phase was gradually consumed by amphibole and biotite forming reactions. Since H_2O was preferentially extracted from the fluid, the Cl-content of the fluid and the activity ratio a_{Cl^-}/a_{OH^-} gradually increased. The consumption of the free fluid phase resulted in the development of fluid-absent domains of the rock. Equilibrium between fluid and minerals was maintained as long as the free fluid was present along the mineral grain boundaries, however, when the fluid was consumed the reactions stopped. Thus, the Cl-content of a mineral grain reflects the composition of the last fluid present along its grain boundaries. By this model, low-Cl amphibole and biotite formed in domains where the free fluid phase was consumed at an early stage, while Cl-rich minerals occur in domains of extended fluid saturation and evolved fluid compositions. Thus, the large compositional gradients of amphibole and biotite that can be observed on thin-section scale reflects, most likely, temporal rather than spatial variations in fluid composition.

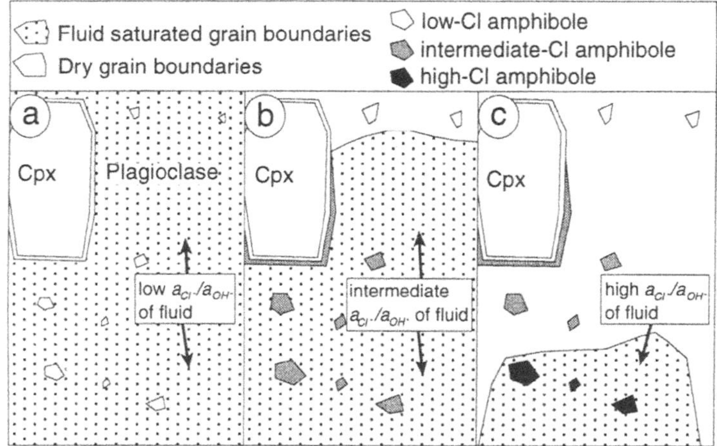

Figure 5. Schematic model for growth of amphibole and consumption of the free fluid phase. a: Initial fluid saturation of grain-boundaries. b: Fluid-absent domains are developed. c: Only small domains of the rock are fluid saturated. The remaining fluid is highly saline.

6.2 Fluid immiscibility

Separation of one fluid phase into two chemically distinct phases can occur at all levels in the crust. Evaporation on the surface and boiling at near surface levels of the crust are well known mechanisms for the generation of saline solutions. At lower levels, unmixing of fluid phases has been demonstrated for several chemical systems. The region of fluid immiscibility in the system $NaCl-H_2O-CO_2$ at 2 kbar and 550°C (from data of Bowers & Helgeson 1983) is shown in Fig. 6. The field of fluid immiscibility of this system is strongly dependent on P and T, thus, gradual changes in the P and T conditions during metamorphism, will lead to gradual variations in the Cl-content of a Cl-rich fluid in equilibrium with a CO_2-rich fluid.

Figures 6a-c illustrate changes that may occur in the composition of a $NaCl-H_2O-CO_2$ fluid at isobaric and isothermal conditions. A fluid with composition A in Fig. 6a which experiences an increase in CO_2-content (e.g. during infiltrating a carbonate-rich rock), will move its composition towards the CO_2 corner. At B′ the fluid will unmix in two phases, a CO_2-rich (B′′) and a Cl-rich (B′). If CO_2 is continuously supplied, and the two immiscible fluids maintain equilibrium, the composition of the Cl-rich fluid will move towards C′ and the composition of the CO_2-rich fluid will move towards C′′, as the bulk composition moves towards C. Figure 6b illustrates the evolution if the Cl-rich fluid is repeatedly isolated from its equilibrium CO_2-rich fluid. If a fluid of composition D experiences an increase in CO_2-content (E) two immiscible fluids will form (E′ and E′′). Recurring mixing between the Cl-rich fluid (E′) with pure CO_2 will move the bulk composition to F which unmix to the fluids F′ and F′′. Thus, under isothermal and isobaric conditions, Cl-bearing fluids which mix with CO_2 may subsequently unmix and form highly saline solutions.

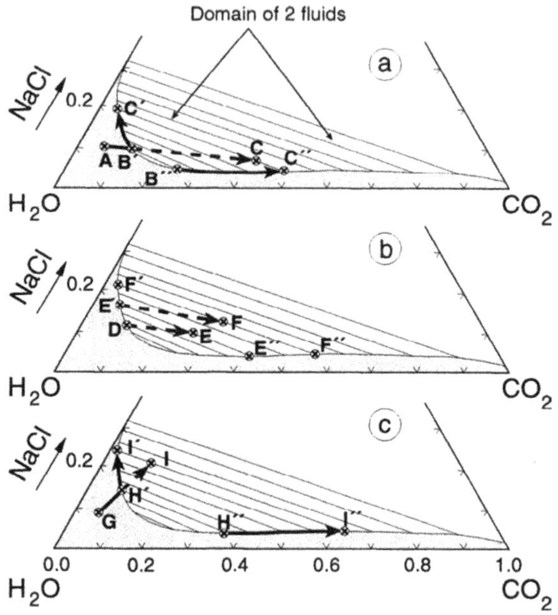

Figure 6. Fluid immiscibility in the system $NaCl$-H_2O-CO_2 at 2 kbar and 550°C (from data of Bowers & Helgeson, 1983). See text for discussion

Figure 6c shows the evolution of a $NaCl$-H_2O-CO_2 fluid (G) during extraction of H_2O (e.g. during hydration reactions). At H′ the fluid will unmix. The composition of the Cl-rich fluid will move towards I′ and the composition of the CO_2-rich fluid will move towards I′′ as the bulk composition moves towards I. Thus, extraction of H_2O from a $NaCl$-H_2O-CO_2 fluid will have an effect similar to addition of CO_2; the fluid phase may unmix into a Cl-rich fluid and a CO_2-rich fluid.

6.3 Fluid filtration (differential transport)

If the water molecules and the dissolved species of a fluid phase are transported at different velocities along the fluid path, the composition of a specific volume of the fluid will change with time. If the differences in the velocities of the various fluid components are sufficiently large, the fluid path may act as a filter, effectively retarding the transport ability of some components. Thus, some components will be enriched in the original volume of fluid, while other components escape through the natural filter.

Russell (1933) suggested that a brine could be formed by preferentially forcing water molecules of a solution through a membrane and leaving the dissolved salts behind. Experiments have shown that the mechanism work on membranes of shale and compacted clay minerals (Hanshaw & Coplen 1973, Kharaka & Berry 1973). The basic principle of the mechanism is that neutral water molecules are permitted to flow through the membrane of net negatively charged clay minerals, while the flow of anions are retarded. Because the solution attempts to maintain electrical neutrality, the flow of cations is also retarded. Land (1995) however argues that although the mechanism may work, it is of no importance during the generation of saline formation waters in sedimentary rocks.

The differentiation in the transport rates of the fluid components through a membrane of shale or clay minerals occur during interactions between the fluid components and the surfaces of the very small pores of the membrane. Differential transport of fluid species may also occur along grain boundaries and crystal defects in crystalline rocks. The Cl-rich amphibole and biotite described by Kullerud (1995, 1996) occur in 3-4 m wide shear zones in a gabbroic host rock. In a 1-2 m wide zone adjacent to the shear zones, Cl-poor amphibole occur in the undeformed rock. Kullerud (1997) suggested that gradients in fluid pressure between the fluid saturated shear zone rock and the dry igneous host rock initiated the diffusion of fluid species into the undeformed rock, along grain boundaries and crystal defects. The low Cl-content of amphibole in the undeformed rock suggests that Cl was transported at much lower velocities than H_2O. Thus, the undeformed gabbroic rock may have acted as a filter, retarding the diffusion of Cl relative to H_2O. The importance of this mechanism during the evolution of fluids in crystalline rocks, however, still remains to be evaluated.

7. SUMMARY

Cl-bearing amphiboles and biotites from various geological environments around the world show large, but systematic compositional variations. For several occurrences of amphibole and biotite, high correlations can be observed between the Cl-content and the contents of cations of the mineral (e.g. Fe and Mg). Most likely, the systematic compositional trends observed at individual localities reflect local gradients in the fluid activity ratio $a_{Cl^-}/_{OH^-}$ at constant P and T during mineral growth. Differences in the compositional trends between localities, on the other hand, were probably related to marked differences in P and T during the formation of the different occurrences.

Cl-rich amphibole and biotite form in equilibrium with very saline solutions. Several sources are possible for such fluids. In some cases, Cl is redistributed locally, during the breakdown of halides. In many cases, however, the Cl-bearing fluid responsible for amphibole and biotite formation is externally derived. Saline solutions may evolve from surface waters (e.g. marine seawater), and magmatic and metamorphic fluids by several mechanisms. During its passage through the crust, a Cl-bearing fluid continuously interacts with its surroundings. H_2O is preferentially incorporated into hydrous minerals during fluid-rock interaction, while Cl remain in the fluid phase, resulting in a gradual increase in the salinity of the fluid. An other mechanism responsible for the formation of saline solutions involves unmixing of the fluid phase. Addition of CO_2 to a $NaCl-H_2O$ fluid may lead to immiscibility between a CO_2-rich hydrous fluid and a saline solution. The saline solution which results from the unmixing will be more saline than the original $NaCl-H_2O$ fluid. Thus, CO_2 may be responsible for dehydration of the saline solution. Probably, both of these mechanisms are important for the formation of saline solutions in the crust. Differential transport rates of Cl and H_2O, for example through natural filter membranes of clay minerals and along grain boundaries and crystal defects may also result in increased fluid salinity. Such mechanisms are, however, probably of minor importance, due to the slow rates of the differentiation processes, especially in crystalline rocks.

ACKNOWLEDGEMENTS

Constructive reviews by Kurt Bucher, Ingrid Stober and Kjell P. Skjerlie are greatly appreciated.

REFERENCES

Boudreau, A.E., Mathez, E.A. & McCallum, I.S. (1986) Halogen geochemistry of the Stillwater and Bushweld Complexes: Evidence for transport of the platinum-group elements by Cl-rich fluids. *J. Petrol.* **27,** 967-986.

Boudreau, A.E., Stewart, M.A. & Spivack, A.J. (1997) Stable Cl isotopes and origin of high-Cl magmas of the Stillwater Complex, Montana. *Geology* **25,** 791-794.

Bowers, T.S. & Helgeson, H.C. (1983) Calculation of the thermodynamic and geochemical consequences of nonideal mixing in the system H_2O-CO_2-NaCl on phase relations in geologic systems: Equation of state for H_2O-CO_2-NaCl fluids at high pressures and temperatures. *Geochim. Cosmochim. Acta* **47,** 1247-1275.

Castelli, D. (1988) Chloropotassium ferro-pargasite from Sesia-Lanzo marbles (Western Italian Alps): a record of highly saline fluids. *Rend. Soc. Ital. Mineral. Petrol.* **43,** 129-138.

Dick, L.A. & Robinson, G.W. (1979) Chlorine-bearing potassian hastingsite from a sphalerite skarn in southern Yukon. *Cand. Mineral.* **17,** 25-26.

Enami, M., Liou, J.G. & Bird, D.K. (1992) Cl-bearing amphibole in the Salton Sea geothermal system, California. *Cand. Mineral.* **30,** 1077-1092

Fyfe, W.S. (1987) The fluid inventory of the crust and its influences on crustal dynamics. *Geol. Assoc. Cand. Spec. Pap,* **33,** 1-3.

Hanshaw, B.B. & Coplen, T.B. (1973) Ultrafiltration by a compacted clay membrane - II. Oxygen and hydrogen isotopic fractionation. *Geochim. Cosmochim. Acta.* **37,** 2311-2327.

Ito, E. & Anderson, A.T.Jr. (1983) Submarine metamorphism of gabbros from the Mid-Cayman rise: petrographic and mineralogic constraints on hydrothermal processes at slow-spreading ridges. *Contrib. Mineral. Petrol.* **82,** 371-388.

Jacobsen, S.S. (1975) Dashkesanite: High-chlorine amphibole from St. Paul's rocks, Equatorial Atlantic, and Transcaucasia, U.S.S.R. *Min. Sci. Invest., Contrib. Earth Sci., Smithsonian Inst.* **14,** 17-20.

Jacobs, D.C. (1976) Geochemistry of biotite in the Santa Rita and Hanover-Fierro stocks. Central mining district, Grant County, New Mexico. Ph. D. thesis (unpublished), Univ. of Utah, Salt Lake City, Utah, 212p.

Jacobs, D.C. & Parry,, W.T. (1979) Geochemistry of biotite in the Santa Rita porphyry copper deposit, New Mexico. *Econ. Geol.* **74,** 860-887.

Jiang, S.-Y., Palmer, M.R., Xue, C.-J. & Li, Y.-H. (1994) Halogen-rich scapolite-biotite rocks from the Tongmugou Pb-Zn deposit, Qinling, north-western China: implicacations for the ore-forming processes. *Mineral. Mag.* **58,** 543-552.

Jiang, S.-Y., Palmer, M.R., Li, Y.-H. & Xue, C.-J. (1996) Ba-rich micas from the Yindongzi-Daxigou Pb-Zn-Ag and Fe deposits, Quinling, northwestern China. *Mineral. Mag.* **60,** 433-445.

Kamineni, D.C., Bonardi, M. & Rao, A.T. (1982) Halogen-bearing minerals from Airport Hill, Visakpatnam, India. *Amer. Mineral.* **67,** 1001-1004.

Kharaka, Y.K. & Berry, F.A.F. (1973) Simultaneous flow of water and solutes through geological membranes - I. Experimental investigation. *Geochim. Cosmochim. Acta.* **37,** 2577-2603.

Krutov, G.A. (1936) Dashkesanite: a new chlorine amphibole of the hastingsite group. *Mineral. Abstr.* **6,** 438.

Kullerud, K. (1995) Chlorine, titanium and barium-rich biotites: factors controlling biotite composition and implications for garnet-biotite geothermometry. *Contrib. Mineral. Petrol.* **120,** 42-59.

Kullerud, K. (1996) Chlorine-rich amphiboles: interplay between amphibole composition and an evolving fluid. *Eur. J. Mineral.* **8,** 355-370.

Kullerud, K. (1997) Evolution and differentiation of fluids in the lower crust. *Geonytt* **24 I,** 57-58.
Land, L.S. (1995) Na-Ca-Cl saline formation waters, Frio Formation (Oligocene), south Texas, USA: Products of diagenesis. *Geochim. Cosmochim. Acta.* **59,** 2163-2174.
Leake, B.E., Wooley, A.R. Arps, C.E.S., Birch, W.D., Gilbert, M.C., Grice, J.D., Hawthorne, F., Kato, A., Kisch, H.J., Krivovichev, V.G., Linthout, K., Laird, J., Mandarino, J.A., Maresch, W.V., Nickel, E.H., Rock, N.M.S., Schumacher, J.C., Smith, D.C., Stephenson, N.C.N., Ungaretti, L, Whittaker, E.J.W. & Youzhi, G. (1997) Nomenclature of amphiboles: Report from the Subcommittee on Amphiboles of the International Mineralogical Association, Commision on New Minerals and Mineral Names. *Amer. Mineral.* **82,** 1019-1037.
Léger, A., Rebbert, C. & Webster, J. (1996) Cl-rich biotite and amphibole from Black Rock Forest, Cornwall, New York. *Amer. Mineral.* **81,** 495-504.
Magenheim, A.J., Spivack, A.J., Michael, P.J. & Gieskes, J.M. (1995) Chlorine stable isotope composition of the oceanic crust: Implications for Earth´s distribution of chlorine. *Earth Planet. Sci. Lett.* **131,** 427-432.
Makino, K., Tomita, K. & Suwa, K. (1993) Effect of chlorine on the crystal structure of a chlorine-rich hastingsite. *Mineral. Mag.* **57,** 677-685.
Markl, G., Musashi, M. & Bucher, K. (1997) Chlorine stable isotope composition of granulites from Lofoten, Norway: Implications for the Cl isotopic composition and for the source of Cl enrichment in the lower crust. *Earth Planet. Sci. Lettr.* **150,** 95-102.
Markl, G. & Bucher, K. (1998) Composition of fluids in the lower crust inferred from metamorphic salt in lower crustal rocks. *Nature* **391,** 781-783.
Markl, G., Ferry, J. & Bucher, K. (in press) Formation of saline brines and salt in the lower crust by hydration reactions in partially retrogressed granulites from the Lofoten Islands, Norway. *Amer. J. Sci.*
Mora, C.I. & Valley, J.W. (1989) Halogen-rich scapolite and biotite: Implications for metamorphic fluid-rock interactions. *Amer. Mineral.* **74,** 721-737.
Morrison, J. (1991) Compositional constraints on the incorporation of Cl into amphiboles. *Amer. Mineral.* **76,** 1920-1930.
Munoz, J.L. (1984) Fe-OH and Cl-OH exchange in micas with application to hydrothermal ore deposits. *Rev. in Mineral.* **13,** 469-493
Munoz, J.L. & Swenson, A. (1981): Chlorine-hydroxyl exchange in biotite and estimation of relative HCl/HF activities in hydrothermal fluids. *Econ. Geol.* **76,** 2212-2221.
Oberti, R., Ungaretti, L. Cannillo, E. & Hawthorne, F.C. (1993) The mechanism of Cl incorporation in amphibole. *Amer. Mineral.* **78,** 746-752.
Oen, I.S. & Lustenhouwer, W.J. (1992) Cl-rich biotite, Cl-rich hornblende, and Cl-rich scapolite in meta-exhalites: Nora, Bergslagen, Sweden. *Econ. Geol.* **87,** 1638-1648.
Ramberg, H. (1952) Chemical bonds and the distribution of cations in silicates. *J. Geol.* **60,** 331-355.
Rao, A.T. & Rao, P.C.S. (1987) Halogen bearing potassium hastingsite from the charnockites of Eastern Ghats. *Indian Mineral.* **28,** 1-5.
Rosenberg, P.E. & Foit, F.F.Jr. (1976) Fe2+-F avoidance in silicates. *Geochim. Cosmochim. Acta* **41,** 345-346.
Russel, W.L. (1933) Subsurface concentration of chloride brines. *AAPG Bull.* **17,** 1213-1228.

Sato, H., Yamaguchi, Y. & Makino, K. (1997) Cl incorporation into successively zoned amphiboles from the Ramnes cauldron, Norway. *Amer. Mineral.* **82,** 316-324.

Sharma, R.S. (1981) Mineralogy of a scapolite-bearing rock from Rajastan, northwest peninsular India. *Lithos* **14,** 165-172.

Sonnenthal, E.L. (1992) Geochemistry of dendritic anorthosites and associated pegmatites in the Skaergaard Intrusion, East Greenland: Evidence for metasomatism by a chlorine-rich fluid. *J. Volc. Geotherm. Res.* **52,** 209-230.

Suwa, K., Enami, M. & Horiuchi, T. (1987) Chlorine-rich potassium hastingsite from West Ongul Island, Lützow-Holm Bay, East Antarctica. *Mineral. Mag.* **51,** 709-714.

Tracy, R.J. (1991) Ba-rich micas from the Franklin marble, Lime Crest and Sterling Hill, New Jersey. *Amer. Mineral.* **76,** 1683-1693.

Trommsdorff, V., Skippen, G. & Ulmer, P. (1985) Halite and sylvite as solid inclusions in high-grade metamorphic rocks. *Contrib. Mineral. Petrol.* **89,** 24-29.

Vanko, D. (1986) High Chlorine amphiboles from oceanic rocks: product of highly-saline hydrothermal fluids? *Amer. Mineral.* **71,** 51-59.

Vielzeuf, D. (1982) The retrogressive breakdown of orthopyroxene in an intermediate charnockite from Saleix (French Pyrénées) *Bull. Mineral.* **105,** 681-690.

Volfinger, M., Robert, J.-L., Vielzeuf, D. & Neiva, A.M.R. (1985) Structural control of the chlorine content of OH-bearing silicates (micas and amphiboles). *Geochim. Cosmochim. Acta* **49,** 37-48.

Zhu, C. & Sverjensky, D.A. (1992): F-Cl-OH partitioning between biotite and apatite. *Geochim. Cosmochim. Acta* **56,** 3435-3467.

RARE EARTH ELEMENTS AND YTTRIUM AS GEOCHEMICAL INDICATORS OF THE SOURCE OF MINERAL AND THERMAL WATERS

P. MÖLLER
GeoForschungsZentrum Potsdam
Telegrafenberg , D-14473 Potsdam

Abstract

Abundances of rare earth elements REE and Y (combined to REY) in mineral and geothermal waters are compared with those in aquifer rocks and leachates from these rocks. The study includes mineral waters from Kyselka, Czech Republic and the Black Forest, Germany, geothermal waters from Kizildere, Turkey, a brine from the Continental Deed Drilling Project, Germany, an iron-rich spring in Nishiki-numa, Hokkaido, Japan, and vent fluids from the oceanic floor. All REY/Ca patterns of waters from felsic rocks are similar, whereas those from mafic rocks show a wide spread which is attributed to different modes of crystallisation. The source-rock-normalised patterns of REY of leachates of magmatic rocks plot closely together. Their metamorphic equivalents scatter widely. The source-rock-normalised leachates show that in mafic rocks REY/Ca are considerably more accessible than in felsic rocks. The retention of REY in less altered granites and basalt decrease from La to Lu, whereas in strongly altered rocks and most metamorphites REY patterns are either horizontal or increasing from La to Lu. Negative Y and variable Eu anomalies indicated that the accessibility and/or chemical behaviour of these two elements during weathering of rocks and migration of fluids is different from the other REE. REY are retained by factors up to 10000 in crystalline rocks but are leached like Ca (and other elements) from glassy material. Anomalous Eu, Y and Ce yield information on the fluid-rock interaction. Eu is sensitive to temperature, whereas Y is not. Anomalous Eu is inherited and may be enhanced at temperatures above 250°C. Ce is sensitive to oxygen fugacity and pH. Y seems to be sensitive to pH and to ligands dominating REY complexation in solution and on surfaces. In general, Y is released more easily from the rocks than REE and is less retained by sorption onto mineral surfaces.

1. Introduction

The chemical composition of groundwater and geothermal water varies depending on the chemical and mineralogical composition and crystallisation history of the dominantly controlling source rocks (Garrels and Mackenzie, 1967; Humphris et al., 1978). Following the suggestions of Garrels (1967) and Drever (1988), the main alteration reactions are the decomposition of plagioclase to kaolinite and smectite. Although this might explain the molar ratios of Na^+/Ca^{2+} and HCO_3^-/H_4SiO_4 of most waters, these reactions are not relevant to trace elements in the waters (Möller et al., 1997a), because high fractions of the latter originate from dissolution of accessory minerals. Amongst all trace elements, the group of the rare earth elements (REE) and yttrium are of particular interest in the study of the trace element fractionation processes

I. Stober and K. Bucher (eds.), Hydrogeology of Crystalline Rocks, 227–246.

during dissolution and fluid transport.

The relevant characteristics of REE and Y are: (i) They are rare in nature (but not so rare as generally believed) which reduces the possibility to form minerals of their own; (ii) they behave chemically very similarly; (iii) their tendency to form chemical complexes on mineral surfaces and in solution varies systematically from La to Lu; (iv) they are enriched in all Ca minerals; in non-Ca minerals they are strongly fractionated depending on the radius (volume) of the lattice site at which they substitute major elements (Morgan and Wandless, 1980; Saunders, 1984; Möller, 1988; Blundy and Wood, 1991).

One of the obstacles in studying the distribution of REE and Y in natural water-rock interactions is that it is often difficult to sample (i) the representative rock that controls the REE and Y abundances in the water, and (ii) the water under its in-situ PT conditions and uncontaminated by water from the surroundings.

Water sampling is problematic because it is mostly done at 1 atm and outflow temperatures at the surface. For instance, if CO_2 is present, it may escape and subsequently, pH increases and carbonates precipitate. Since REE and Y are co-precipitated with calcite but less with aragonite, the changes in PTX conditions during ascent of natural waters might influence the original composition. Most aquifers are petrologically inhomogeneous and the residence time of water is locally variable. The REE and Y contents in the water will never represent equilibria with the sampled host rock because of the formation of metastable components, surface coatings on minerals and the kinetics of ion exchange. At best, a steady state is reached. With respect to the country rocks, one can only attempt (i) to make the best choice with respect to the reacted rocks and (ii) to sample the water as close to the source as possible. Having all these constraints in mind, REY patterns of waters (natural leachates) and experimental leachates from corresponding aquifer rocks will be compared in the following sections with the aim to study the systematics of REE and Y fractionation during weathering or alteration of rocks in the temperature range from 10 to 400°C.

2. Rare earth element systematics in hydrous systems

It is convenient to normalise the zigzag abundance distribution of REE in rocks and minerals by that of a reference material such as C1 chondrites (Coryell et al., 1963; Anders and Grevesse, 1989) or different local "shales" (McLennan, 1989). Bau and Dulski (1995) suggested to include Y into the series of REE because Y has the same radius (Shannon, 1976) and charge as Ho. Henceforth, when Y is included, the resulting patterns will be signified as REY patterns. The normalisation generally leads to smooth patterns except for those samples in which individual elements (Ce, Eu, Y) behave in an anomalous manner. The chondrite-normalised REY patterns of waters vary over 6 orders of magnitude (Fig. 1) in the temperature range of 10 to 400°C (Table 1). They either increase or decrease from La to Lu, but with a tendency of a flat trend for the heavy REE and Y. Many patterns show Ce, Eu, and Y anomalies.

The rocks, which the waters are in contact with (Table 1), clearly split into two groups (Fig. 2). The felsic rocks are marked by strongly negative Eu anomalies, whereas the mafic rocks show none. The marble (Saz/Kiz) has a positive one. In all rocks studied,

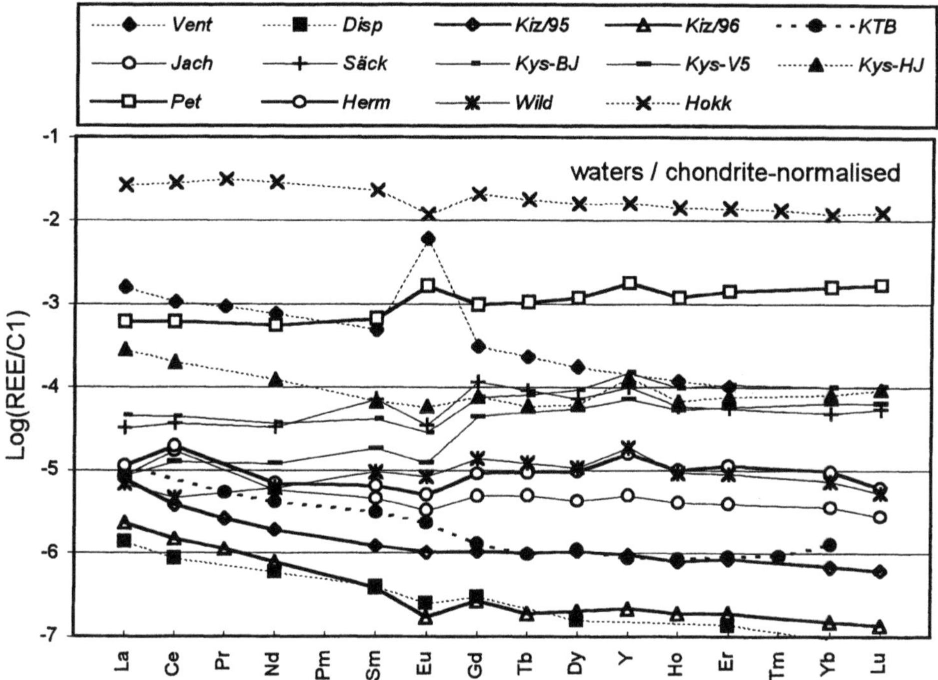

Fig. 1: Chondrite-normalised REY patterns of selected groundwaters and thermal waters from felsic (solid lines) and basic rocks (broken lines). Igneous and metamorphic rocks are indicated by thin and bold lines, respectively. Data are taken from Table 2. ***Vent:*** Mean of 22 vent fluids from the East Pacific rise (11°, 13°, 21°N) (Klinkhammer et al., 1994). ***Disp:*** Mean of 7 dispersed flows from Teahitia Seamount (Michard et al., 1993). ***Hokk:*** Iron-spring, Nishiki-numa, Hokkaido, Japan (Bau et al., 1998). ***Kiz:*** Kizildere, Menderes rift valley, Turkey; means of geothermal waters sampled in 1995 and 1996 from various wells at the pilot plant of Kizildere (unpubl). ***KTB***: 4000 m brine (Möller et al., 1994) from the pilot bore hole of the Continental Deep Drilling Project/ Eastern Bavaria, Germany. The water drained from a sequence of amphibolites (Giese, 1993). ***Jach***: Mean of 2 analyses of thermal water from underground wells of the Svornost shaft at Jachymov in the Kruzne hory granite, Bohemia, Czech Republic. ***Kys***: Waters from the wells at Kyselka (mean of the wells BJ10 and BJ13; well V5) all drilled into the granite of the Slavkosky-les, and mean of two analyses of waters from two wells (HJ4 and HJ5) in the basalt of the Doupovske hory near Kyselka, Bohemia, Czech Republic (Möller et al., 1998). ***Pet***: Mean of two analyses of water from the well A7 at Peterstal and mean of two analyses of water from a well at Hermersberg, Central Black Forest, Germany; both wells are drilled into granite-veined gneisses. ***Säck***: Thermal water from a well at Bad Säckingen, Southern Black Forest,Germany. ***Wild***: Mean of thermal water recovered from two wells (I and IV) in the Kegelbach granite at Bad Wildbad, Northern Black Forest, Germany (Möller et al., 1997a).

the respective normalised Y and Ho values are the same. The overall variability of REY patterns is less than 2.5 orders of magnitude, which is less than in the waters derived from them.

In the study of water-rock interaction, however, it is more appropriate to normalise the REE in the waters to those of the source rocks (Fig. 4). In this way, the *inherited* and/or *acquired* features can more easily be evaluated than by comparison of the separately chondrite-normalised REY patterns of waters and rocks. *Inherited* are features that are derived from the source rocks, and which, therefore, directly refer to the naturally leached rock. *Acquired* are features that are gained during fluid migration due to

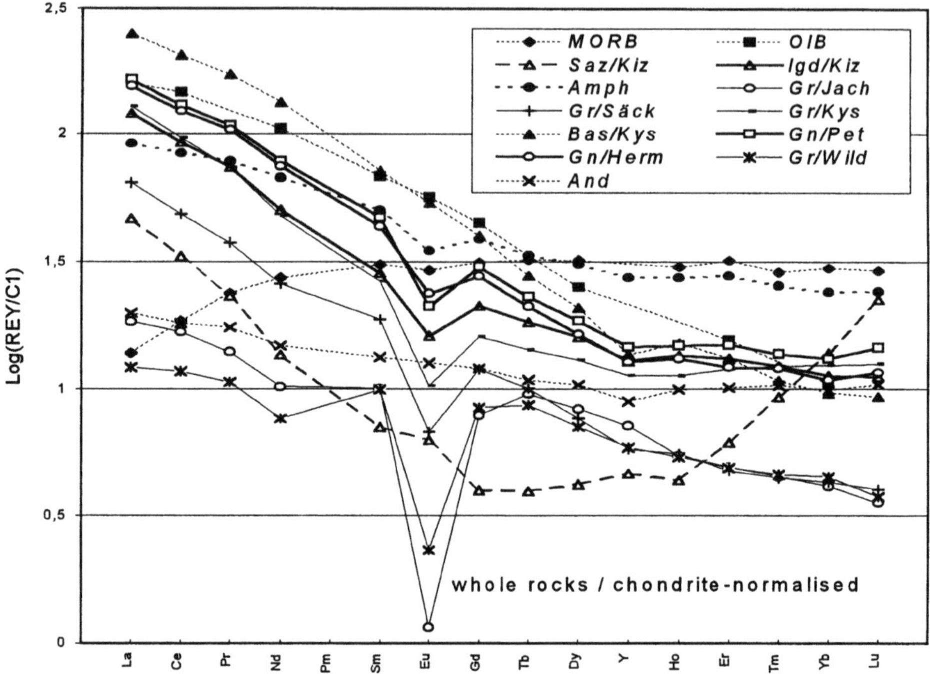

Fig. 2: Chondrite-normalised REY patterns of felsic (solid lines) and basic (broken lines) aquifer rocks. Data are taken from Table 2. ***MORB***: dredged mid-ocean ridge basalt from the East Pacific rise (21°S) (M. Bau, unpubl.). ***OIB***: ocean island basalt, mean of 16 analyses from Teahitia (Hemond et al., 1994). ***And***: andesite from Nishiki-numa, Japan (Bau et al., 1998). ***Saz/Kiz***: outcropping marble of the Sazak Formation, Kizildere, Turkey; ***Igd/Kiz***: outcropping mica schist from Igdecik Formation, Kizildere, Turkey (unpubl.). ***Amph***: amphibolite from the Deep Drill Hole at 3850m, Oberpfalz, Germany (Möller et al., 1994). ***Gr/Jach*** and ***Gr/Kys***: drill core samples of granites from Jachymov/ Krusne hory and Kyselka/Slavkovsky-les, Czech Republic. ***Bas/Kys***: alkali-basalt from Doupovske hory, Kyselka, Czech Republik (Möller et al., 1998). ***Gn/Pet*** and ***Gn/Herm***: drill core samples of the gneisses from Bad Peterstal and Hermersberg, Central Black Forest, Germany. ***Gr/Wild***: drill core sample of the Kegelbach granite, Bad Widbach, Northern Black Forest, Germany. ***Gr/Säck***: granite from Bad Säckingen, Southern Black Forest (Möller et al., 1997a).

sorption processes at mineral surfaces forming the pores in the source rocks or due to selective coprecipitation and scavenging (Möller, 1998). *Inherited* are the typical trends of REY patterns and, for instance, the negative Eu anomalies in waters from felsic rocks. The *acquired* characteristics necessitate intensive interaction of water with minerals, for which a high ratio of mineral surfaces to volume of water is a prerequisite. In contrast, the interaction of water with wall rock minerals in fractures is minor, because stationary conditions are achieved more rapidly.

For comparison of natural waters with experimental leachates from the aquifer rocks it is appropriate to relate the REY abundances of the natural water to those of the experimental leachate, and to apply Ca as an internal major element for additional normalisation. Source-rock- and Ca-normalisation group the waters from the igneous rocks and lead to much more coherent trends of REY patterns. In particular, inherited and *acquired* anomalies can be distinguished because inherited anomalies are

2.1. ANOMALOUS BEHAVIOUR OF Ce, Eu AND Y

Among the REY, Ce, Eu, and Y often behave in an anomalous manner relative to the REE series. Ce and Eu anomalies are defined by the ratio of actual to expected values interpolated between the neighbouring REE. Thus, the anomalies are given as the deviation from the smooth normalised patterns. Because Y plots at the place of Ho, the ratio of Y/Ho is preferred to any type of interpolation method.

2.1.1. *Europium*

Calculations (Sverjensky, 1984) and experiments (Bilal, 1991) showed that at relevant fO_2 the divalent fraction of Eu increases with temperature, particulary above 250°C (Bau and Möller, 1992). Thus, Eu behaves differently from the other REE at high temperatures. Due to the large ionic radius of Eu^{2+} (0.139^{VIII} nm compared to 0.121 nm of Eu^{3+}; Shannon, 1976) it is not incorporated into Ca minerals at the same level as the other REE (Ca^{2+}: 0.126^{VIII}; 0.114^{VI}; Shannon, 1976). Since rocks rarely consist of large-ion-dominated minerals, into which Eu^{2+} would fit, Eu is enriched in residual fluids or is sorbed onto mineral surfaces. Its concentration in the range of pM/kg is definitely too low to form an Eu- accessory minerals. The fraction of "loosely-bound" Eu on mineral surfaces, often referred to as excess Eu, is more easily leachable from rocks by percolating water than the structurally incorporated REE, although at low temperatures all released Eu is trivalent (Möller et al., 1993). This fraction of Eu, contributes to any *inherited* positive anomaly. Another contribution to an *inherited* anomaly is the alteration of feldspars, particularly plagioclase. Since most magmatic feldspars have a positive Eu anomaly, their dissolution or alteration yields fluids with enhanced Eu. For instance, high temperature alteration of plagioclase (>250°C) is considered as the source of the strong positive Eu anomalies in black smoker fluids (Fig. 1) at mid-oceanic ridges (Klinkhammer et al., 1994) which may be further enlarged due to sorption effects at high temperatures. In contrast to plagioclase, alkali feldspars are much more stable in contact with hydrothermal fluids and, therefore, their contribution of Eu is less than that of plagioclase. Separated biotite from felsic rocks, which hosts numerous tiny solid inclusions of accessory minerals, shows strongly negative Eu anomalies, which result from the many inclusions. Pure biotite itself has low REY abundances and negligible Eu anomalies (Bea et al., 1994). When biotite in felsic rocks is chloritized (low temperature reactions), the fluid inherits the high abundance levels of REY and the negative Eu anomaly from the tiny inclusions (Möller and Giese, 1997).

Because the excess Eu from the intergranular space of rocks is leached faster than REY and thereby Eu is released from minerals with progress of weathering, the source-rock-normalised REY patterns of water loose the initially positive Eu anomaly with time. This may be the reason, why the thermal waters from Kyselka, Bad Säckingen, Hermersberg, and Kizildere only show insignificant Eu anomalies (Fig. 4).

Chemically similarly behaving species may be fractionated by different sorption kinetics, when tri- and divalent ions or ions with different electron configurations (such as Y and Ho) interact with mineral surfaces forming the walls of the pores. Considering Coulomb forces only, the divalent Eu is less strongly sorbed onto surfaces than the trivalent species. For instance, the divalent Eu moves faster through pores than the trivalent REY (Möller and Holzbecher, 1998). Fig. 3 shows calculated breakthrough

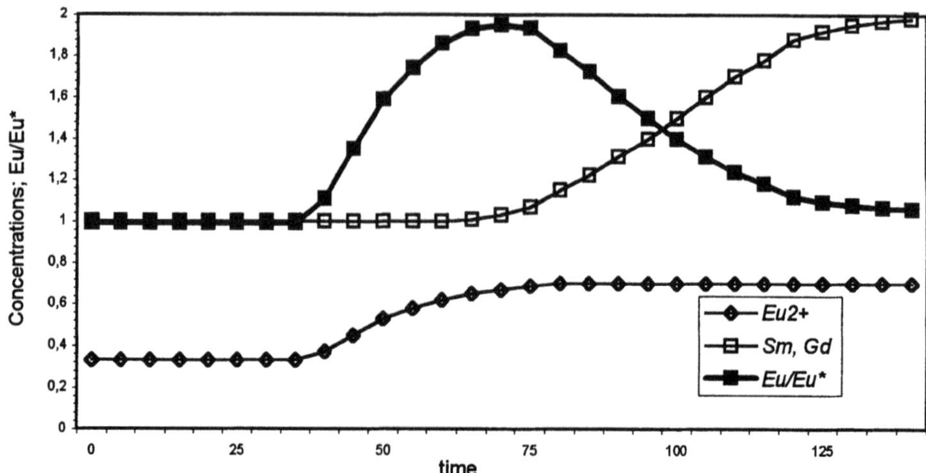

Fig. 3: Calculated breakthrough curves of two differently behaving chemical species and the Eu-anomaly Eu/Eu*. For instance, this figure illustrates the fractionation of Eu^{2+} and trivalent REE due to sorption onto mineral surfaces. It is assumed that one third of total dissolved Eu is divalent and moves faster than trivalent Sm and Gd, i.e. it is less adsorbed, when the pore fluid is replaced by another one which is higher in REE concentration by a factor of two (redrawn after calculations in Möller and Holzbecher, (1998).

curves which illustrate the time-dependent changes in element ratios and the development of an Eu anomaly. If the primary pore fluid is replaced by one with higher REE abundances, which is commonly the case if the replacing fluid has considerably higher temperature than the original pore fluid, a positive Eu anomaly must develop with time in the effluent. Such an Eu anomaly is *acquired* during migration of fluids at high temperatures only, whilst some Eu is divalent. At the temperature below about 200°C an Eu anomaly can only be *inherited* from the source rocks (Bau and Möller, 1992).

2.1.2. *Yttrium*

Due to the close physical similarity in size and charge, Y behaves often similar to Ho, although Y is not a 4*f* element. In fresh to poorly altered igneous rocks, Y and Ho behave alike (Fig. 2). In alteration processes, however, the small dissimilarity of sorption onto mineral surfaces, controlled by surface complexation (Bau et al., 1996; Diakonov et al., 1998), is multiplied in migrating fluids, and Y-Ho fractionation occurs. Different from anomalous Eu, the Y/Ho ratio is not principally dependent on temperature. In some groundwaters and thermal waters Y is enriched (see Y and Ho in Fig. 1 and 3). This anomaly is only *acquired*, whilst the rock is not in a transient equilibrium with the migrating water. If leaching lasts long enough, a steady state equilibrium might be reached and Y as well as Eu anomalies may vanish. As shown in Fig. 4, many waters exhibit *acquired* anomalies of Y. Although the waters from the granites of Kyselka and Bad Säckingen do not show significant Eu anomalies, they still show recognisable anomalous Y/Ho ratios.

2.1.3. *Cerium*

Ce anomalies are quite common in surface waters. For instance, oxic seawater is characterised world-wide by a negative Ce anomaly (Elderfield, 1988). It is assumed that this anomaly is caused by accumulation of Ce in ferromanganese nodules and crusts (Addy, 1979; Elderfield et al., 1981). In general, positive Ce anomalies form on Fe-Mn colloids and surface coatings (Braun et al., 1990) or by bacterial activity (Moffet, 1990). A similar, strongly negative Ce anomaly is observed in the thermal water of Pamukkale, Turkey (unpubl.) and groundwater from the Calcareous Alps near Trento/Italy (unpubl.). It is common to all these waters that they are below pH 7, bicarbonate-rich, and high in Eh. It is assumed that the precipitation of FeOOH along the migration paths of the infiltrating meteoric waters also induce oxidation to Ce^{4+} which is preferentially sorbed onto freshly precipitated oxyhydroxides. Because most igneous and metamorphic rocks yield waters with low Eh values, negative Ce anomalies are absent from waters derived from such rocks. In contrast, Ce behaviour in alkaline lakes is considerably different as shown by strong positive Ce anomalies (Möller and Bau, 1993). This is best explained by carbonate complexation of Ce^{4+} (unpublished results). Among the studied waters from felsic rocks, positive Ce anomalies have been found in the thermal water from Jachymov (Möller et al., 1998) and the mineral water from Hermersberg (Möller et al., 1997a). This might be explained by the type of alteration in these areas. For instance, the granite of Jachymov has been altered along with the post-Variscan uranium mineralization in this area. Since U is only mobilised by oxidising fluids, it may be supposed that also Ce was partly oxidised and fixed at mineral surfaces of the altered granite. The present-day water is chemically reducing and contains Fe^{2+}. If this type of water passes the Jachymov granite, some of the surfacially fixed Ce^{4+} is leached as Ce^{3+} and added to the Ce fraction from dissolved minerals of the leached rocks.

Because fresh igneous and metamorphic rocks do not show Ce anomalies, all positive Ce anomalies in rocks are achieved by alteration reactions, whereas the negative ones (Bad Wildbad) might be explained by oxidation of Fe together with Ce. In general, Ce anomalies are not typical features of primary rocks but the result of alteration reactions.

2.2. ACCESSIBILITY OF REY IN ROCKS

Depending on the distinct conditions during crystallisation of igneous and metamorphic rocks the accessibility of REY from these rocks may be different. Solidification of igneous rocks proceeds by cooling, i.e. minerals formed from melts at high temperatures. Many of them become unstable at low temperatures and, therefore, recrystallise or are altered in the presence of fluids. The trace elements, often hosted by the major minerals, are thereafter partly released from hydrated accessory minerals. In metamorphic rocks, crystallisation of alteration minerals follows increasing temperature and pressure, and the incompatible trace elements are rejected from major minerals. Thus, the distribution of REY is influenced by the composition of the fluid and the mode of crystallisation history.

The fraction of any solid that is dissolved by fluid-rock interaction, either naturally or experimentally, does not necessarily represent the average composition of the rock. The

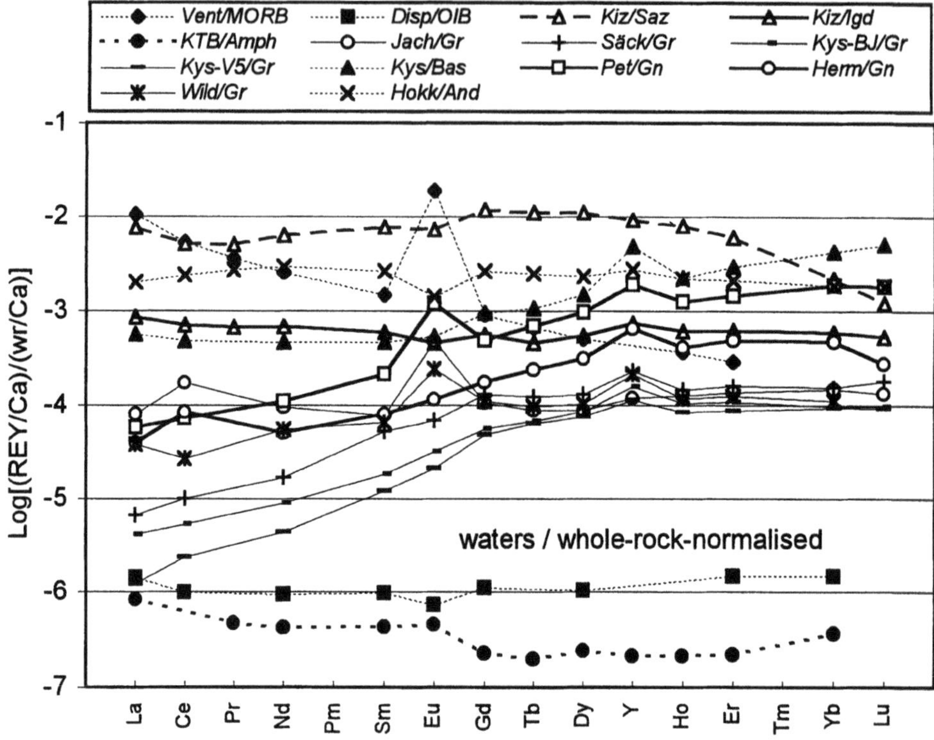

Fig. 4: Source-rock-normalised REY/Ca patterns of waters from felsic (solid lines) and basic (broken lines) aquifer rocks. Data are given in Table 2; abbreviations are explained in Table 1.

REY in the solute are significantly controlled by soluble tracer phases with REY patterns that may be totally different from those of the major components. Since varying fractions of REY are bound to accessory minerals, detailed knowledge of the distribution of REY among the major minerals is not sufficient for deriving the REY patterns of waters that have interacted with the rocks. It is the solubility of the minor phases (particularly of the phosphates; Irber, 1996) that control the behaviour of REY in water-rock interactions.

2.3 LEACHING PROCEDURE

In order to study the REY contributions of minor phases, a leaching procedure was applied that allows the determination of the easily soluble fraction of elements at 70°C and self-adjusting pH values in range of 3 to 4 depending on the type of rock and its alteration (Möller and Giese, 1997). Four aliquots of 1 g of powdered rock samples (grain size <100 μm) are weighted into polyethylene bottles, where they are mixed with 4 g of ion exchange resin in H^+ form (BIORAD AG50W-X8) and 100 ml bidistilled water. All bottles were kept in a shaking water bath at 70°C. After 1, 4, 10, and 20

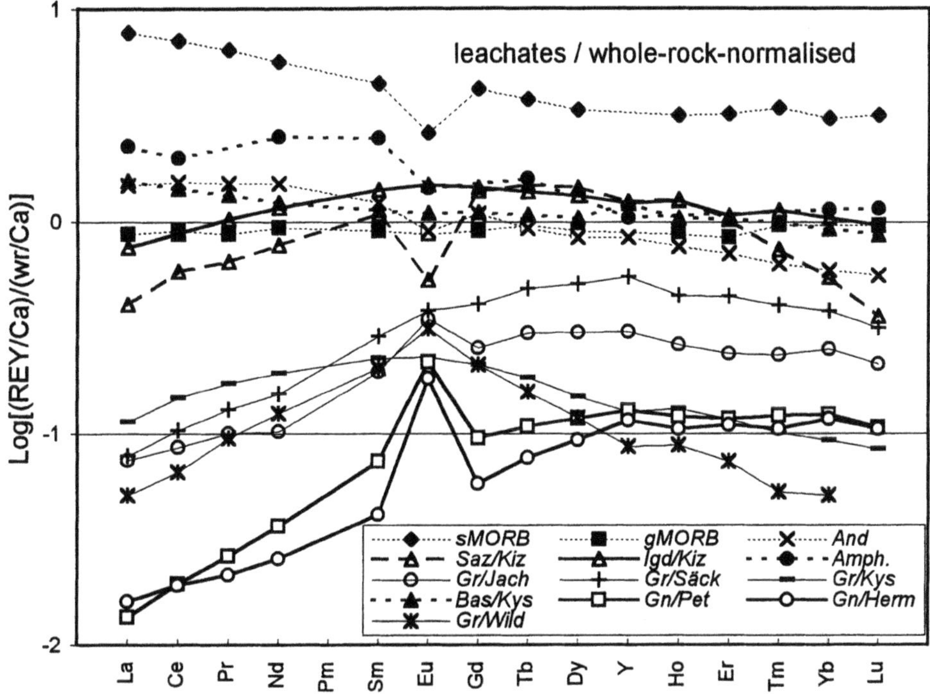

Fig. 5: Source-rock-normalised REY/Ca pattern of leachates of the sampled aquifer rocks from localities from which the thermal and mineral waters were collected. Data are given in Table 2. Solid and broken lines signify felsic and mafic environments. Abbreviations are explained in Table 1 excepting ***sMORB*** and ***gMORB***, which represent leachates of the semicrystalline interior and the glassy skin of a dredged pillow.

hours, one batch was stopped by decanting the solution and separating the rock powder and resin by wet sieving. The resin was transferred into a chromatography column and the collected ions were eluted with 40 ml of 4 M HNO_3. In the eluate major and trace elements were determined by ICP-AES and ICP-MS, respectively.

The following aspects were studied by this leaching method:

- The quantities of the easily soluble fractions of elements in rocks under controlled acidic conditions in the pH range of 3 to 4.
- The distribution of REY between easily or less soluble groups of minerals in rocks. Although the specific minerals are not known, the leached fraction is equivalent to a bulk analysis of the soluble components, which cannot be obtained by any other method.
- The change of the soluble fractions with alteration of the rocks.
- The distribution of REY among the soluble mineral phases controls the REY abundances in natural waters. Under the applied conditions of pH 3 many minerals are more soluble than at pH 6, some are even unstable such as calcite and apatite. Thus, the results at pH 3 represent the maximum accessible REY fractions which can be leached from a rock in a long run under acidic condition, i.e. in the presence of CO_2.

Where possible, drill core samples of the aquifer rocks were chosen for the leaching experiments. Although the rocks are never the exact equivalents of the inhomogeneous aquifer rocks, they may be considered as the best material which the mineral waters can be related to. The REY analyses of these rocks were used for the subsequent normalisation of REY values of the corresponding waters and leachates.

3. Results

3.1. WATER FROM FELSIC ROCKS

3.1.1. *Sampling areas*

In Bohemia, Czech Republic, the sparkling mineral water (Na-Ca-HCO_3) from Kyselka is produced from the altered Carboniferous Slavkovsky-les granite that underlies a Tertiary basalt of the Doupovske hory. The water was sampled at the filling station of the Mattoni Company in Kyselka. At Jachymov, a thermal Ca-Na-HCO_3 water was recovered from wells drilled from underground about 100 m into the muscovitised Krusne hory granite. This low-CO_2, thermal water is used for cures. From all these rocks drill core samples were available. A detailed study of water from these localities has been published elsewhere (Möller et al., 1998).

The sparkling Ca-Na-HCO_3 thermal water from Bad Wildbad, northern Black Forest, Germany, is considered to be derived from the Kegelbach granite (Stober, 1995) from which a drill core sample was available. The sparkling mineral water from Bad Peterstal (Na-Ca-HCO_3) and the low-CO_2 water from nearby Hermersberg (Na-Ca-HCO_3) originate from altered gneisses with granitic veinlets. Samples for leaching studies were available from drill cores. The thermal water from Bad Säckingen (Na-Ca-Cl) is produced from a granitic aquifer. Here, the granite of Bad Säckingen had to be sampled at a nearby outcrop. Detailed descriptions of samples from the Black Forest area are given by Möller et al. (1997a).

The CO_2-rich, Na-HCO_3-SO_4 waters from the geothermal pilot plant at Kizildere, W-Anatolia, Turkey, was recovered from the silencers at 1 bar and 100°C. Temperature and pressure at the drill heads are about 190°C and 13 atm, respectively. The recalculated bottom hole temperatures are about 220°C (Giese, 1997). In the two years of sampling REY patterns of the same type but differing in the REY levels by factors of 3 were obtained (unpublished data). For that reason, the samples from 1995 and 1996 are presented separately. The assumed aquifer is either a Tertiary marble (Sazak Formation) and/or the underlying mica schists that are in part rich in marble (Igdecik Formation). Since drill core material was not available, the country rocks were collected from nearby outcrops. The REY patterns of mica schists and marble from Kizildere are different (Fig. 2). Since it is unknown which of the two types of rocks controls REY abundances in the geothermal water, Kiz95 and Kiz96 are related to the marble (Kiz/Saz) and mica schist (Kiz/Idg), respectively (Fig. 4).

3.1.2. *REE patterns of waters*

The source-rock-normalised REY/Ca patterns of all waters from granites (Fig. 4) represent a rather small field compared with the wide scatter of chondrite-normalised patterns (Fig. 1). The light REE (LREE) show a somewhat wider spread than the heavy

REE (HREE) and Y. The very narrow spread of HREE and Y indicates that the distribution of these elements between soluble Ca- and REY-bearing minerals is similar too, probably because of similar surface complexes. The gneiss-related waters from Peterstal deviate from the granitic trend as defined by the samples Kys-BJ/Gr, Kys-V5/Gr, Säck/Gr, Wild/Gr, and Jach/Gr in Fig. 4. The mica-schist-related REY pattern of the geothermal water from Kizildere is more horizontal than the gneiss-related ones. The marble-related water from Kizildere has the highest REY level of all samples shown in Fig. 4. Where determined, many patterns show enhanced Y abundances in water which indicate that these aquifer rocks are either in disequilibrium with the percolating water, or fresh rocks are constantly involved in the water-rock interaction due to progress of weathering or tectonic events. Four out of seven waters from felsic rocks *inherited* no Eu anomaly, and two a positive one (Fig. 4), although all granites are characterised by a negative Eu anomaly (Fig. 2), which indicates that the excess Eu from the intergranular space has not been leached yet. Only if the process of leaching lasted long enough, a stationary equilibrium for Eu distribution is established and the Eu anomalies vanish (Fig. 1: Kys-BJ; Kys-V5; Säck; Herm).

Ce anomalies are created by earlier alteration processes. If the infiltrating fluid was oxidising, Fe released by weathering was precipitated as FeOOH which scavenges REY. It might be assumed that part of the Ce was oxidised to Ce^{4+} and was much more retained by of the precipitate during aging than the trivalent REY (Bau et al., 1998). Thus, the surface coatings of minerals in rocks are slightly enriched by Ce. The present-day water is chemically reducing and contains Fe^{2+}. If this type of water passes the Jachymov granite, some of the surface-coating with Ce^{4+} is reduced to Ce^{3+}, leached and contributes to the amount of the Ce fraction from dissolved minerals of the weathered rocks. Along such lines of arguments, the positive Ce anomaly in the water from Jachymov and Hermersberg might be explainable.

3.1.3. *Leachates*

The quantities of experimentally leached REY, obtained in this study, exceed by far those in natural waters (Table 1). The REY/Ca ratios are larger by 2 to 5 orders of magnitude. Thus, both processes tap different volumes of minerals and probably different solid phases. For instance, the experimental leachate completely dissolves carbonates, apatite, and hydrous phosphates within 20 hours, whereas the natural waters interact mainly with the altered surfaces of minerals. This has to be considered, when comparing Figs. 4 and 5. Since in batch-leaching experiments anomalies cannot be *acquired* but only be *inherited*, the absence of any significant Y-Ho fractionation during leaching evidence that the enhanced Y/Ho ratios in waters are due to sorption processes during migration.

The leachates of the granites show a coherent trend, which is different from that of the corresponding waters. Here, LREE plot more closely together than the HREE and Y. This suggests that the accessible LREE are bound to minerals of very similar solubility, whereas HREE and Y originate from a greater variety of minerals.

The leachates of the gneisses from Bad Peterstal Gn/Pet and of the basement rocks of Kizildere Igd/Kiz are distinctly different from those of the granites (Fig. 5). The leachates of gneisses exhibit strongly positive Eu anomalies, whereas those of the rocks from Kizildere show none (Igd/Kiz) or even negative ones (Saz/Kiz). Remarkable is the

positive Eu anomaly of the leachate of the gneisses Gn/Pet and Gn/Herm. Because the water from Hermersberg (Herm/Gn in Fig. 4) shows no Eu anomaly but the leachate does, it is assumed that the latter tap Eu-enriched minerals such as K-feldspars.
Comparing the behaviour of Eu in Figs. 5 and 4 it is obvious that the waters and leachates from the granites of Kyselka and Bad Säckingen do not contain any excess Eu (at least not in the studied samples). In contrast, the natural waters and corresponding leachates of the granites from Bad Wildbad and Jachymov and the gneiss from Bad Peterstal still leach minerals or surface coatings that are enriched in Eu relative to the bulk composition. In other places, the expected Eu anomaly of the water is much smaller. A reason for the absence of Eu anomalies may be that the excess Eu has previously been leached from the rocks. This may happen, if the process was ongoing over geologic time scales.
The trends of source-rock-normalised REY patterns of the leachates of both types of rocks from Kizildere are rather similar but with significant differences in Eu anomalies (Fig. 5). Although the marble shows a small positive Eu anomaly (Fig. 2), the leachate is characterised by a negative one (Fig. 5). This is only possible, if the minerals accessible to leaching have a deficit in Eu, although the whole rock is slightly enriched. On the other hand, the water from the mica schist is characterised by a strongly negative Eu anomaly, whereas the leachate has none. This shows that some Eu is present in easily leachable solids, which, just by chance, lead to absence of an Eu anomaly in the leachate. Since the geothermal water from Kizildere only has a negligible tendency to negative Eu anomalies (Fig. 1), the latter can only be *inherited* from the mica schist. The leachates of the mica schists indicate the absence of excess of Eu, whereas the leachates of the marble indicate the presence of soluble Eu-deficient mineral phases. If the marble dominated the REY pattern of the geothermal water, the Eu anomaly should be much more negative than determined in the water. Summarising, the mica schist or its equivalents is the most probable source that controls REY patterns of the geothermal fluids in Kizildere. The high-permeability of the enclosed marble in the Idgecik Fm can, of course, be part of the aquifer system. On the other hand fluid interaction with the marble of the Sazak Fm cannot completely be ruled out, because the low REY contents in the marble compared to the high ones in mica schist would be difficult to recognise in mixtures.
None of the REY patterns of the leachates (Fig. 5) and whole rocks (Fig. 2) show Ce anomalies, although they are quite common in the respective water patterns (Fig. 4). This is due to the fact that the experimental leaching dissolves larger parts of the accessible minerals, whereas natural leaching just only interacts with surfaces of minerals, which are obviously composed differently from the bulk. The excess Ce in the waters from the Jachymov granite and the gneiss from Hermersdorf must originate from minor solid phases such as surface coatings that only insignificantly contribute to the experimental leaching because their fraction is negligible.

3.2. WATER FROM MAFIC ROCKS

3.2.1 *Sampling areas*

The basalt from the Doupovske hory overlies the Carboniferous Slavkovsky-les granite.

The CO_2-rich (about 2300 mg/kg; 11°C), Ca-Na-HCO_3 water from wells in the alkalibasalt were sampled at the filling station of the Mattoni Company, Kyselka, Czech Republic. The rock samples were taken from drill cores which show strong hematitisation and precipitation of calcite in fractures.

The Continental Deep Drilling Project (KTB) produced about 270 m^3 of a highly saline (70 g TDS/kg) Ca-Na-Cl brine from the open hole section between 3850-4000 m depth at temperatures of 129°C (Maiwald and Lodemann, 1994). The brine drained from fissures in the amphibolite into the borehole (Möller et al., 1997b). The temperature of the sampled brine was about 30°C at 250 m below surface under N_2/CH_4 cover. The "4000 m brine" was sampled at the end of the pumping test, i.e., after 3 months. Thus, it may also have drawn water from the overlying paragneisses, the only different geological unit at this location.

Data for vent fluids, dispersed flows, and the Hokkaido iron-spring are taken from the literature. The analyses of black smoker fluids (Na-Ca-Cl) from the East Pacific Rise (11°, 13°, 21°N) are taken from Klinkhammer et al. (1994). These ca. 400°C vent fluids were sampled by submercibles. The mid-ocean-ridge basalt (MORB) were dredged at the East Pacific Rise (21° S). This pillow had a glassy rim which was separately leached (gMORB). The interior of the pillow was semicrystalline (sMORB).

The dispersed flows with temperatures of about 30°C are from Teahitia, Society Islands, and were sampled during the Cyana cruises (Michard et al., 1993). REE analyses of ocean island basalts (OIB) of Teahitia as well as Meahitia are reported by Hemond et al., 1994. The main problem with these waters is that the thermal regime is still unclear: the water may have derived (i) from interaction of basalt with limited amounts of low-temperature fluids containing only small fractions of seawater, (ii) from high-temperature interaction with basalt followed by significant mixing with seawater, or (iii) both processes (Michard et al., 1993).

The cold (8°C) Ca-Na-SO_4 water from the mildly acidic (pH 4) iron-spring at Nishiki-numa, Hokkaido/Japan, is compared with the basaltic andesite from which it is derived (Bau et al., 1998). The pH value of the water is low because of the weathering of sulphide minerals in the andesite indicating previous alteration.

3.2.2. *Water*

Different from waters from felsic rocks, REY patterns of waters from the basic rocks show extreme scatter (Figs. 1 and 3). Even when Ca- and whole-rock-normalised, they split into two groups: (i) waters from basalts and andesite with high levels of REY, and (ii) waters from amphibolite and OIB with low contents of REY. They differ by 3 orders of magnitude. With exception of the high-temperature vent fluids and the low-temperature mineral water from Kyselka, all other patterns are relatively flat. Only the water from the alkalibasalt of Kyselka, Czech Republic, shows enhanced Y, the iron-rich spring of Nishiki-numa, Japan, has a just visible Y anomaly, whereas for the remaining samples Y was not available in the whole rocks. A strongly positive Eu anomaly is only observed in the high temperature, acidic, black smoker fluids, whereas the water from the andesite exhibits a negative one. The vent fluids with temperatures above 400°C show steep patterns with strongly positive Eu anomalies, which is typical for all black smoker fluids (Klinkhammer et al., 1994; Michard, 1989), but they are not anomalous in Y abundances (Bau et al., in press). Eu is *inherited* to a large extent

(Klinkhammer et al., 1994; Michard, 1989) but might also be acquired by sorption processes during migration due to the high temperature of these systems at which Eu is partly divalent (Möller and Holzbecher, 1998). The dispersed flows are flat with only slightly negative Eu anomalies. The REY trends of the vent fluids and the cold mineral waters from Kyselka are oppositely directed. This might be due to high acidity of the vent fluids, by which OH^- complexation at surfaces and in solution is circumvented, and chemical complexation by HCO_3^- in the mineral waters occurs. The absence of a significant Eu anomaly in the 4000 m fluid of KTB indicates that the country rocks do not contain excess Eu. The published enhanced Ce is an artefact caused by drilling (Möller et al. 1994) and is, therefore, not shown in Figs. 1, 4, 5, and 6.

3.2.3. *Leachates*

The leachates of the amphibolite, MOR basalts, alkalibasalt, and andesite resemble much more each other (Fig. 5) than the corresponding waters (Fig. 4). The source-rock-normalised REY pattern of the leachate of the glassy MORB is around unity, indicating (i) homogeneous distribution of REY and Ca in the solid, and (ii) congruent bulk dissolution. The semicrystalline MORB, however, shows distinct distribution of REY in mineral phases that are easily soluble (Irber et al., 1996). The accessibility of REY in MORB is high compared with the studied alkalibasalt and amphibolite. In general, all the studied mafic rocks show significantly higher accessibility of REY than the felsic rocks, and Eu and Y behave in a normal fashion.

The low-temperature, dispersed flows from the Teahitia Seamount (Michard et al., 1993) show patterns that are very similar to leachates of the semicrystalline MORB (Giese and Bau, 1994). In the semicrystalline MORB, OIB and amphibolite REY are present in solid phases that are more soluble than plagioclase, which usually has a positive Eu anomaly. This explains the deficit of Eu in the leaching patterns.

3.3. RETENTION OF REY BY CRYSTALLINE ROCKS

Relating the REY/Ca patterns of the 20 hours leachates to the corresponding REY/Ca patterns of the waters (eq.1) yields insight into the tendency of the aquifer rocks to retain REE and Y (Fig. 6).

$$R = \frac{(REY/Ca)_{leachate}}{(REY/Ca)_{water}} \quad (1)$$

Although the pH of the interacting fluids as well as the temperature in the natural leaching process are not comparable with those in the leaching experiments, the resulting retention patterns (R patterns) visualise that for most of the rocks the retention is in the range of 100 to 10000. In any case, Ca is more severely leached than REY which, therefore, become enriched in the alteration minerals or surface coatings. In granite-water systems, R values systematically decrease from La to Lu, whereas they are very variable in mafic rock-water systems, where R values are lowest when the rock is in a glassy state, i.e. REY are homogeneously distributed and are leached to the same amount as the major elements. Under such conditions, solubility of the major elements

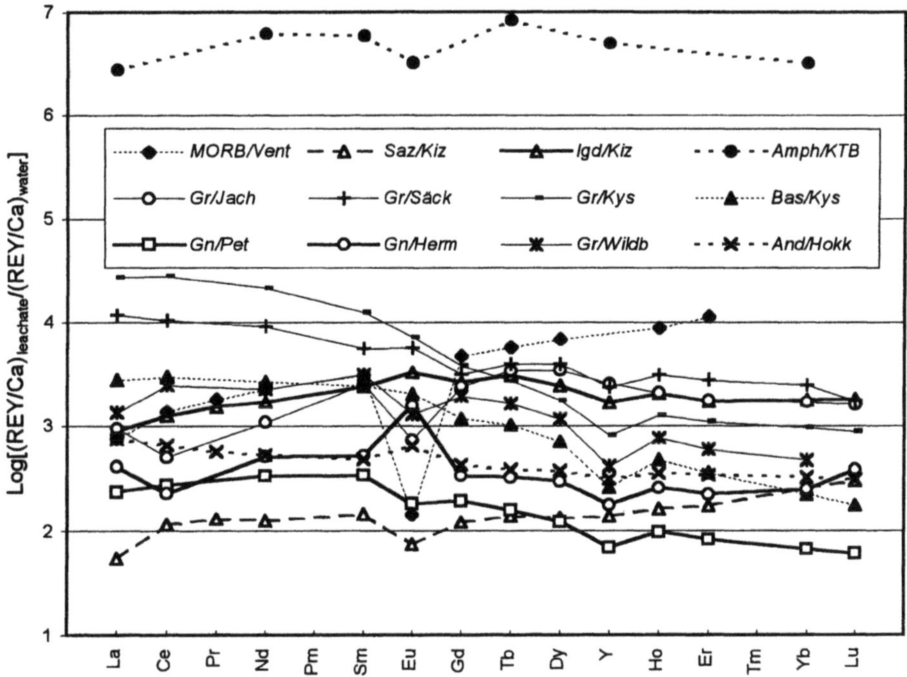

Fig. 6: REY patterns of R-values as defined by the ratio of the composition of 20-hours leachates of rocks to that of the corresponding waters (eq.1). Where several analyses were available, means are used for calculations. ***MORB/Vent***: smoker systems. ***Saz/Kiz***: leachates of marly limestone of the Sazak Formation related to geothermal water from Kizildere (analyses from 1995). ***Igd/Kiz***: leachates of marly limestone of the Igdecik Formation related to geothermal water from Kizildere (analyses from 1996). ***Amph/KTB***: leachate of amphibolite from 3800 to 4000 m related to the brine from 4000 m, KTB drill hole in the Oberpfalz, Germany. ***Gr/Jach***: leachates of altered granites from drill cores related to the thermal water from underground drill holes in the mine of Jachymov; both from the same locality. ***Gr/Säck***: leachate of outcropping granite related to thermal water of Bad Säckingen, Southern Black Forest, Germany. ***Gr/Kys*** and ***Bas/Kys***: leachates of granites and basalt from drill cores from Kyselka, Bohemia, Czech Republic, related to the mineral water recovered from these rocks. ***Gn/Pet*** and ***Gn/Herm***: leachates of gneisses from Bad Peterstal and Hermersberg, Central Black Forest; Germany related to mineral water from the corresponding drill holes. ***Gr/Wild***: leachates of granites from drill cores of granites from Bad Wildbad, Northern Black Forest related to thermal waters of this area. ***And/Hokk***: leachates of andesites related to waters of the iron-spring Nishiki-numa, Japan.

also control the solubility of the minor elements. If the REY form own minerals, their solubility controls the abundance in solutions that interacted with the crystalline rocks.

The quite narrow band in Fig. 6 indicates that essentially the same processes act on REY in water-rock interactions. The products may be not the same, but behave in a similar way.

The decreased Y values indicate that Y is often less strongly bound in rocks than its REE neighbours. This is not astonishing because Y is not a 4*f* element. Only under magmatic conditions Y behaves precisely like Ho. In aqueous systems, Y displays its own behaviour, which may be related to small differences in solubility products of the hydroxides (Diakonov et al., 1998).

The R values should depend on the chemical complexation at surfaces and in solution and formation of secondary minerals during the natural alteration processes. No systematic trend is recognisable for the effect of HCO_3^- in solution, which varies between less than 10 to more than 1000 mg/kg in the pH range from 5 to 8 at about 30°C (Tab. 1). Tentatively, it might be assumed that the trends of R values are largely controlled by surface complexation as represented by hydrolysis reactions. Only the vent fluids with pH values between 3 and 4 show R values increasing from La to Lu, thereby indicating that at these temperatures other processes control the REY distribution than at lower temperatures and higher pH values.
Positive Eu and negative Ce anomalies in R patterns are due to reduction and oxidation these elements prior to leaching, respectively. During leaching these elements are either more or less easily accessible than the trivalent REE. The high retention of Eu in the gneiss from Bad Peterstal is most probably due to the stability of alkalifeldspar, whereas the low retention in MORB is due to the rapid decomposition of plagioclase under the acidic conditions in the smoker fluids.

4. Conclusions

Although waters from magmatic and metamorphic rocks exhibit widely varying chondrite-normalised REY patterns, the source-rock-normalised REY/Ca patterns of water show much closer similarities, i.e. host rocks chemically control the REY patterns of waters. All waters from felsic rocks are similar, whereas those from mafic rocks show a wide spread. This might be attributed to different modes of crystallisation of the aquifer rocks. The studied felsic rocks are all coarse grained, whereas the basaltic and andesitic rocks are fine grained with variable amounts of glass. The minor phosphate contents of the less crystalline basic rocks, for instance, may give rise to the formation of the more easily soluble accessories than in the slowly crystallising phosphate-enriched felsic rocks.
The source-rock-normalised, 20-hours leachates show some similarities with the correspondingly normalised waters. The waters and leachates of the felsic and mafic magmatic rocks both form distinctive groups of source-rock-normalised REY/Ca patterns, whereas those of the metamorphic or strongly altered equivalents differ considerably in REY levels. For instance, REY patterns of leachates of gneisses are lower than those of granites, but in natural waters they are inverse; those of the leachates of basalts are all in the same range, whereas the waters derived from the amphibolite and the semicrystalline MORB exhibit considerably higher REY levels than the less altered basalts. The leached REY/Ca ratio from the glassy MORB is about unity, whereas it increases up to one order of magnitude after partial crystallisation.
When relating the leachates of rocks to the corresponding water composition a rather narrow band of retention patterns of REY are obtained covering only 2 orders of magnitude with only one exception. Within this band different types of REY patterns are recognised. The retention of REY in the less altered granites Gr/Kys and Gr/Säck and the basalt Bas/Kys decrease from La to Lu, whereas the strongly altered rocks and most metamorphites show either horizontal or increasing patterns from La to Lu. Most of the pattern show negative Y and variable Eu anomalies. Thereby, it is indicated that the accessibility and/or chemical behaviour of these two elements behave different from

the other REE during weathering of rocks and migration of fluids.
Anomalies of Eu, Y and Ce reveal the history of the fluid-rock interaction. Eu is sensitive to temperature, whereas Y is not. Anomalous Eu is *inherited* and may be enhanced at temperatures above 250°C. Ce is sensitive to oxygen fugacity and pH. Y seems to be sensitive to pH and to ligands dominating REY complexation in solution and on surfaces. Some waters *acquired* enhanced Y/Ho ratios irrespective of their source rocks. Using the anomalies of REY patterns of waters temperature-dependent reactions can be discussed. Metamorphic waters are geologically old and could have interacted intensively with the aquifer rocks, leading to total removal of excess Eu. Such waters would not show Eu anomalies source-rock-normalised patterns. Ce anomalies may still occur depending on the development of Eh with time. Although groundwater is geologically very young, it might be in a steady state equilibrium with the aquifer rocks. Under such conditions, an acquired positive Y and *inherited* Eu anomaly indicate temperatures of water-rock interaction below 200°C. Since in mildly acidic to neutral waters no or only positive Y anomalies are observed, this indicates that Y is released more easily from the rocks than REE and is less retained by sorption onto mineral surfaces.
In summary, REY represent a unique tool to study the behaviour of trace elements in water-rock interaction with time.

5. Acknowledgement

This contribution benefitted from the thorough analytical work of P. Dulski and the constructive criticism of two anonymous reviewers.

6. References

Addy S.K. (1979) Rare earth element patterns in manganese nodules and micronodules from northwest Atlantic. *Geochim. Cosmochim. Acta* **43**, 1105-1115.

Anders, E. and Grevesse, N. (1989) Abundance of elements: Meteoric and solar. *Geochim. Cosmochim. Acta* **53**, 197-214.

Bau, M. and Dulski, P. (1996) Antropogenic origin of posititve gadolinium anomalies in river water. *Earth Planet. Sci. Lett.* **143**, 245-255.

Bau, M. and Möller, P. (1992) Rare earth element fractionation in metamorphogenic hydrothermal calcite, magnesite and siderite. *Mineral. Petrol.* **45**, 231-246.

Bau, M., Koschinsky, A., Dulski, P. and Hein, H.J. (1996) Comparison of the partitioning behaviours of yttrium, rare-earth elements, and titanium betwen hydrogenetic marine ferromanganese crusts and seawater. *Geochim. Cosmochim. Acta* **60**, 1709-1725.

Bau, M., Usui, A., Pracejus, B., Mita, N., Kanai, Y., Irber, W., and Dulski, P. (1998) Geochemistry of low-temperature water-rock interaction: Evidence from natural waters, andesites and Fe-oxyhydroxide precipitates at Nishiki-numa iron-spring, Kokkaido, Japan. *Chem. Geol,* **15**, 293-307.

Bea, F., Pereira, M.D., Corretge, L.G., and Fershitater, G.B. (1994) Differentiation of strongly peraluminous, perphosporous granites: The Pedrobernardo pluton, Central Spain. *Geochim. Cosmochim. Acta* **58**, 2609-2627.

Bilal, B.A. (1991) Thermodynamic study of Eu^{3+}/Eu^{2+} redox reaction in aqueous solutions at elevated temperatures and pressures by means of cyclic voltammetry. *Z. Naturforsch.*, **46a**, 1108-1116.

Blundy, J.D. and Wood B.J. (1991) Crystal-chemical controls on the partitioning of Sr and Ba between plagioclase feldspar, silicate melts, and hydrothermal solutions. *Geochim. Cosmochim. Acta* **55**, 193-209.

Braun J.J., Pagel M., Muller J.P., Bilong P., Michard A., and Guillet B. (1990) Cerium anomalies in lateritic profiles. *Geochim. Cosmochim. Acta* **54**, 781-795.

Coryell, C.D., Chase, J.W., and Winchester, J.W. (1963) A procedure for geochemical interpretation of terrestrial rare earth abundance patterns, *J. Geophys. Res.* **68**, 559-566.

Diakonov, I. I., Ragnarsdottir K.V. and Tagirov, B. R. (1998) Standard thermodynamic properties and heat capacity equations of rare earth hydroxides: I. Ce(III)-, Pr-, Sm-, Eu(III)-, Gd-, Tb-, Dy-, Ho-, Er-, Tm-, Yb-, and Y-hydroxides. Comparison of thermochemical and solubility data. *Chem. Geol.* **151**, 327-347

Drever, J.I. (1988) The geochemistry of natural waters. 2nd edt. Prentice Hall, Englewood Cliffs, New Jersey, 437p.

Dulski P. (1994) Interferences of oxide, hydroxide, and chloride analyte species in the determination of rare earth elements in geological samples by inductively coupled plasma-mass spectrometry. *Fresenius J.Anal. Chem.* **304**, 193-203.

Elderfield H. (1988) The oceanic chemistry of the rare earth elements. *Philos Trans. R.* Soc. *London* Ser. A **325**, 105-126.

Elderfield H., Hawkesworth C.J., and Greaves M.J. (1981) Rare earth element geochemistry of oceanic ferromanganese nodules and associated sediments. *Geochim. Cosmochim. Acta* **45**, 513-528.

Garrels R. M. (1967) Genesis of some ground waters from igneous rocks. In P. H. Abelson (ed.) *Researches in Geochemistry*, **2**, 405-420.

Garrels R. M. and Mackenzie F. T. (1967) Origin of the chemical compositions of some springs and lakes. In R. F. Gould (ed) Equilibrium concepts in natural water systems. *Am. Chem. Soc. Adv. Chem. Ser.* **67**, 222-242.

Giese, L. (1997) *Geotechnische und umweltgeologische Aspekte bei der Förderung und Reinjektion von Thermalfluiden zur Nutzung geothermischer Energie am Beispiel des Geothermalfeldes Kizildere und des Umfeldes, W-Anatolien/Türkei.* PhD Thesis, Free University Berlin, 250p.

Giese, U. (1993) *Bestimmung des leicht mobilisierbaren Ionenanteils in Gesteinen.* PhD Thesis, Free University Berlin, 155p.

Giese, U. and Bau, M. (1994) Trace element accessibility in mid-ocean ridge and ocean island basalt: an experimental approach. *Min. Mag.* **58A**, 329-330.

Hemond, C., Devey, C.W., and Chauvel, C. (1994) Source compositions and melting processes in the Society and Austral plumes (South Pacific Ocean): Element and isotope (Sr, Nd, Pb, Th) geochemistry. *Chem. Geol.* **115**, 7-45.

Humphris S.E., Morrison M.A., and Thompson R.N. (1978) Influence of rock crystallisation history upon subsequent lanthanide mobility during hydrothermal alteration of basalt. *Chem. Geol.* **23**, 125-137.

Irber, W. (1996) *Laugungsexperimente an peraluminischen Graniten als Sonde für Alterationsprozesse im finalen Stadium der Granitkristallisation mit Anwendung auf das Rb-Sr-Isotopensystem.* PhD thesis, 319pp, Free Univ. Berlin

Irber, W., Bau, M. and Möller, P. (1996) Experimental leaching with cation exchange resin: a method to estimate element availabilities in geological samples. *J. Conf. Abstr.* **1**, 280.

Klinkhammer, G.P., Elderfield, H., Edmond, J.M., and Mitra, A. (1994) Geochemical implications of rare earth element patterns in hydrothermal fluids from mid-ocean ridges. *Geochim. Cosmochim. Acta* **58**, 5105-5113.

McLennan S.M. ((1989) Rare earth elements in sedimentary rocks: Influence of provenance and sedimentary processes. In B.R. Lipin and G.A. McKray (eds) *Geochemistry and mineralogy of rare earth elements.* Mineral. Soc.Amer. 169-200.

Maiwald, U. and Lodemann, M. (1994) Continuing recordings of physicochemical and hydraulic parameters during the pumping test 1991 at KTB pilot borehole (KTB-VB) *Sci. Drill.* **4**, 95-99.

Michard, A. (1989) Rare earth systematics in hydrothermal fluids. *Geochim. Cosmochim. Acta* **53**, 745-750.

Michard, A., Michard, G., Stüben, D., Stoffers, P., Cheminee, J.-L., and Binard, N. (1993) Submarine thermal springs associated with young volcanoes: The Teahitia vents, Society Island, Pacific Ocean. *Geochim. Cosmochim. Acta* **57**, 4977-4986.

Moffet, J.W. (1990) Microbially mediated cerium oxidation in seawater. *Nature* **345**, 421-423.

Möller P. (1988) The dependence of partition coefficients on differences on ionic volumes in crystal-melt systems. *Contrib. Mineral. Petrol.* **99**, 62-69

Möller P. (1998) Eu anomalies in hydrothermal minerals: Kinetic versus thermodynamic interpretation. Proc. 9th IAGOD Symp. Peking 1994, Schweizerbart Verlag, 239-246.

Möller, P. and Bau, M. (1993) Rare-earth patterns with positive cerium anomaly in alkaline waters from Lake Van, Turkey. *Earth Planet. Sci. Lett.* **117**, 671-676.

Möller, P., Dulski, P., and Giese, U. (1994) Rare earth elements in KTB-VB fluids. *Sci. Drill.* **4**, 113-122.

Möller P., Dulski P., and Giese U. (1993) REE distribution in granites of the Fichtelgebirge. *Z. geol. Wiss.* **21** 193-206.

Möller, P. and Giese, U. (1997) Determination of easily accessible metal fractions in rocks by batch leaching with acid cation-exchange resin. *Chem. Geol.* **137**, 41-55.

Möller P. and Holzbecher E. (1998) Eu anomalies in hydrothermal fluids and minerals: A combined thermochemical and dynamic phenomenon. *Freib. Forsch.-H.* **C475**, 73-84.

Möller, P., Morteani, G., Fuganti, A., Dulski, P., and Gerstenberger, H. (1998) Rare earth elements, yttrium and H, O, C, Sr, Nd, and Pb isotope studies in mineral waters and corresponding rocks from NW Bohemia, Czech Republic, *Appl. Geochem.*, **13**, 975-994.

Möller P., Stober I., and Dulski P. (1997a) Seltenerdelement-, Yttrium-Gehalte und Bleiisotope in Thermal- und Mineralwässern des Schwarzwaldes. *Grundwasser* **2**,118-132

Möller, P., and 26 authors (1997b) Paleofluids and recent fluids in the upper continental crust: Results from the German Continental Deep Drilling program (KTB) *J. Geophys. Res.* **102 B8**, 18233-18254.

Morgan, J.W. and Wandless, G.A. (1980) Rare earth element distribution in some hydrothermal minerals: Evidence for crystallographic control. *Geochim. Cosmochim. Acta* **44**: 973-980.

Saunders, A.D. (1984) *The rare earth element characteristics of igneous rocks from the ocean basins.* In P. Henderson (ed): Rare earth element geochemistry, Elsevier, Amsterdam Oxford, New York, Tokyo, 205-236.

Shannon, R.D. (1976) Revised effective ionic radii and systematic studies of interatomic distances in halides and chalcogenides. *Acta Crystallogr.* **A32**, 751-767.

Stober, I. (1995) *Die Wasserführung des kristallinen Grundgebirges.* Stuttgart, 191p.

Sverjensky, D.A. (1984) Europium redox equilibria in aqueous solution. *Earth Plant. Sci. Lett.* **67**, 70-78.

Table 1: Compilation of waters and corresponding aquifer rocks

Locality	Abbrev.	Temp. °C	Type of water	CO_2+HCO_3 mg/kg	pH	Source rock	Abbrev.
Vent, East Pacific Rise	Vent	400	Na-Cl	Low	4	MOR basalt	MORB
Dispersed Flow, Teahitia, Society Islands	Disp	30	Na-Cl	Low	5-6	ocean island basalt	OIB
Hokkaido, Japan	Hokk	8	Ca-SO_4	Low	4	basaltic andesite	And
Kizildere, Turkey	Kiz	220/100	Na-HCO_3	13 bars	6.8	mica schist	Igd
Kizildere, Turkey	Kiz	220/100	Na-HCO_3	13 bars	6.8	marble	Saz
Continental Deep Drilling Project, Germany	KTB	119/30	Ca-Na-Cl	<100	5.8-8.3	amphibolite	Amph
Jachymov, Czech Republic	Jach	34	Ca-Na-HCO_3	14	7.5	granite	Gr
Säckingen, Germany	Säck	25	Na-Ca-Cl	450	6.8	granite	Gr
Kyselka, Czech Rebublic	Kys/BJ	17	Na-Ca-Cl	2000	5.8	granite	Gr
Kyselka, Czech Rebublic	Kys/V5	14	Na-Ca-Cl	1800	5.7	granite	Gr
Kyselka, Czech Rebublic	Kys/HJ	11	Na-Ca-Cl	2300	6.2	alkalibasalt	Bas
Bad Peterstal, Germany	Pet	17	Na-Ca-Cl	1610	6.2	gneiss	Gn
Hermersberg, Germany	Herm	17	Na-Ca-Cl	<5	8.2	gneiss	Gn
Bad Wildbad, Germany	Wild	37	Ca-Na-HCO_3	19	7	granite	Gr

Table 2: Compilation of Ca and REY abundances in waters, aquifer rocks and 20 hours leachates from the latter. The asterix* indicates values that are not given as concentrations but % of leached fraction (Giese and Bau, 1994). The elements in waters were determined after preconcentration (Bau and Dulski, 1996) in waters, and directly in the digestion solutions of the rocks and the leachates by ICP-MS (Dulski, 1994).

Waters

	Vent	*Disp*	*Hokk*	*Kiz/95*	*Kiz/96*	*KTB*	*Jach*	*Säck*	*Kys/BJ*	*Kys/V5*	*Kys/HJ*	*Pet*	*Herm*	*Wild*
Ca (mM/kg)	25.2	10.5	2.19	0.12	0.03	393	0.45	7.43	1.93	1.28	5.45	10.2	0.35	0.90
La (pM/Kg)	2643	2.32	44100	13.4	3.92	19	14.7	54	76.8	14.7	482.9	1032	19.2	11.6
Ce	4491	3.73	121000	16.5	6.42		73.6	156	188.2	54.5	858.3	2605	83.1	20.0
Pr	585		19400	1.65	0.71	3.40								
Nd	2350	1.86	88400	5.99	2.45	13	18.0	102	114.4	37.3	390.7	1730	21.9	19.4
Sm	478	0.39	22200	1.19	0.37	3.09	4.51	70	40.1	17.7	65.8	650	6.32	9.29
Eu	2194	0.09	4320	0.38	0.06	0.86	1.22	12.6	10.3	4.52	21.5	601	1.87	3.08
Gd	387	0.38	25700	1.32	0.33	1.62	6.14	145	94.9	54.4	95.4	1209	11.3	17.1
Tb	53		3950	0.22	0.04	0.22	1.12	21	18.3	11.4	13.4	233	2.14	2.77
Dy	262	0.23	23400	1.57	0.30	1.65	6.46	110	140.1	81.8	92.8	1730	14.5	16.0
Y			281000	16.8	3.77	15	88	1747	2636	1252	2281	31052	272	325
Ho	41		4780	0.27	0.06	0.29	1.39	20	33.6	18.0	23.0	403	3.42	3.13
Er	96	0.13	13200	0.79	0.18	0.86	3.76	53	103.6	54.9	73.46	1316	10.62	8.47
Tm			1870			0.13								
Yb		0.09	11200	0.63	0.14	1.20	3.38	46	94.8	60.6	75.5	1472	8.95	6.84
Lu			1700	0.08	0.02		0.38	7.39	14.0	8.83	12.8	230	0.83	0.73

Source rocks

	MORB	*OIB*	*And*	*Saz/Kiz*	*Igd/Kiz*	*Amph.*	*Gr/Jach*	*Gr/Säck*	*Gr/Kys*	*Gr/Kys*	*Bas/Kys*	*Gn/Pet*	*Gn/Herm*	*Gr/Wild*
Ca ppm	93000	69600	133	206511	59116	107000	3024	4000	913	913	108500	6285	7500	2430
La ppm	3.25	37.5	4.66	11.0	28.4	21.5	4.32	15.1	30.2	30.2	59.0	38.7	36.5	2.86
Ce	11.1	88.3	10.9	20.1	55.9	50.6	10.1	29.3	58.2	58.2	124.2	78.5	74.3	7.08
Pr	2.12		1.55	2.07	6.60	6.93	1.24	3.35	6.53	6.53	15.4	9.64	9.26	0.95
Nd	12.4	47.4	6.69	6.16	22.8	30.5	4.61	11.7	21.6	21.6	60.3	35.2	33.7	3.45
Sm	4.54	10	1.96	1.04	4.20	7.38	1.47	2.75	3.90	3.90	10.49	6.95	6.40	1.46
Eu	1.64	3.16	0.71	0.35	0.91	1.96	0.06	0.38	0.58	0.58	3.01	1.18	1.33	0.13
Gd	6.2	8.85	2.37	0.78	4.17	7.66	1.54	2.36	3.17	3.17	7.90	5.97	5.46	1.66
Tb	1.16		0.39	0.14	0.66	1.21	0.34	0.36	0.51	0.51	1.00	0.83	0.76	0.31
Dy	7.82	6.08	2.53	1.02	3.89	7.54	2.02	1.85	3.16	3.16	5.06	4.49	4.00	1.72
Y			13.9	7.20	20.1	42.7	11.1	8.99	17.7	17.7	21.3	22.8	19.9	9.12
Ho	1.7		0.56	0.24	0.76	1.53	0.30	0.31	0.63	0.63	0.85	0.83	0.74	0.30
Er	5.1	2.47	1.61	0.97	2.09	4.43	0.78	0.75	1.91	1.91	2.11	2.38	1.94	0.77
Tm	0.69		0.25	0.22	0.30	0.61	0.11	0.11	0.30	0.30	0.26	0.33	0.29	0.11
Yb	4.86	1.71	1.61	2.26	1.84	3.91	0.67	0.70	2.02	2.02	1.57	2.14	1.77	0.73
Lu	0.7		0.25	0.54	0.27	0.58	0.08	0.10	0.30	0.30	0.22	0.35	0.28	0.09

Leachate/20h

	sMORB	*gMORB*	*And*	*Saz/Kiz*	*Igd/Kiz*	*Amph.*	*Gr/Säck*	*Gr/Jach*	*Gr/Kys*	*Gr/Kys*	*Bas/Kys*	*Gn/Pet*	*Gn/Herm*	*Gr/Wild*
Ca ppm	2.7*	3.2*	318*	532	238	10.2	42	69	5.03	5.03	229	85	89	29
La ppb	21*	2.8*	16.7*	11.4	86.1	23.4	4.46	20.4	18.8	18.8	195	7.05	6.92	1.75
Ce	19.3*	2.9*	40.3*	30.0	199	20.6	12.0	52.2	47.2	47.2	377	20.6	16.8	5.61
Pr	17.4*	2.8*	5.66*	3.43	27.3		1.72	7.49	6.17	6.17	43.8	3.40	2.35	1.08
Nd	15.4*	3*	24.4*	12.31	107	26.1	6.5	31.1	22.9	22.9	158	17.3	10.2	5.15
Sm	12.2*	2.9*	5.81*	2.90	24	25.6	3.97	13.6	4.87	4.87	25.2	6.90	3.14	3.55
Eu	7.2*	2.8*	1.54*	0.48	5.47	14.8	0.31	2.48	0.73	0.73	6.98	3.45	2.86	0.49
Gd	11.5*	2.9*	6.21*	2.82	24.5		5.41	16.6	3.69	3.69	18.5	7.63	3.76	4.21
Tb	10.3*	3.1*	0.87*	0.55	3.69	16.5	1.41	3.02	0.52	0.52	2.28	1.21	0.69	0.58
Dy	9.2*	2.9*	5.12*	3.84	20.9		8.39	16.3	2.60	2.60	11.1	7.11	4.43	2.44
Y			28.1*	23.2	99.1	10.7	46.2	85.4	12.30	12.3	55.3	39.2	27.2	9.50
Ho	8.7*	2.8*	1.02*	0.80	3.85		1.10	2.38	0.45	0.45	1.87	1.35	0.92	0.32
Er	8.8*	2.7*	2.73*	2.57	9.0		2.56	5.75	1.23	1.23	4.59	3.75	2.51	0.68
Tm	9.4*	3.1*	0.38*	0.42	1.35		0.35	0.74	0.16	0.16	0.54	0.54	0.36	0.07
Yb	8.4*	3.2*	2.27*	3.17	7.71	11.7	2.32	4.55	1.03	1.03	3.07	3.54	2.45	0.45
Lu	8.7*	3.1*	0.33*	0.50	1.04	11.8	0.25	0.52	0.14	0.14	0.41	0.50	0.35	0.05

Chapter 4

Microbial Processes in Crystalline Rocks

THE HYDROGEN DRIVEN INTRA-TERRESTRIAL BIOSPHERE AND ITS INFLUENCE ON THE HYDROCHEMICAL CONDITIONS IN CRYSTALLINE BEDROCK AQUIFERS

K. PEDERSEN
Göteborg University, Department of Cell and Molecular Biology, Section Microbiology
Box 462, SE-405 30 Göteborg, Sweden

1. Evidence for an intra-terrestrial bioshpere

Diverse and active populations of microorganisms have been observed in most subterranean and sub-seafloor environments investigated (Bachofen 1997), including granitic rock aquifers at depths ranging down to 1240 m (Pedersen 1997a). Documentation of *in situ* activity of microbial populations in deep granitic rock groundwaters suggests that the microbes present are active at low, but significant levels (Ekendahl and Pedersen 1994, Pedersen and Ekendahl 1990, 1992a-b). Igneous rocks are too hot when formed to host life of any kind. Therefore, observed life in crystalline bedrock must have entered after cooling and fracturing of the rock mass.

The drilling and excavation to access these microbial ecosystems are vigorous operations and it can be argued that the observed life is an artefact of the access operations (Pedersen 1993a) and not a true deep biosphere. The risk for microbial contamination of the aquifers by the drill water used to cool the drill and transport the drill cuttings out of the borehole during drilling is obvious and investigations have, therefore, been undertaken to study this risk. During the end of the construction phase of the Äspö hard rock laboratory (HRL) (Pedersen 1997a), samples were collected from boreholes in the Äspö HRL tunnel concomitant with geological, hydrological and hydrogeochemical characterisation of designated experimental rock volumes (Winberg et al. 1996). 16S rRNA gene sequencing and culturing methods were used to investigate whether a lasting microbial contamination due to the drilling occurred. Samples were collected from the drill water source, the drilling equipment and from the drilled boreholes. The results showed that total and viable counts of bacteria in drilled boreholes were higher than what could be explained by introduction with the drillwater, indicating that microbes were present before drilling. A total of 158 bacterial 16S rRNA genes that were cloned from the drill water source, the drilling equipment and the drilled boreholes were partially sequenced. The drilled boreholes generally had a 16S rRNA diversity that was entirely different from what was found in samples from the drilling equipment, implying that unique intrinsic microorganisms were present in the sampled groundwater.

I. Stober and K. Bucher (eds.), Hydrogeology of Crystalline Rocks, 249–259.

Proving that certain species of microorganisms found in drilled boreholes are intrinsic and not introduced during drilling is extremely difficult. Instead, as was shown by Pedersen et al. (1997b), the opposite situation is easier to investigate, i.e. testing if a known contaminating microbial population establishes or not in deep granitic aquifers during drilling. The 600 m long tubing used for drill water supply constituted a source of bacterial contamination to the rest of the drilling equipment and the boreholes. Nevertheless, using molecular and culturing methods, it was shown that although large numbers of contaminating bacteria were introduced in the boreholes during drilling, they did not become established in the aquifers at detectable levels. Therefore, it seems reasonable to conclude that we find no evidence for lasting microbial contamination of boreholes drilled in granitic rock. The reason for this is the inability of foreign microbes to adapt to the prevailing oligotrophic, reducing, anaerobic and low temperature environmental conditions in deep granitic aquifers.

We have recently described two new species isolated from fractures deep under Äspö HRL, *Desulfovibrio aespoeensis* (Motamedi and Pedersen 1998) and *Metanobacterium subterraneum* (Kotelnikova et al. 1998). These two species are adapted to life under the conditions prevailing where they were isolated from, i.e. they are most probably intrinsic. Recent findings of bacteria-like fossils in a granitic aquifer 207 m below ground at Äspö (Pedersen et al. 1997a) support the hypothesis of a deep and intrinsic subterranean biosphere. It can, based on the results arrayed above, be concluded that the absolute majority of the microorganisms found in the new boreholes were present in the intersected aquifers before drilling.

2. The hydrogen driven intra-terrestrial biosphere hypothesis

Groundwater at depths of 500 m can be very old and ages of 10,000 years are not unusual. This poses a conceptual problem for the deep intra-terrestrial biosphere: What ultimate energy source is it being using? Organic carbon from sun driven surface ecosystem would not last for a long time because documentation of *in situ* activity of microbial populations in deep granitic rock environments suggests that the microbes present are actively degrading organic carbon at low but significant rates (Ekendahl and Pedersen 1994, Pedersen and Ekendahl 1990, 1992a-b). Typical values at depth are one to two mg dissolved organic carbon per litre groundwater (Pedersen and Karlsson 1995). Any energy source at this depth must be renewable. Throughout our work results have indicated the presence of autotrophic microorganisms in the studied deep granitic rock environments that utilise hydrogen as a source of energy. Therefore, a hypothesis of a hydrogen-driven biosphere in deep granitic aquifers has been suggested (Pedersen 1993b, Pedersen 1997a, Pedersen and Albinsson 1992). The organism base for this biosphere is suggested to be composed of acetogenic bacteria that have the capability of reacting hydrogen with carbon dioxide to produce acetate (homoacetogens) and methanogens that yield methane from hydrogen and carbon dioxide (autotrophic methanogens) or from acetate produced by homoacetogens (acetoclastic methanogens) (Fig. 1). A similar hypothesis has recently been published for deep basaltic rock aquifers (Stevens and McKinley 1995) as have arguments against it (Anderson et al. 1998). One of the aims of our recent studies has, therefore, been to collect evidence for a hydrogen-driven deep biosphere in deep granitic aquifers and has been focused on acetogenic bacteria and methanogens as the autotrophic base for such a biosphere. Distribution, numbers and physiological diversity of homoacetogens and methanogens in deep granitic rock aquifers at the Äspö HRL were investigated using a variety of methods. The results showed that methanogens and homoacetogens are present and are metabolically active in the Äspö HRL groundwaters at all investigated depths down to 450 m (Kotelnikova and Pedersen 1998). Pure cultures of autotrophic, rod-shaped methanogens were isolated and one of them could be described as a new species, *Methanobacterium subterraneum* (Kotelnikova et al. 1998).

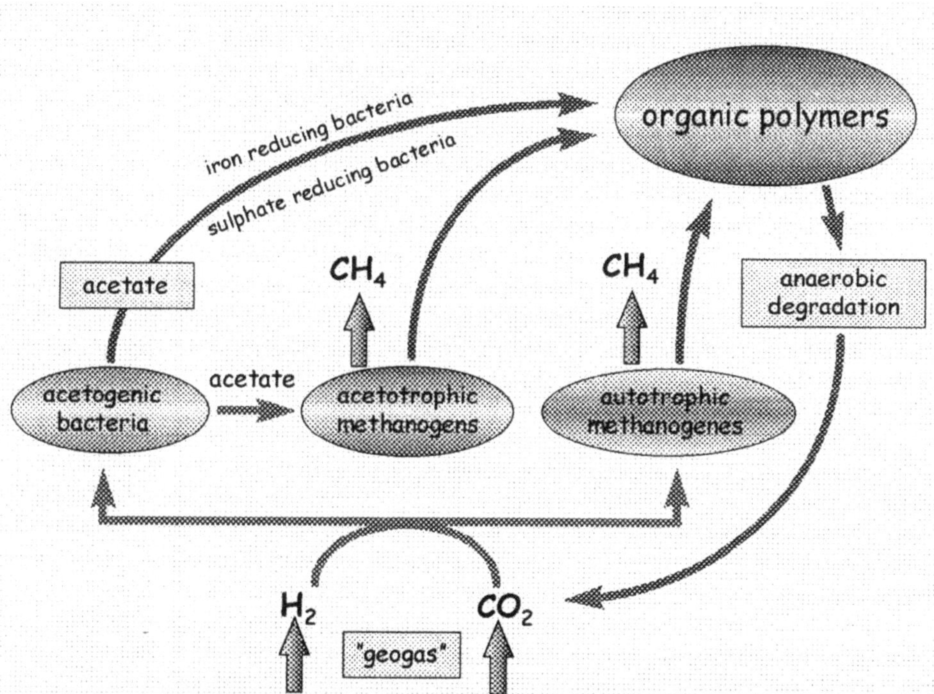

Figure 1. The deep hydrogen driven biosphere hypothesis, illustrated by its carbon cycle. At relevant temperature and water availability conditions, intra-terrestrial microorganisms are capable of performing a life cycle that is independent of sun-driven ecosystems. Hydrogen and carbon dioxide from the deep crust of earth or organic carbon from sedimentary deposits can be used as energy and carbon sources. Phosphorus is available in minerals like apatite and nitrogen for proteins, nucleic acids etc. can be obtained via fixation of nitrogen which predominates in most ground waters (Table 1).

Continuous support of hydrogen to the populations in the investigated granitic rock environment can be delivered from radiolysis of water. The Baltic Shield granites contain uranium that in its decay series among other species produces radium and radon. The radioactivity during this decay generate heat in the rock, e.g. 7 μWm^{-3} in Malingsbo granite (Malmquist et al. 1983) and explains a significant part of the heat flow from deep Swedish rock. Radium, and also thorium, migrate from the rock matrix to the aquifers were it tend to sorb on the aquifer surfaces. Additionally, radon gas will be generated and dissolved in the groundwater. The alpha radiation produced during decay of these, and other species as well (Landström et al, 1971), may be enough to explain a continuous hydrogen production from radiolysis of water in deep granitic aquifers to the intra-terrestrial biosphere. This possibility is presently being investigated in more detail.

Until recently, it has been a general concept that all life on Earth depends on the sun via photosynthesis, including most of the geothermal life forms found in deep sea trenches as they use oxygen for the oxidation of reduced inorganic compounds (almost all oxygen on earth is produced via photosynthesis). Our results now suggest that a deep subterranean granitic biosphere exists, driven by the energy available in hydrogen formed through radiolysis, mineral reactions or by volcanic activity (Fig. 1). Knowledge about this biosphere has just begun to emerge and is expanding the spatial borders for life from a thin layer on the surface of the planet Earth and in the seas to a several kilometre thick biosphere reaching deep below the ground surface and the sea floor (Whitman et al. 1998). If this theory holds, life may have been present and active deep down in Earth for a very long time and it cannot be excluded that the place for the origin of life was a deep subterranean igneous rock aquifer environment (probably hot with a high pressure) rather than a surface environment.

3. Intra-terrestrial microbes and biogeochemical processes

The intra-terrestrial biosphere, concluded above to exist, may influence the geochemical situation in many different ways. A full understanding of deep subterranean environments can, therefore, not be achieved until microbial processes are included in models, theories and interpretation of results. This is because microbes catalyse many reactions that, for kinetic reasons, are very slow or not possible at low temperature and pressure. One obvious example is the bacterial reduction of sulphate to sulphide in anoxic waters. Another example is that bacterial activity usually influences the redox potential in the environment. If, for instance, the environment is rich in Fe(III) and organic matter, iron-reducing bacteria will dominate, produce Fe(II) and carbon dioxide in large quantities and the resulting redox potential will be controlled by Fe(II) at or below approximately -100 mV.

Microbial decomposition and production of organic material depend on the sources of energy and electron-acceptors present. Organic carbon, reduced inorganic molecules or hydrogen are possible energy sources in subterranean environments. During microbial oxidation of these energy sources the microbes use electron acceptors in a certain order according to Fig. 2. First oxygen is used, thereafter follows the utilisation of nitrate, manganese, iron, sulphate, sulphur and carbon dioxide.

Simultaneously, fermentative processes supply the respiring microbes with hydrogen and short organic acids. As the solubility of oxygen in water is low and because oxygen is the preferred electron acceptor by many microbes utilising organic compounds in shallow groundwater, anaerobic reduced environments and processes usually dominate at depth in the subterranean environment.

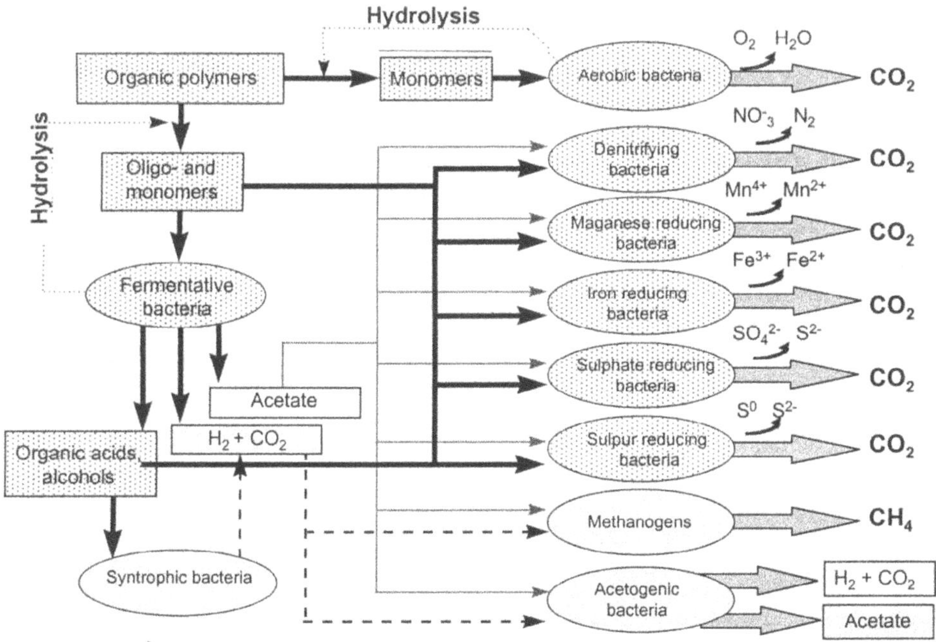

Figure 2. The degradation of organic carbon can occur via a number of different metabolic pathways, characterised by the principal electron acceptor in the carbon oxidation reaction. A range of significant compounds occurring in groundwater are formed or consumed during this process.

3.1 IRON REDUCING BACTERIA

Iron reducing bacteria were discovered to be of major biogeochemical importance in granitic rock during a block scale redox experiment at the Äspö HRL. The unexpected redox stability of the studied system could only be explained by the mobilization of solid phase ferric iron oxi-hydroxides to liquid phase ferrous iron by iron reducing bacteria with organic carbon as electron donor (Banwart 1995, Banwart et al. 1994, 1996). We have isolated several different bacteria from this habitat able to reduce ferric iron to ferrous iron, including *Shewanella putrefaciens* (Pedersen et al. 1996). The 16S-rRNA gene sequences show that several of the dominating species sampled had a 95% or more identity with known iron reducing bacteria like *Pseudomonas medosina* (Pedersen and Karlsson 1995). Our results imply that much of the ferrous iron found in anoxic ground water from deep crystalline rock aquifers may be a product of microbial iron reduction and not only due to non-biological reduction of ferric iron.

3.2 SULPHATE REDUCING BACTERIA

Sulphate reducing bacteria frequently appear in the Äspö HRL environments at depths greater than approximately 100 m; isolates as well as 16S-rRNA genes related to sulphate reducing bacteria have been found (Pedersen et al. 1996, 1997b). Evidence and indications of sulphate reduction based on geological, hydrological, ground water, isotope and microbial data in and around the Äspö HRL tunnel were evaluated by a multi-disciplinary research group (Laaksoharju et al. 1995) and the most important conclusions are given below.

Geological data were evaluated to find the amount of sulphide which could be calculated to result from the sulphate reduction. The conclusion is that the amount of pyrite normally occurring in the fracture coatings could explain the amount of sulphate reduced. However, there are other processes in the geological time span which have also produced pyrite. Therefore the existence of pyrite is not a conclusive evidence for sulphate reduction.

The hydrogeological conditions were evaluated in order to describe possible transport phenomena related to the observed sulphate reduction. The questions to be answered were: Can sulphate reduction take place in the sea bottom sediments and the resulting sulphide be transported with ground water to the tunnel? Could the ground water flow conditions in the tunnel either increase or decrease the effect of biological sulphate reduction? The answer to the first questions is yes, the process can occur in the sea bed sediments and the effect on hydrochemistry can be observed in the water inflow in the tunnel. Hydrogeological calculations imply a transport time of approximately 100-400 days for the water passing through the sediments to reach the tunnel in a proportion of 25%. The second question answer is that the ground water flow conditions around the tunnel would not affect the biological process directly. However, if the sulphate reduction had been an ancient process, then the effects would soon be washed out, which has not been the case. In addition, the existence of high bicarbonate and low sulphate concentrations in the probing holes on the very first sampling occasion after the tunnel was excavated strongly imply that the process is ongoing.

The ground water chemistry was evaluated by multivariate mixing and mass balance calculations. The calculations demonstrated that an understanding of the fluxes of compounds, rather than measurements of concentrations only, is necessary for modelling sulphate consumption and bicarbonate production by sulphate reducing bacteria. These calculations defined the specific conditions where the process could be ongoing. The results show that the salinity range of 4000-6000 mg/l of chloride is the optimal one. Sulphate reduction seems to occur in anaerobic brackish ground water with access to dissolved sulphate and organic carbon. These conditions are mainly found in the sea bed sediments, in the tunnel section under the Baltic Sea and in some deep ground water.

Isotope data were expected to give a definite answer to where the sulphate reduction takes place, since the bacterial processes always result in an enrichment of the lighter isotopes in their metabolic products. Concerning both the $\delta^{13}C$ and the $\delta^{34}S$ isotopes, the results generally point towards the existence of bacterial sulphate reduction (see Tullborg, this book). However, there are several processes in the geological evolution which could have given the same isotopic signatures as well.

Microbiological data were collected in boreholes where the hydrochemistry indicated an ongoing or previously ongoing sulphate reduction. The results showed that sulphate reducing bacteria were occurring, sometimes at large numbers, and that they could be correlated to a ground water composition with high bicarbonate and low sulphate concentrations.

4. Microbial production and consumption of gases

The content of gases that can be produced or consumed by microorganisms in deep groundwater have been analysed by us earlier, but the method used did not separate hydrogen from helium (Table 1). The deep biosphere hypothesis (Fig. 1) depends totally on the presence of hydrogen in deep groundwater. We have, therefore, recently invested significant work in the measurement of dissolved gases in deep granitic rock environments, now including hydrogen. The technique developed is described in detail by Pedersen (1997). Recent results for dissolved gas, analysed with this method, have been produced for the boreholes denoted KA3005, KA3010 and KA3110 and the results are presented in Table 1. It can be seen that the total amount of gas found differs at most 3 times between boreholes and sites. The sensitivity and reproducibility of the gas extraction and analysis methods are good and hydrogen could be detected. Each of the found gases are discussed in more detail below, from a microbiological perspective.

Nitrogen is by far the most dominant gas in all samples analysed (Table 1). Some of the nitrogen may have been dissolved from air in rain and surface waters that become groundwater with time, but the solubility of nitrogen at 10°C and atmospheric pressure is 19.6 ml l^{-1}. Most of the nitrogen values in Table 1 exceeds this solubility limit and other sources of dissolved nitrogen to groundwater must exist as well. Nitrogen can be used by nitrogen fixing bacteria as a source of nitrogen and many produce nitrogen from nitrate during an anaerobic respiration process called denitrification (Pedersen and Karlsson 1995). Microbial processes could contribute to the pool of dissolved nitrogen in groundwater through denitrification processes, but it is unknown if this occurs in enough amounts to explain the excess of dissolved nitrogen. Helium is a noble gas and it is not produced or consumed by microorganisms.

All living and active organisms expel carbon dioxide from their degradation of organic material and many microorganisms and all plants and algae can transform carbon dioxide to organic carbon (Fig. 2). The concentration of this gas may, therefore, be influenced by microorganisms with subsequent effects on the carbonate system, pH and precipitation and dissolution of carbonate minerals.

TABLE 1. The content of nitrogen, hydrogen, helium and carbon-containing gases and the total volumes of gas extracted from groundwater samples of the Stripa borehole V2, the Laxemar borehole KLX01 and the Äspö boreholes KR0012, 13 and 15 (Pedersen 1993b, Pedersen and Ekendahl 1992a-b) and the Äspö boreholes KA3005, KA3010 and KA3110 (Pedersen 1997b).

Boreholes	Sampling depth (m)	N_2 (μL L^{-1})	H_2 (μL L^{-1})	He (μL L^{-1})	CO (μL L^{-1})	CO_2 (μL L^{-1})	CH_4 (μL L^{-1})	C_2H_6 (μL L^{-1})	[a] $C_2H_{2\text{-}4}$ (μL L^{-1})	Total gas (μL L^{-1})
Stripa										
V2	799-807	25000	n.a[b].	<10	<1	32	245	0.3	<0.1	25277
V2	812-821	31000	n.a	<10	<1	11	170	0.6	<0.1	31181
V2	970-1240	24500	n.a	<10	<1	10	290	2.9	<0.1	24803
Laxemar										
KLX01	830-841	46500	n.a	4600	0.5	460	26	<0.1	<0.1	51586
KLX01	910-921	37000	n.a	3500	0.1	500	27	<0.1	<0.1	41027
KLX01	999-1078	18000	n.a	2450	0.7	1600	31	<0.1	<0.1	22082
Äspö HRL										
KR0012	68	22000	n.a	40	0.1	6050	1030	<0.1	0.1	29120
KR0013	68	25000	n.a	110	0.2	9640	1970	<0.1	0.1	36720
KR0015	68	22000	n.a	64	0.1	15037	4070	<0.1	0.1	41171
KA3005/2[c]	400	25930	1.68	1757	<1	1082	1715	<0.1	<0.1	32300
KA3005/4	400	26661	0.11	3809	<1	2100	1849	<0.1	<0.1	34419
KA3010/2	400	40626	30.96	7946	1.4	142	55	<0.1	<0.1	48801
KA3110/1	414	14861	14.50	448	<1	1832	925	<0.1	<0.1	18080

[a] The content of $C_2H_2 + C_2H_4$
[b] not analysed
c number after slash denotes sampled borehole section

The suggested deep biosphere hypothesis requires hydrogen as its energy base. Hydrogen is expected to act as an inert gas in most geochemical systems and it is therefore usually overlooked and not analysed for. Some data on hydrogen in hard rock were published earlier (Sherwood Lollar et al. 1993a-b). From 2.2 up to 1574 μM hydrogen in groundwater from Canadian shield and Fennoscandian shield rocks were found. The origin of such hydrogen can vary. Most granitic rocks show a low but significant radioactivity which can generate hydrogen by radiolysis of water. Anaerobic mineral reactions (e.g. anaerobic corrosion of iron) will also create hydrogen (Stevens and McKinley 1995). Finally, deep volcanic gases contain hydrogen. Screening the Äspö HRL groundwater for hydrogen with a simple "closed bottle head space" method revealed significant amounts of hydrogen in most samples analysed (Kotelnikova and Pedersen 1997). The sampling and extraction method used confirmed that hydrogen is present (Table 1) and consequently, there is an energy base available for the deep biosphere.

Methane occurs frequently in subterranean environments all over the globe. Evidence for an ongoing methane generating process in deep Swedish granite has been published (Flodén and Söderberg 1994, Söderberg and Flodén 1991, 1992). Pockmarks in Baltic sea sediments were found, indicating gas eruption from fracture systems in the underlying granite, mainly of methane. Values of 1.3 up to 18576 μM of methane in groundwater from Canadian shield and Fennoscandian shield rocks have been published (Sherwood Lollar et al. 1993a-b). Recent data indicate up to 720 μM of methane down to 440 m depth at Äspö HRL (Kotelnikova and Pedersen 1997). The stable carbon isotope profile is commonly used as an indication of a biogenic origin of the methane. Some results on the $^{13}C/^{12}C$ signatures indicate biogenic origin of the Äspö methane (Banwart et al. 1996).

5. Life in hard rock tunnels

Excavation for tunnels, mining etc. introduces several changes in the subterranean environment that will induce activities in the tunnel by microorganisms other than those present in the fractured rock. Oxygen is normally introduced in tunnels by ventilation which makes growth of aerobic bacteria possible. As the crystalline rock ground water at depth usually is anoxic with a low redox potential, marked redox and oxygen gradients will develop when such ground water reaches the oxygenated tunnel atmosphere. Typical redox pairs participating in these gradients are manganese(II) oxidizing to manganese(IV), ferrous iron to ferric iron, sulphide to sulphate and methane to carbon dioxide. Such gradients are the habitats for many different lithotrophic and also heterotrophic bacteria. Among them are the iron, manganese, sulphur and methane oxidizing bacteria that generate chemical energy for anabolic reactions through the oxidation of reduced inorganic compounds and methane with oxygen. The energy gained by the lithotrophs is used to reduce carbon from CO_2 to organic carbon and this is the first step in an environmental succession that eventually ends as a reduced environment again.

Commonly, seeps of ground water from fractures intersected by the Äspö tunnel or flows of ground water from boreholes turn light brown to dark brown from precipitates that sometimes can be very voluminous. They usually appear within some weeks after excavation/drilling and have in some cases reached a thickness of 10 cm or more. The most frequently occurring inhabitant in these precipitates is the lithotrophic iron-oxidizing bacterium *Gallionella ferruginea* (Hallbeck and Pedersen 1990, 1991, 1995, Hallbeck et al. 1993). It forms moss-like covers on rocks and sediments in ponds in tunnels and is very abundant close to the outflow of ground water from rock wall fractures (Pedersen and Karlsson 1995). At many such outflows, white, threadlike structures are observed. Microscopic observation have revealed them as being sulphur oxidizing bacteria of different types; both extracellular and intracellular deposition of sulphur have been observed. Especially tunnel sections below the sea bed with ongoing sulphate reduction harbour this type of bacteria. Sequencing the 16S rRNA gene from one of these sites has indicated the genus *Thiothrix* to be present (not published).

6. Conclusion

Altogether, our results show that there is a very high probability for the existence of an intra-terrestrial biosphere that is driven by hydrogen from the interior of the earth (Fig. 1) and, therefore, independent of photosynthesis. Our planet obviously has two biospheres, the sun driven biosphere that is well known and accepted by everybody, and the new, unexplored earth driven intra-terrestrial biosphere. Prospective research will aim at exploration of distribution, diversity, *in situ* activity and biogeochemistry of the intra-terrestrial biosphere.

7. Acknowledgements

Everett Shock kindly reviewed this paper.

8. References

Andersson, R.T. Chapelle, F.H. and Lovley, D.L. (1998) Evidence against hydrogen-based microbial ecosystems in basalt aquifers. Science 281, 976-977.

Bachofen, R. (1997) Proceedings of the 1996 International Symposium on Subsurface Microbiology (ISSM-96) 15-21 September 1996 in Davos Switzerland. Guest editor: Reinhard Bechofen. FEMS Microbiol. Rev. 20, 179-638.

Banwart, S. (1995) The Äspö redox investigations in block scale. Project summary and implications for repository performance assessment. SKB Technical Report 95-26. 47 pp. Swedish Nuclear Fuel and Waste Management Co, Stockholm.

Banwart, S., Gustafsson, E., Laaksoharju, M., Nilsson, A.-C., Tullborg, E.-L. and Wallin, B. (1994) Large-scale intrusion of shallow water into a granite aquifer. Water Resour. Res. 30, 1747-1763.

Banwart, S., Tullborg, E.-L., Pedersen, K., Gustafsson, E., Laaksoharju, M., Nilsson, A.-C., Wallin, B. and Wikberg, P. (1996) Organic carbon oxidation induced by largescale shallow water intrusion into a vertical fracture zone at the Äspö Hard Rock Laboratory (Sweden). J. Contam. Hydrol. 21, 115-125.

Ekendahl, S. and Pedersen, K. (1994) Carbon transformations by attached bacterial populations in granitic ground water from deep crystalline bed-rock of the Stripa research mine. Microbiology 140, 1565-1573.

Flodén, T. and Söderberg, P. (1994) Shallow gas traps and gas migrations models in crystalline bedrock areas offshore Sweden. Baltica 8, 50-56.

Hallbeck, L. and Pedersen, K. (1990) Culture parameters regulating stalk formation and growth rate of *Gallionella ferruginea*. J. Gen. Microbiol. 136, 1675-1680.

Hallbeck, L. and Pedersen, K. (1991) Autotrophic and mixotrophic growth of *Gallionella ferruginea*. J. Gen. Microbiol. 137, 2657-2661.

Hallbeck, L. and Pedersen, K. (1995) Benefits associated with the stalk of *Gallionella ferruginea*, evaluated by comparison of a stalk-forming and a non-stalk-forming strain and biofilm studies in *situ*. Microb. Ecol. 30, 257-268.

Hallbeck, L., Ståhl, F. and Pedersen, K. (1993) Phylogeny and phenotypic characterization of the stalk-forming and iron-oxidizing bacterium *Gallionella ferruginea*. J. Gen. Microbiol. 139, 1531-1535.

Kotelnikova, S., Macario, A.J.L. and Pedersen, K. (1998) *Methanobacterium subterraneum*, a new species pf Archaea isolated from deep groundwater at the Äspö Hard Rock Laboratory, Sweden. Int. J. Syst. Bacteriol. 48, 357-367.

Kotelnikova, S. and Pedersen, K. (1997) Evidence for methanogenic *Archaea* and homoacetogenic *Bacteria* in deep granitic rock aquifers. FEMS Microbiol. Rev. 20, 339-349.

Kotelnikova, S. and Pedersen, K. (1998) Distribution and activity of methanogens and homoacetogens in deep granitic aquifers at Äspö Hard Rock Laboratory, Sweden. FEMS Microbiol. Ecol. 26, 121-134.

Laaksoharju, M., Pedersen, K., Rhen, I., Skårman, C., Tullborg, E.-L., Wallin, B. and Wikberg, W. (1995) Sulphate reduction in the Äspö HRL tunnel. SKB Technical Report 95-25. 87 pp. Swedish Nuclear Fuel and Waste Management Co., Stockholm.

Landström, O. Christell, R. And Koski, K. (1971) Field experiments on the application of neutron activation techniques to in situ borehole analysis. Geoexploration, 10, 23-39

Malmqvist, D. Larson, S.Å., Landström, O. And Lind, G. (1983) Heat flow and heat production from the Malingsbo granite, central Sweden. Bull. Geol. Inst. Univ. Uppsala, N.S. 9, 137-152.

Motamedi, M. and Pedersen, K. (1998) Isolation and characterisation of a mesophilic sulphate-reducing bacterium, *Desulfovibrio aespoeensis* sp. nov. from deep ground water at Äspö Hard Rock Laboratory, Sweden. Int. J. Syst. Bacteriol. 48, 311-315.

Pedersen, K. (1993a) The deep subterranean biosphere. Earth Sci. Rev. 34, 243-260.

Pedersen, K. (1993b) Bacterial processes in nuclear waste disposal. Microbiology Europe 1, 18-23.

Pedersen, K. (1997a) Microbial life in granitic rock. FEMS Microbiol. Rev. 20, 399-414.

Pedersen, K. (1997b) Investigations of subterranean microorganisms and their importance for performance assessment of radioactive waste disposal. Results and conclusions achieved during the period 1995 to 1997. Technical Report 97-22. 283 pp. Swedish Nuclear Fuel and Waste Management Co., Stockholm.

Pedersen, K. and Albinsson, Y. (1992) Possible effects of bacteria on trace element migration in crystalline bed-rock. Radiochim. Acta 58/59, 365-369.

Pedersen, K., Arlinger, J., Ekendahl, S. and Hallbeck, L. (1996) 16S rRNA gene diversity of attached and unattached groundwater bacteria along the Access tunnel to the Äspö Hard Rock Laboratory, Sweden. FEMS Microbiol. Ecol. 19, 249-262.

Pedersen, K. and Ekendahl, S. (1990) Distribution and activity of bacteria in deep granitic groundwaters of southeastern Sweden. Microb. Ecol. 20, 37-52.

Pedersen, K. and Ekendahl, S. (1992a) Incorporation of CO_2 and introduced organic compounds by bacterial populations in groundwater from the deep crystalline bedrock of the Stripa mine. J. Gen. Microbiol. 138, 369-376.

Pedersen, K. and Ekendahl, S. (1992b) Assimilation of CO_2 and introduced organic compounds by bacterial communities in ground water from Southeastern Sweden deep crystalline bedrock. Microb. Ecol. 23, 1-14.

Pedersen, K., Ekendahl, S., Tullborg, E.-L., Furnes, H., Thorseth, I.-G. and Tumyr, O. (1997a) Evidence of ancient life at 207 m depth in a granitic aquifer. Geology 25, 827-830.

Pedersen, K., Hallbeck, L., Arlinger, J., Erlandson, A.-C. and Jahromi, N. (1997b) Investigation of the potential for microbial contamination of deep granitic aquifers during drilling using 16S rRNA gene sequencing and culturing methods. J. Microbiol. Meth. 30, 179-192.

Pedersen, K. and Karlsson, F. (1995) Investigations of subterranean microorganisms - Their importance for performance assessment of radioactive waste disposal. SKB Technical Report 95-10. 222 pp. Swedish Nuclear Fuel and Waste Management Co., Stockholm.

Sherwood Lollar, B., Frape, S.K., Fritz, P., Macko, S.A., Welhan, J.A., Blomqvist, R. and Lahermo, P.W. (1993a) Evidence for bacterially generated hydrocarbon gas in Canadian shield and Fennoscandian shield rocks. Geochim. Cosmochim. Acta 57, 5073-5085.

Sherwood Lollar, B., Frape, S.K., Weise, S.M., Fritz, P., Macko, S.A. and Welhan, J.A. (1993b) Abiogenic methanogenesis in crystalline rocks. Geochim. Cosmochim. Acta 57, 5087-5097.

Söderberg, P. and Flodén, T. (1991) Pockmark development along a deep crustal structure in the northern Stockholm Archipelago, Baltic Sea. Beitr. Meereskd. 62, 79-102.

Söderberg, P. and Flodén, T. (1992) Gas seepages, gas eruptions and degassing structures in the seafloor along the Strömma tectonic lineament in the crystalline Stockholm Archipelago, east Sweden. Continent. Shelf Res. 12, 1157-1171.

Stevens, T.O. and McKinley, J.P. (1995) Lithoautotrophic microbial ecosystem in deep basalt aquifers. Science 270, 450-453.

Whitman, W.B., Coleman, D.C. and Wiebe, W.J. (1998) Prokaryotes: The unseen majority. Proc. Natl. Acad. Sci. U. S. A. 95, 6578-6583.

Winberg, A., Andersson, P., Hermanson, J. and Stenberg, L. (1996) Investigation programme for selection of experimental sites for the operational phase. Äspö Progress Report HRL 96-01. Results from the SELECT project. 78 pp. Can be ordered from: Swedish Nuclear Fuel and Waste Management Co, Box 5864, S-102 40 Stockholm, Stockholm.

ANCIENT MICROBIAL ACTIVITY IN CRYSTALLINE BEDROCK – RESULTS FROM STABLE ISOTOPE ANALYSES OF FRACTURE CALCITES

EVA-LENA TULLBORG
Terralogica AB
Gråbo, Sweden

1. Introduction

Water, energy and a carbon source are the fundamental requests for microbial activity. Can these requirements be met in a crystalline aquifer at depth? Examples of groundwater showing presence of bacteria at large depth in crystalline environment are numerous and different types and species of underground bacteria have been successfully identified (e.g. Pedersen and Karlsson, 1995; Pedersen, this volume). However, sampling of groundwater in boreholes always involves a number of question-marks concerning representativity, undisturbed conditions and actual depth of the water and consequently also the microbes sampled.

In order to proof the existence of microbial activity in situ, it should be documented before the penetration of the rock by drilling or excavation. Such evidences can, in best case, be fossilised bacteria, but can also be other traces left behind. It is for example known that bacteria can cause extreme fractionation in stable isotope ratios of carbon and sulphur since they usually prefer the light isotopes. Locally this leads to extremely low $\delta^{13}C$-values in CO_2 and HCO_3^-, if bacteria mediated production takes place at closed conditions. In contrast, very high values (positive $\delta^{13}C$-values in HCO_3^-) can be found, if methane forming microbes have been active, since the remaining CO_2 then can be enriched in ^{13}C. Sampling of gases and groundwater for isotope analyses however, still involves the problem with representativity mentioned above. One possibility to avoid such difficulties is therefore to study fracture coatings from open water conducting fractures in drill cores. The purpose of such studies is e.g. to find out if HCO_3^- with extreme isotopic signature, is incorporated in calcite preserved on the fracture walls or sulphide produced by sulphate reducing bacteria preserved in pyrite. Concerning stable isotope analyses of calcite the $\delta^{18}O$-values can provide additional information about the groundwater environment favourable for bacteria mediated processes.

This paper is a compilation of stable isotope results from calcite fracture infillings in different crystalline bedrocks in Sweden. The analyses have been previously reported (mainly in technical reports, see reference list) but are here compiled and discussed with focus on indications of microbial activity at depth in the crystalline bedrock. Such indications are the findings of bacteria like fossils in a calcite coated fracture at 200 metres depth in a crystalline bedrock aquifer at Äspö, south central Sweden (cf. section 2).

2. Findings of bacteria like fossils in fracture calcite at 200 metres depth in a crystalline aquifer

Samples from a calcite coated fracture from 207 metres depth on the island of Äspö, south-east coast of Sweden (Fig.1), yielded enough material to allow for a combined geochemical and microbiological study. Results of electron-microscopy investigations could be compared with carbon, oxygen and sulphur isotope compositions of the calcite and embedded pyrite (Pedersen et al., 1997; Tullborg et al., 1999). Twenty $\delta^{13}C/\delta^{18}O$ analyses from different spots of the fracture filling showed that calcite has precipitated on the fracture walls during at least four different low-temperature hydro-chemical regimes as interpreted from the $\delta^{18}O$-values and mineralogy. One of these events corresponds to a possible recent origin (approximately in equilibrium with present groundwater at ambient temperatures). The $\delta^{13}C$-values vary from -7 to -46.5 o/oo (Fig. 2a) and three $\delta^{34}S$ analyses carried out on pyrite crystals from the coating, showed

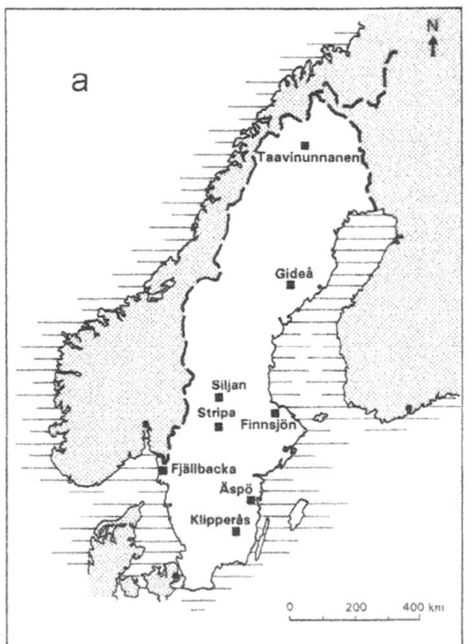

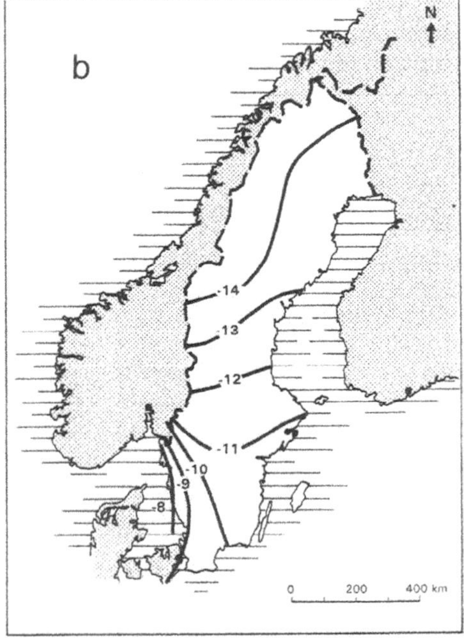

Figure 1a) Location of the sampled sites in Sweden *b)* Mean annual $\delta^{18}O$ values (SMOW) of precipitation in Sweden, (from Burgman et al., (1987).

values of -16.3, -3.4 and -2.8 o/oo (CDT standard). Bacteria-like fossils occurring in colonies and as typical biofilms, identified by electron-microscopy and X-ray microanalyses, demonstrated these fossils to be enriched in carbon (Fig 2b). From the $\delta^{13}C/\delta^{18}O$ diagram (Fig 2a), it can be seen that, (a) the low $\delta^{13}C$-values are associated with a range of $\delta^{18}O$-values, and (b) the $\delta^{13}C$-values show large variations for samples with similar $\delta^{18}O$-values.
This is not the pattern expected if a water with extreme carbon isotope composition had been transported into the fracture. In contrast, these data and the observations of bacteria fossils, support a local production within this fracture of HCO_3^- caused by microbial activity during different periods of time and at different hydrochemical environments.

The occurrence of pyrite precipitates embedded in the calcite and their variation in sulphur isotope ratios suggests that the microbial activity may at least partly have been caused by sulphur reducing bacteria. Such bacteria have been identified in sediments outside Äspö and in groundwater in boreholes as deep as 600 metres (Pedersen et al., 1996). Groundwater in sediment porewater and also in the Äspö tunnel passing 80 to 180 m under the Baltic sea, are significantly modified by sulphate reducing bacteria (low $SO_4^{2-}$and high HCO_3^-), whereas the deep groundwaters are not seriously modified. Present day sulphate reduction is mainly taking place in water of brackish composition characterised by $\delta^{18}O$-values between -6 to -8 o/oo.

The microbial activity and resulting mineralisations of calcite and pyrite in fracture KAS 02:207m may originate from relatively recent events to events that may be several hundreds of million years old, since low temperature conditions have prevailed at least since the Mesozoic (Tullborg et al., 1996). The present groundwater sampled in this section of the bedrock is a Na-Ca-Cl water with TDS of 6500mg/l, pH of 7.4, Cl^-/SO_4^{2-} weight ratio of 35.5 (compared with 8 for the Baltic sea outside Äspö). The water is saturated in respect of calcite and has a TOC value of 3-6 mg/l (Smellie et al., 1995).

The findings of bacteria fossils together with extremely low carbon isotope calcite, support the idea of intrinsic microbial activity at depth. However, this was a very exceptional and exclusive study since the amount of material from calcite coated fractures in drill core samples from crystalline rocks is usually very limited, and such detailed studies can not be carried out frequently. However, if we assume that in situ microbial activity at least locally causes extreme carbon isotope values, then a compilation of the stable isotope values available from fracture calcites from different parts of the Fennoscandian shield in Sweden, can provide information about the extent and distribution of the underground bacterial activity in crystalline bedrock at depth.

3. Fracture calcites

In order to investigate fracture minerals at depth there is a need of drill cores if tunnels and rock caverns are missing. Relatively little interest was paid to the fracture minerals in non-mineralised rocks in Sweden before the start of the Swedish Nuclear Fuel and Waste Management Co (SKB) site investigations and only a small number of drillcores

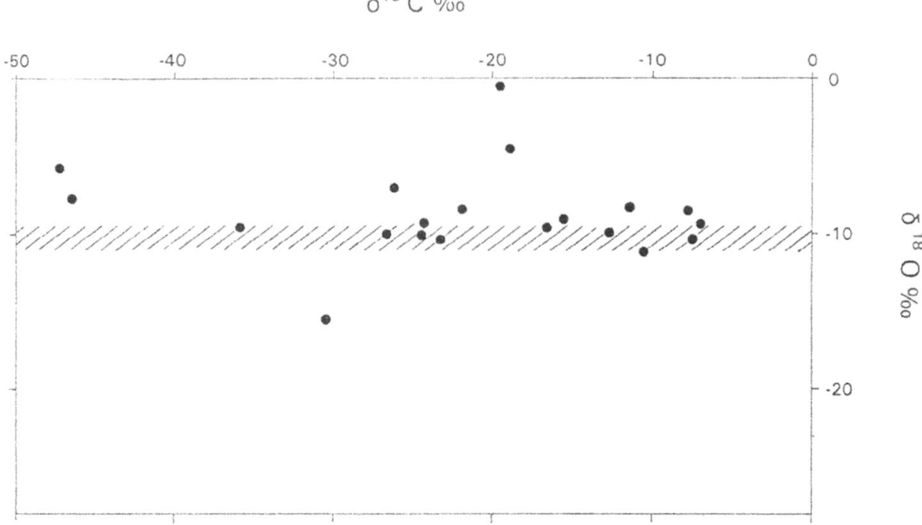

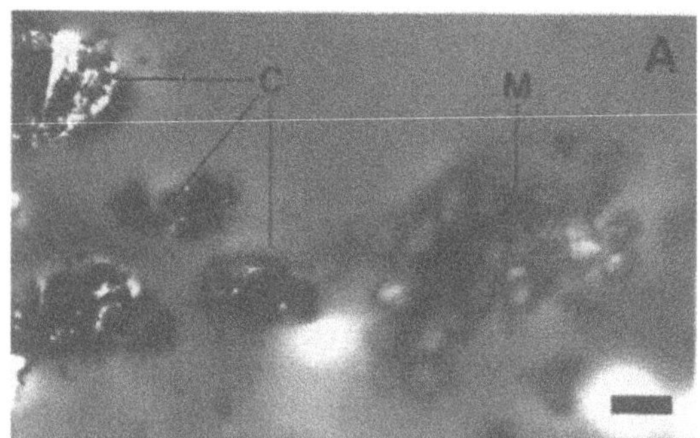

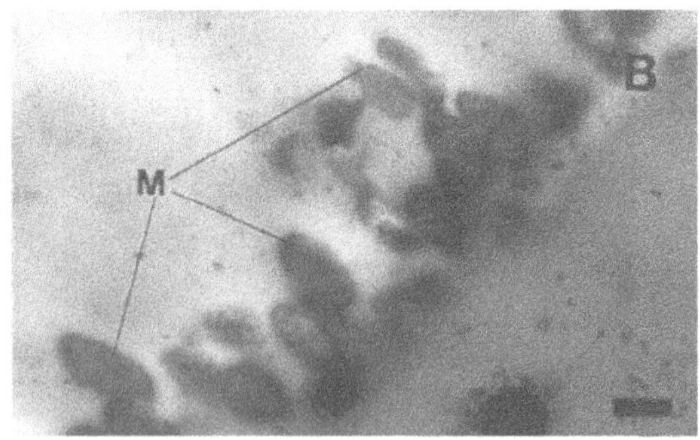

Figure 2(a) $\delta^{18}O$ vs. $\delta^{13}C$ (PDB) in calcite samples from one fracture at 207 m depth at Äspö, southeastern Sweden. Hatched area represents calcite precipitated in equilibrium with present groundwater at ambient temperaturesf fractionation factor according to 0'Neil et al. (1969). *(b)* Thin-section transmission electron microscopy of fracture calcite from the same fracture as above (KAS 02:207 m). A: Microcolony of fossil microorganisms (M) and calcite grains (C). Scale bar represents 1 μm. B: fossil microorganisms (M) arranged in typical biofilm microfilm formation. X-ray microanalysis was performed on these biofilm microfossils. Scale bar represents 1 μm. From Pedersen et al., 1997.

were available from crystalline basement rocks. Since the knowledge about fracture mineral types and their variation with e.g. depth was limited the initial SKB investigations were concentrated on identification and mapping of the fracture minerals. From these investigations it was clear that calcite was a common fracture mineral in most areas.
The studies hitherto have mainly concentrated on giving an overview and a relative chrono-stratigraphy of the events that can be traced in the fractures. Another purpose has been to trace reactivation of the fractures. Special attention has been given to low temperature processes such as the possibility to trace stable isotope redistribution caused by different ground waters, the depth penetration of oxygenated water, and the extent of microbial activity. A major problem with fracture fillling studies on drill cores is the usually limited amounts of sample available. It is also difficult to decide what fractures or part of fractures have been originally open and exposed to the present groundwater.

Results of $\delta^{18}O$ and $\delta^{13}C$ analyses of fracture calcites, from nine different (Fig. 1) areas in Sweden, have been compiled in order to study the responses to different groundwater systems as recorded in their isotope ratios (Tullborg, 1997). All areas are situated within the crystalline basement of the Fennoscandian shield. Most of them belong to the Palaeo- to Meso-Proterozoic provinces in the Svecofennian Domain that have suffered metamorphism and reactivation of fractures. However, the Fjällbacka site , (Fig.1) is located in the Late Meso-Proterozoic Bohus Granite of the Southwest Scandinavian Domain and is post-kinematic in relation to the Sveconorwegian (Grenvillian) orogeny. Only one of the sites is located in a gabbroic rock (Taavinunnanen in the very north of Sweden) whereas the other sites are located in rocks of granitic to quartz monzodioritic composition.

The data compiled have previously been reported in a number of reports and papers. Analyses of carbon and oxygen isotopes have been carried out on fracture calcites from, Finnsjön (Larson and Tullborg 1984a) Gideå (Tullborg and Larson, 1983), Taavinunnanen (Larson and Tullborg, 1984), Klipperås (Tullborg, 1989,) and Äspö (Tullborg and Wallin, 1991; Banwart et al., 1994, Winberg, 1996; Tullborg, 1997 and Tullborg et al., 1999) as part of the SKB programme concerning radioactive waste disposal in crystalline rocks. Fracture calcites analysed within the international Stripa Project (Frape et al., 1992a) are also included.

Within the Deep Earth Gas project in the Siljan impact structure (first phase) $\delta^{18}O$ and $\delta^{13}C$ analyses of fracture calcite were carried out in order to trace different carbon sources (Smellie and Tullborg, 1985). At Fjällbacka similar analyses were used to trace water flow paths in a study concerning the potential of crystalline rocks for use in geothermal heat extraction (Eliasson et al., 1990).

The above mentioned references include descriptions of the sites, geological settings and fracture filling histories. The areas also represent different hydrological regimes with respect to regional groundwater flow, discharge/recharge conditions, etc. Calcites have been sampled from drillcores corresponding to depths of 1100 metres below the present surface from both open and sealed fractures.

More than 400 analyses are included in this compilation. Results are given in Tullborg (1997). The Stripa and Äspö sites are the best-documented sites and the numbers of calcite analyses are approximately 100 from each of these sites.

The fracture calcites analysed precipitated during different periods of time representing various groundwater compositions, ranging from old hydrothermal fillings to calcites precipitated during conditions similar to those present. Concerning ages hydrothermal conditions (>150°C) are generally referred to the Precambrian, whereas the Phanerozoic is regarded as a period of low temperature conditions (<150 °C) in the Fennoscandian shield.

The present groundwater shows large variations in composition at different depths as well as in different areas. Modern fresh water, with $\delta^{18}O$-values in agreement with the annual mean of the precipitation of the area, is found at a depth ranging from 0-50 m and in some areas down to more than 500 metres. However, a careful evaluation of the groundwater data is necessary since the penetration depth of modern fresh water in recharge areas especially may be modified due to the drilling. At greater depths, older waters with various salinity are found. Some areas like Finnsjön, Stripa and Äspö have saline water with low $\delta^{18}O$-values, which exclude a simple marine origin. In contrast a complex evolution involving interaction with glacial meltwater (or cold climate recharge water) and a deep "brine" type of saline water is suggested (cf. Moser et al., 1989; Laaksoharju and Skårman, 1995 and Laaksoharju and Wallin, 1997). One hypothesis is that the hydraulic head beneath a land ice is high enough to significantly enhance the penetration depth of a glacial meltwater into the bedrock (Svensson, 1996a; 1996b). In addition, brackish water from different stages of the Baltic Sea was involved in the formation of groundwater at Äspö (Laaksoharju and Wallin, 1997).

Despite the different ages and origins of the groundwaters at different sites, the waters below 50 to 100 metres depth are generally reducing as shown by negative Eh, and dissolved Fe^{2+} and HS^- in the waters. These waters are also saturated in respect of calcite (cf. Laaksoharju et al., 1993).

3.1 SELECTION OF SAMPLES

The analyses of fracture fillings included in this compilation were carried out during a period of approximately 15 years. In the early 1980s relatively little was known about the $\delta^{18}O$ and $\delta^{13}C$ values of calcites from fractures in the crystalline basement except for in mineralised rocks and geothermal areas. For this reason all kinds of fractures; thin and broad, open and sealed, as well as fractures with calcite coexisting with hydrothermal and with low temperature minerals etc. were sampled. Furthermore, at that time, we were not quite aware of the small scale isotope variations (zoning, etc.), and the sampling volumes were usually somewhat larger than those used today (around 10 mg instead of 2-5 mg). The main purpose of the studies has been to describe past and present water-rock interaction using $\delta^{18}O$ and $\delta^{13}C$ values. The later studies focused on tracing relatively late and recent water-circulation and therefore the sampling has concentrated on water conducting fractures and fracture zones.

3.2 SAMPLING AND ANALYTICAL PROCEDURE

$\delta^{18}O$ and $\delta^{13}C$ analyses of the carbonates (except for those from Stripa) have been carried out at three laboratories; Institutt for Energiteknikk (IFE), Kjeller, Norway, Department of Marine Geology, Göteborg University, and Institute of Geology and Geochemistry, Stockholm University. They used similar conventional techniques and the results are comparable. The analytical technique used at IFE is given below: The samples were dried for four hours at 400°C, put into glass bottles together with 2 ml 100% H_3PO_4 and evacuated to $5x10^{-3}$mbar. They reacted with the acid for 2 hours at 25°C. The CO_2 gas produced was cleaned by freezing and analysed in a VG Optima gas mass spectrometer for measurements of the isotope ratio of stable carbon and oxygen isotopes. The results were related to a standard as follows:

$$\delta^{13}C_{sample} = \left[({}^{13}C/{}^{12}C_{sample}/{}^{13}C/{}^{12}C_{stand}) - 1 \right] \times 10^3$$

The same equation is valid for the $^{18}O/^{16}O$ fractionation expressed as $\delta^{18}O$. The standard used for the carbon and oxygen analyses was related to the PDB standard. Some of the analyses from Äspö (Tullborg and Wallin 1991; Banwart et al., 1994) were related to SMOW but converted to PDB values. The accuracy of the $\delta^{13}C$ and the $\delta^{18}O$ analyses is ±0.1 o/oo.

The samples have been prepared using a small knife or a dental drill. Only a few mg of carbonate were needed and in some fractures it was possible to sample different generations of calcites from a single fracture. The sampling was usually based on information from optical microscopy of the fillings. However, from the detailed studies at Äspö (Tullborg et al., 1999 and Pedersen et al., 1997) it is obvious that large variations in isotopic composition can occur due to microscale zoning. This zoning is difficult, and mostly impossible to detect when using an optical microscope.

3.3 RESULTS

A plot of $\delta^{18}O$ versus $\delta^{13}C$ for all fracture calcites is shown in Figure 3. It is difficult to distinguish between originally open and sealed fractures in the drillcores. Some fractures can be partly open and water conducting although they are sealed in the intersection represented by the drillcore. In figure 3 all types of fractures are represented and the analyses from different depth intervals are categorised as 0-250 m, 250-500 m, and more than 500 metres. Taking all the data into account the $\delta^{18}O$-values vary within the range + 1.1 to -27 o/oo whereas the $\delta^{13}C$-values vary within an even larger range of +18 to -74 o/oo. Most of the samples however, have $\delta^{13}C$-values between -2 to -20 o/oo. The largest variation in $\delta^{13}C$-values was recorded in samples with $\delta^{18}O$-values in the range of -5 to -15 o/oo. The samples with high positive $\delta^{13}C$-values (+1 to +18 o/oo) were found at depths of between 5 m and 819 m and the calcites with extremely low $\delta^{13}C$-values (-74 to -26 o/oo) were found at depths of between 74 m and 904 m (Fig 4a). Most of the calcites with $\delta^{18}O$-values below -20 o/oo are hydrothermal in origin and have $\delta^{13}C$-values between -7 and -2 o/oo.

$\delta^{13}C$-values of HCO_3^- in the present groundwaters (65 analyses from SKB's database SICADA) range from -22.3 to -6.5 o/oo of which more than 90 % are within the interval -10 to -21 o/oo. From Stripa generally lower $\delta^{13}C(HCO_3^-)$ values are reported (-14.0 to -27 o/oo) in addition to two extremely low values (-35,6 and -32.6 o/oo) detected at 900 metres depth in one of the boreholes (Murphy and Davis, 1992; Fritz et al., 1989). Some extremely low $\delta^{13}C$-values of calcites from this depth are also reported from this site (Frape et al., 1992), indicating correspondence between the

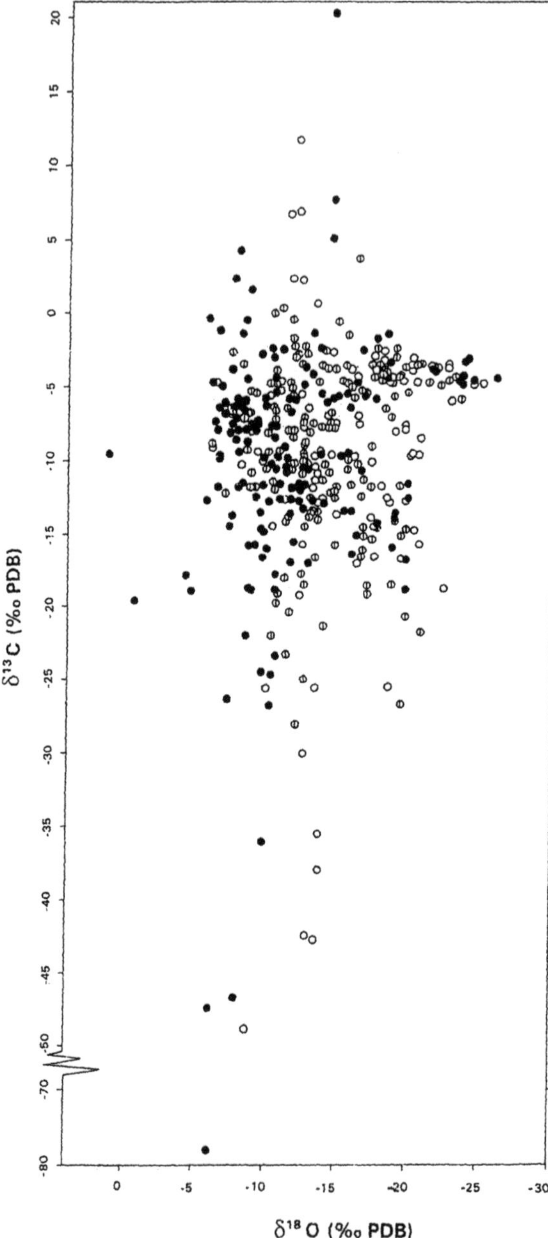

Figure 3. $\delta^{18}O$ versus $\delta^{13}C$ for all the fracture calcites analysed (cf. refs). Filled circle = fracture calcites from 0-250 m depth below the surface (N=162), divided circle = 250-500 m depth (N=176) and open circle = samples from > 500 m depth (N=93). From Tullborg (1997).

composition of the calcite and the present groundwater. This is commented by Fritz et al., (1989) as a possible indication of a biologically active system at depth approaching 1000 m. Taken the $\delta^{13}C(HCO_3^-)$-values from all sites into account (Fig 4b), there is a tendency towards more negative $\delta^{13}C$-values with depth. The near-surface samples (upper 100 metres) show mostly values around -15 ± 3 o/oo, indicative of contributions from soil and atmospheric CO_2 together with calcite dissolution (Fig 4b). The relatively small variations in values indicate open conditions. The HCO_3^- contents in the waters range between 400 to 32 mg/l with a general decrease with depth. The TOC concentrations for the waters sampled below 100 m usually do not exceed a few mg. Higher values reported from Stripa were regarded as the result of contamination (Murphy and Davis, 1992).

Two main processes may explain the decrease with depth, (a) Successive depletion of ^{13}C in the water caused by precipitation of calcite, which is enriched in ^{13}C compared to the water (b) subsurface production of HCO_3^- by microbial mediated breakdown of organic material. As the general trend is decreasing HCO_3^-content with depth this production is probably low, but may locally be significant.

Comparing the $\delta^{13}C$-values of HCO_3^- with the $\delta^{13}C$-values of the fracture calcites (Figs. 4a and b) it can be concluded that most of the calcite values are largely similar to equilibrium with the present groundwater at ambient temperatures, i.e. an enrichment of c. 3 o/oo in the calcite compared with the water. However, the extreme values in the calcites are not matched with the $\delta^{13}C$-values in the water. The explanation for the extreme $\delta^{13}C$(calcite)-values can be that the calcite is precipitated from a water volume with extreme $\delta^{13}C$ isotope composition transported through the fracture system. In such a case this water existed in the past, not found in the bedrock today. Another explanation is that the extreme values are produced in situ by bacteria mediated processes, causing local disequilibria and large variations in $\delta^{13}C(HCO_3^-)$ on small scale, indicating "closed/semi-closed conditions" (low flow compared to the reactions rate of the bacteria mediated processes). Since we know that the number of fractures carrying calcites with extreme $\delta^{13}C$ are relatively few, the variation in isotopic composition is very large over small distances (microscale), and microbes are identified in the deep groundwaters (Pedersen and Karlsson, 1995) the second explanation is favoured. The amounts of HCO_3^- with extreme $\delta^{13}C$-values are probably very small, and therefore the volume of water with extreme values will be dispersed and the most extreme values not possible to detect in the water samples.

With one exception, the extreme values were found at depths between 62 to 500 m at Äspö, whereas at Stripa there are several observations of extreme values (positive and negative) down to 853 m. Arguments for abiogenic methane from depth as nutrients for bacteria, have been put forward by e.g. Pedersen and Karlsson, 1995. This may be an explanation for the extreme $\delta^{13}C$-values in carbonates at large depths but the processes responsible for these precipitates have not yet been revealed.

$\delta^{13}C$-values in calcites from the different sites are shown in histograms in figure 5. It can be seen that values below -20 o/oo and above +1 o/oo were, with one exception, found at Äspö and Stripa. The reason for this is not yet fully understood.

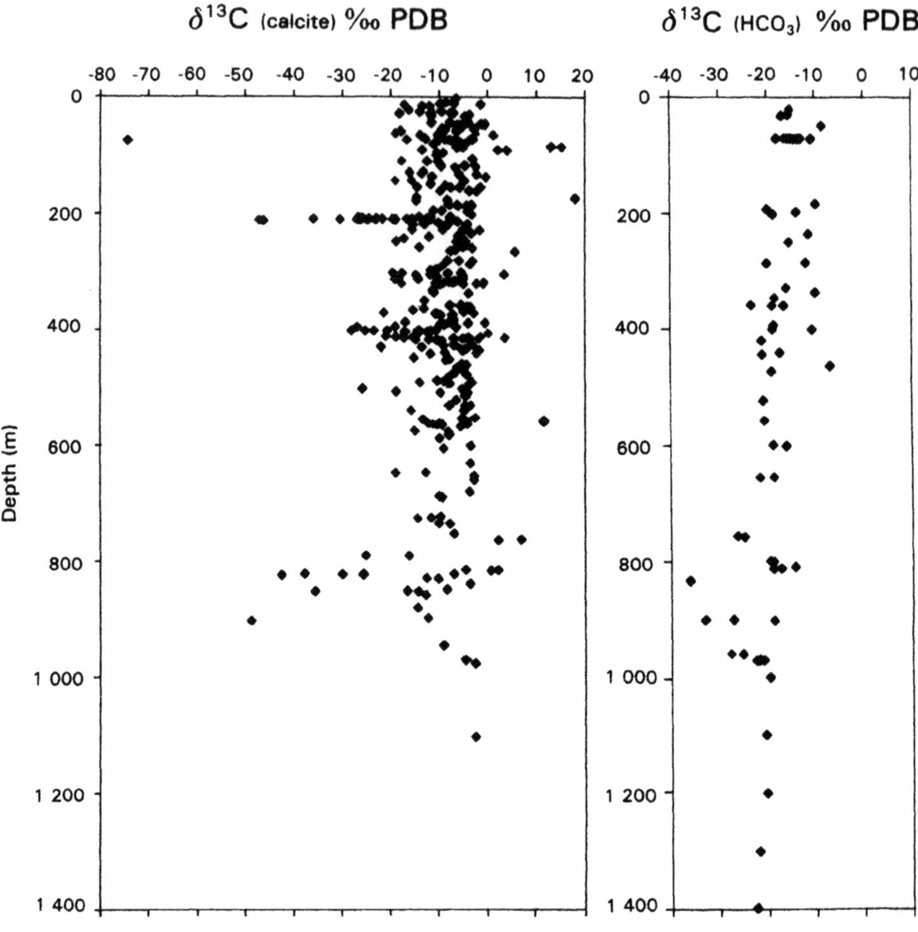

*Figure 4. (*a) δ^{13}C in fracture calcites versus depth. *(b)* δ^{13}C in HCO_3^- in groundwaters from the sampled sites except for Fjällbacka, Siljan and Kråkemåla where no data were available.

Two explanations can be suggested:

(1) Increased addition of organic matter in the groundwater of these areas, e.g. during the downward percolation of meteoric water and or brackish seawater.

(2) An increased flux of abiogenic methane. This is so far not supported by gas analyses.

The first alternative seems, based on the present knowledge, to be the best explanation for most of the extreme values. It is suggested that the Äspö and Stripa belongs to one

of two major types of hydrogeochemical systems distinguished (type B below). However, the hydrogeological conditions at Stripa are severely influenced by the mine which has been operating for more than hundred years.
The two "types" of hydrogeochemical systems are:
A) Typical recharge areas, like Taavinunnanen, Fjällbacka and to some extent Klipperås, with thin soil covers or low organic content in the overburden. These are characterised by calcite dissolution in water conducting fractures down to 50 - 100 m depth. This zone of dissolution is accompanied by precipitation of Fe-oxyhydroxide indicating the penetration depth of oxygenated water. The $\delta^{13}C$-values in the calcites below the dissolution zone are within the range -3 to -14 o/oo with the exceptions of two samples of -17 o/oo. Mean values are -6.5 o/oo for Taavinunnanen, -7.7 o/oo for Fjällbacka and -9.2 o/oo for Klipperås. No extreme values are found indicating that the production of biogenic CO_2 (at closed conditions) in the groundwater is very low. The redox capacities in this type of hydro-geochemical system is usually low in the shallower part of the the fracture system. At greater depth the main redox buffer capacity is suggested to be due to the Fe^{2+} minerals and sulphides present.
B) Areas characterised by low hydraulic gradient and presence of sediments rich in organics material; In these areas calcite dissolution is more or less absent or only observed in the upper 10-20 metres. The $\delta^{13}C$ values in the open fractures are somewhat lower (-11.3 o/oo for Äspö) than in the type A areas (the extreme values <-25 o/oo and >0 being excluded). Extreme carbon values in fracture calcites are present, indicating bacterial activity. This type of system has a much higher redox buffer capacity and there are no evidences for penetration of oxygenated water into these aquifers, except for in the upper tens of metres.

Some of the areas investigated cannot be categorised into any of these types and at some sites both of them can be found, due to large variations in hydrogeology (caused by variations in discharge/recharge, fracture frequency and thickness of soil cover etc.). It is important to note that it is quite possible that microbial activity at depth in the bedrock takes place also in Taavinunnanen, Fjällbacka and Klipperås sites but has probably a much lower activity and thus low production of isotopically extreme $\delta^{13}C(HCO_3^-)$ –values .

4. Conclusion

The combined TEM and stable isotope study of fracture calcite at 207 m from Äspö showed that microbial activity has taken place in situ in the crystalline bed rock before the disturbances caused by drilling and tunnel excavation took place.

At closed conditions the microbial activity can cause extreme fractionation in e.g. the stable carbon and sulphur isotope systems. The resulting extreme values may be preserved in precipitates on the fracture walls e.g. calcite and pyrite. It is expected that sulphate and Fe reducing bacteria cause low $\delta^{13}C(HCO_3)$ whereas methane forming bacteria may cause high $\delta^{13}C(HCO_3)$. Therefore, a compilation of the stable isotope results with focus on extremely high or extremely low $\delta^{13}C$ values is suggested to give an overview of the extent and distribution of microbial activity at depth. However, ideally, this requires that all calcite samples have been sampled in the same way and

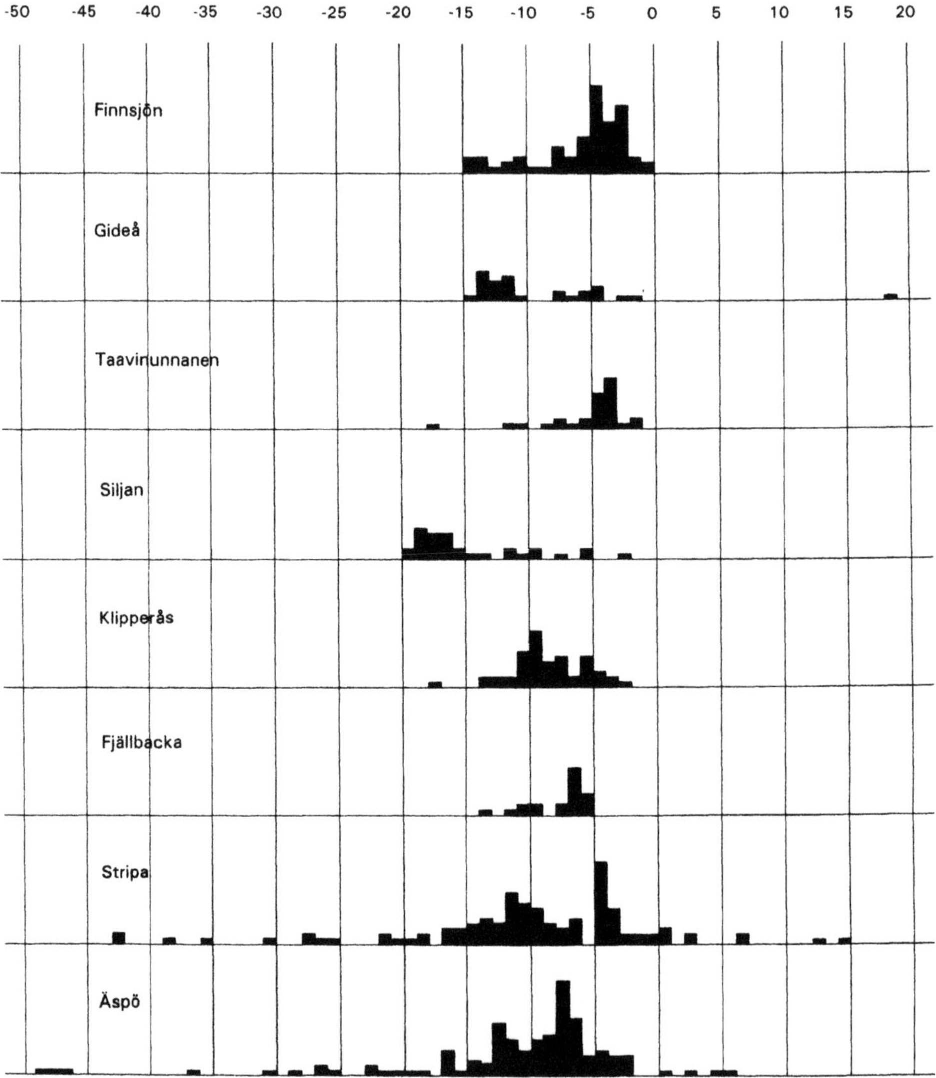

Figure 5. δ^{13}C-values of all fracture calcites analysed, except for five analyses from Kråkemåla close to Äspö (cf. Tullborg, 1997). Note the relatively small δ^{13}C-interval (0 to -20 o/oo) for all sites except for that of Äspö and Stripa..

that only small amounts of sample have been analysed since the calcite samples showing extreme values otherwise can be mixed up and not possible to recognise.

Most of the extreme values have been recorded at depth between 50 and 500 m but some occurrences have been found as deep as 900 metres below see level. From the present material it seems that organic material and gases from the surface are more important than flux of abiogenic gases as nutrients for bacteria. However, the results do not exclude the possibility of gases as nutrients especially not for the deeper samples.

The extreme carbon isotope values are found in calcites with various $\delta^{18}O$ indicating that microbial activity can take place in different water types. However, they are most common in calcites corresponding to Baltic seawater and temperate meteoric water. This observation is in accordance also with data from the Finnish site Oilkiluoto (Blyth et al., 1998). The low $\delta^{13}C$ calcites are so far undated.

In conclusion, bacterial activity at depth in the crystalline bedrock of the Fennoscandian shield can be documented by the extreme $\delta^{13}C$–values, although the amounts of calcite produced in situ at depth seems to be relatively small.

ACKNOWLEDGEMENT

Sven Åke Larson, Bill Wallin, Marcus Laaksoharju and Karsten Pedersen are acknowledged for their valuable comments on the manuscript.

5. References

Banwart, S., Tullborg, E-L., Pedersen, K., Gustafsson, E., Laaksoharju, M., Nilsson, A-C., Wallin, B. and Wikberg, P. (1994) Organic oxidation induced by large scale shallow water intrusion into a vertical fracture zone at the Äspö Hard Rock Laboratory. Fourth International Conference on the chemistry and migration behaviour of actinides and fission products in the geosphere. Charleston, SC USA, *Contaminant hydrology,* 537-542.

Blyth, A., Frape, S., Blomqvist, R., Nissinen, P and McNutt, R. (1998): An isotopic and fluid inclusion study of fracture calcite from borehole OL-KR1 at the Olkiluoto site, Finland. *Posiva Report* **98-04**, ISSN 1239-3096.

Burgman, J. O., Calles, B. and Westman, F. (1987) Results from a ten year study of 18-O in precipitation and run-off in Sweden. Symp. *Use of isotope techniques in water resources development.* **IAEA-SM-299/107**: 579-587.

Eliasson, T., Tullborg, E.L. and Landström, O. (1990) Fracture filling minerals and geochemistry at the Swedish HDR research site. In: *Hot Dry Rock Geothermal energy,* Camborne School of mines international conference, R. Baria (editor). ISBN 1-85365-217-2. 425-435.

Frape, S. K., Fritz, P. Kamineni, D.C. and Gibson, I.L. (1992) Chemical, isotopic and petrographic description of fracture mineralogy, Stripa pluton, Sweden. In

Hydrogeochemical investigations in boreholes at the Stripa Mine. S.N Davis & D.K. Nordstrom (editors), *Stripa Project SKB Technical Report,* **TR 92-13**. ISSN 0349-5698.

Fritz, P., Fontes, J-C., Frape, S.K., Louvat, D., Michelot, J-L. and Balderer, W. (1989) The isotope geochemistry of carbon in groundwater at Stripa. *Geochimica et Cosmochimica Acta* **53,** 1765-1775.

Laaksoharju, M. (ed.), (1995) Sulphate reduction in Äspö HRL tunnel. *SKB Technical Report* **TR 95-25**, ISSN 0284-3757.

Laaksoharju, M. and Skårman, C. (1995) Groundwater sampling and chemical characterization of the Äspö HRL tunnel in Sweden. *SKB Progress Report* **25-95-29.**

Laaksoharju, M. and Wallin, B. (eds.) (1997) Evolution of the groundwater chemicstry at the Swedish Äspö Hard Rock Laboratory site. Proceedings of the second Äspö International Geochemistry Workshop, June 6-7, 1995. *SKB International Report* **ICR-97-04**. ISSN 1104-3210.

Laaksoharju, M., Smellie, J., Tuotsalainen, P. and Snellman, M. (1993) An approach to quality classification of deep groundwaters in Sweden and Finland. *SKB Technical Report* **TR 93-27**, ISSN 0284-3757.

Larson, S.Å. and Tullborg, E-L. (1984a) Stable isotopes of fissure-filling calcite from Finnsjön, Uppland, Sweden. *Lithos,* **17**, 117-125.

Larson, S.Å. and Tullborg, E-L. (1984b) Fracture fillings in the gabbro massif of Taavinunnanen, northern Sweden. *SKB/KBS Technical Report,* **TR 84-05**, ISSN 0348-7504.

Moser, H., Wolf, M., Fritz, P., Fontes, J-Ch, Florowski, T., and Payne, B.R. (1989) Deuterium, oxygen-18, and tritium in Stripa Groundwater. *Geochim Cosmochim. Acta*,**53,**1757-1763.

Murphy, E. and Davis, S.N. (1993) Analyses of organic and inorganic carbon species in Stripa groundwater . In Hydrochemical investigations in boreholes at the Stripa mina. The hydrochemical Advisory group and their Associates: S.N Davis & D.K. Nordstrom (editors), *Stripa Project SKB Technical Report,* **TR 92-19**. ISSN 0349-5698.

O´Neil, J.R., Clayton, R.N. and Mayeda, T.K., 1969: Oxygen isotopefractionation in divalent metal carbon nates. *The Journal of Chemical Pysics*, **51**, 5547.

Pedersen, K. and Karlsson, F. (1995) Investigations of subterranean bacteria - Their influence on performance assessment of radioactive waste disposal. *SKB Technical Report* **TR 95-10,** ISSN 0284-3757.

Pedersen, K., Investigations of subterranian microorganisms in deep crystalline bedrock and their importance for the disposal of nuclear waste. This volume.

Pedersen, K., Ekendahl, S., Tullborg, E-L., Furnes, H., Thorseth, I. and Tumyr, O. (1997) Evidence of ancient life in a granitic aquifer at 200 meters depth. *Geology,* **25**, 827-830.

Smellie, J.A.T. and Tullborg, E-L. (1985) Geochemical investigations in the Siljan area, Sweden. *SGAB report* IRAP 85214.

Svensson, U. (1996a) Regional groundwater flow due to an advancing and retreating glacier - scoping calculations. *SKB Palaeohydrological Programme,* Progress Report **96-35.**

Svensson, U. (1996b) Simulations of regional groundwater flows, as forced by glaciation cycles. *SKB Palaeohydrological Programme,* Progress Report **96-36**.

Tullborg E-L., (1989) The influence of recharge water on fissure-filling minerals – A study from Klipperås, southern Sweden.. *Chem. Geol.* **76**, 309-320.

Tullborg, E-L., (1997) Recognition of low-temperature processes in the Fennoscandian shield. *PhD thesis*, Earth Sciences Centre, Göteborg University, A17, 1997. ISSN 1400-3813.

Tullborg, E-L. and Larson S.Å. (1983) Fissure fillings from Gideå, central Sweden. *SKB/KBS Technical Report* **TR 83-74**. ISSN 0348-7504.

Tullborg, E-L. and Wallin, B. (1991) Stable Isotope studies of calcite fracture fillings ($\delta^{18}O$, $\delta^{13}C$) and groundwaters ($\delta^{18}O$, δD). Part II in: Tullborg, E-L., Wallin, B. and Landström, O. Hydrogeochemical studies of fracture minerals from water conducting fractures and deep groundwaters at Äspö. *SKB Progress Report* **25-90-01.**

Tullborg, E-L., Landström, O., and Wallin, B. (1999) Low temperature, trace element mobility influenced by microbial activity - indications from fracture calcites and pyrite in crystalline basement. *Chemical Geology,* in press.

Winberg, A., (ed.) (1996) First TRUE stage - Tracer Retention Understanding Experiments. Descriptive structural-hydraulic models on block and detailed scales of the TRUE-1 site. *SKB International Cooperation Report* **ICR 96-04.** ISSN 1104-3210.

Water Science and Technology Library

1. A.S. Eikum and R.W. Seabloom (eds.): *Alternative Wastewater Treatment.* Low-Cost Small Systems, Research and Development. Proceedings of the Conference held in Oslo, Norway (7–10 September 1981). 1982 ISBN 90-277-1430-4
2. W. Brutsaert and G.H. Jirka (eds.): *Gas Transfer at Water Surfaces.* 1984 ISBN 90-277-1697-8
3. D.A. Kraijenhoff and J.R. Moll (eds.): *River Flow Modelling and Forecasting.* 1986 ISBN 90-277-2082-7
4. World Meteorological Organization (ed.): *Microprocessors in Operational Hydrology.* Proceedings of a Conference held in Geneva (4–5 September 1984). 1986 ISBN 90-277-2156-4
5. J. Němec: *Hydrological Forecasting.* Design and Operation of Hydrological Forecasting Systems. 1986 ISBN 90-277-2259-5
6. V.K. Gupta, I. Rodríguez-Iturbe and E.F. Wood (eds.): *Scale Problems in Hydrology.* Runoff Generation and Basin Response. 1986 ISBN 90-277-2258-7
7. D.C. Major and H.E. Schwarz: *Large-Scale Regional Water Resources Planning.* The North Atlantic Regional Study. 1990 ISBN 0-7923-0711-9
8. W.H. Hager: *Energy Dissipators and Hydraulic Jump.* 1992 ISBN 0-7923-1508-1
9. V.P. Singh and M. Fiorentino (eds.): *Entropy and Energy Dissipation in Water Resources.* 1992 ISBN 0-7923-1696-7
10. K.W. Hipel (ed.): *Stochastic and Statistical Methods in Hydrology and Environmental Engineering.* A Four Volume Work Resulting from the International Conference in Honour of Professor T. E. Unny (21–23 June 1993). 1994
    10/1: Extreme values: floods and droughts ISBN 0-7923-2756-X
    10/2: Stochastic and statistical modelling with groundwater and surface water applications ISBN 0-7923-2757-8
    10/3: Time series analysis in hydrology and environmental engineering ISBN 0-7923-2758-6
    10/4: Effective environmental management for sustainable development ISBN 0-7923-2759-4
    Set 10/1–10/4: ISBN 0-7923-2760-8
11. S.N. Rodionov: *Global and Regional Climate Interaction: The Caspian Sea Experience.* 1994 ISBN 0-7923-2784-5
12. A. Peters, G. Wittum, B. Herrling, U. Meissner, C.A. Brebbia, W.G. Gray and G.F. Pinder (eds.): *Computational Methods in Water Resources X.* 1994 Set 12/1–12/2: ISBN 0-7923-2937-6
13. C.B. Vreugdenhil: *Numerical Methods for Shallow-Water Flow.* 1994 ISBN 0-7923-3164-8
14. E. Cabrera and A.F. Vela (eds.): *Improving Efficiency and Reliability in Water Distribution Systems.* 1995 ISBN 0-7923-3536-8
15. V.P. Singh (ed.): *Environmental Hydrology.* 1995 ISBN 0-7923-3549-X
16. V.P. Singh and B. Kumar (eds.): *Proceedings of the International Conference on Hydrology and Water Resources* (New Delhi, 1993). 1996
    16/1: Surface-water hydrology ISBN 0-7923-3650-X
    16/2: Subsurface-water hydrology ISBN 0-7923-3651-8

16/3: Water-quality hydrology ISBN 0-7923-3652-6
16/4: Water resources planning and management ISBN 0-7923-3653-4
Set 16/1–16/4 ISBN 0-7923-3654-2

17. V.P. Singh: *Dam Breach Modeling Technology.* 1996 ISBN 0-7923-3925-8
18. Z. Kaczmarek, K.M. Strzepek, L. Somlyódy and V. Priazhinskaya (eds.): *Water Resources Management in the Face of Climatic/Hydrologic Uncertainties.* 1996 ISBN 0-7923-3927-4
19. V.P. Singh and W.H. Hager (eds.): *Environmental Hydraulics.* 1996 ISBN 0-7923-3983-5
20. G.B. Engelen and F.H. Kloosterman: *Hydrological Systems Analysis.* Methods and Applications. 1996 ISBN 0-7923-3986-X
21. A.S. Issar and S.D. Resnick (eds.): *Runoff, Infiltration and Subsurface Flow of Water in Arid and Semi-Arid Regions.* 1996 ISBN 0-7923-4034-5
22. M.B. Abbott and J.C. Refsgaard (eds.): *Distributed Hydrological Modelling.* 1996 ISBN 0-7923-4042-6
23. J. Gottlieb and P. DuChateau (eds.): *Parameter Identification and Inverse Problems in Hydrology, Geology and Ecology.* 1996 ISBN 0-7923-4089-2
24. V.P. Singh (ed.): *Hydrology of Disasters.* 1996 ISBN 0-7923-4092-2
25. A. Gianguzza, E. Pelizzetti and S. Sammartano (eds.): *Marine Chemistry.* An Environmental Analytical Chemistry Approach. 1997 ISBN 0-7923-4622-X
26. V.P. Singh and M. Fiorentino (eds.): *Geographical Information Systems in Hydrology.* 1996 ISBN 0-7923-4226-7
27. N.B. Harmancioglu, V.P. Singh and M.N. Alpaslan (eds.): *Environmental Data Management.* 1998 ISBN 0-7923-4857-5
28. G. Gambolati (ed.): *CENAS. Coastline Evolution of the Upper Adriatic Sea Due to Sea Level Rise and Natural and Anthropogenic Land Subsidence.* 1998 ISBN 0-7923-5119-3
29. D. Stephenson: *Water Supply Management.* 1998 ISBN 0-7923-5136-3
30. V.P. Singh: *Entropy-Based Parameter Estimation in Hydrology.* 1998 ISBN 0-7923-5224-6
31. A.S. Issar and N. Brown (eds.): *Water, Environment and Society in Times of Climatic Change.* 1998 ISBN 0-7923-5282-3
32. E. Cabrera and J. García-Serra (eds.): *Drought Management Planning in Water Supply Systems.* 1999 ISBN 0-7923-5294-7
33. N.B. Harmancioglu, O. Fistikoglu, S.D. Ozkul, V.P. Singh and M.N. Alpaslan: *Water Quality Monitoring Network Design.* 1999 ISBN 0-7923-5506-7
34. I. Stober and K. Bucher (eds): *Hydrogeology of Crystalline Rocks.* 2000 ISBN 0-7923-6082-6

GPSR Compliance
The European Union's (EU) General Product Safety Regulation (GPSR) is a set of rules that requires consumer products to be safe and our obligations to ensure this.

If you have any concerns about our products, you can contact us on

ProductSafety@springernature.com

In case Publisher is established outside the EU, the EU authorized representative is:

Springer Nature Customer Service Center GmbH
Europaplatz 3
69115 Heidelberg, Germany

www.ingramcontent.com/pod-product-compliance
Ingram Content Group UK Ltd.
Pitfield, Milton Keynes, MK11 3LW, UK
UKHW021333070726
13610UKWH00011B/55

* 9 7 8 9 4 0 1 7 1 8 1 7 2 *